SECOND EDIT

Writing Reaction Mechanisms in Organic Chemistry

ADVANCED ORGANIC CHEMISTRY SERIES

SERIES EDITOR

James K. Whitesell

SECOND EDITION

Writing Reaction Mechanisms in Organic Chemistry

ADVANCED ORGANIC CHEMISTRY SERIES

Audrey Miller

Professor Emeritus
Department of Chemistry
University of Connecticut
Storrs, Connecticut

Philippa H. Solomon

Edison, New Jersey

San Diego   London   Boston   New York   Sydney   Tokyo   Toronto

Academic Press
A Harcourt Science and Technology Company
525 B Street, Suite 1900, San Diego, California 92101-4495, U.S.A.
http://www.apnet.com

Academic Press
24-28 Oval Road, London NW1 7DX, UK
http://www.hbuk.co.uk/ap/

Harcourt/Academic Press
A Harcourt Science and Technology Company
200 Wheeler Road, Burlington, Massachusetts 01803
http://www.harcourt-ap.com

Library of Congress Catalog Card Number: 99-61410

International Standard Book Number: 0-12-496712-4

PRINTED IN THE UNITED STATES OF AMERICA
99 00 01 02 03 04 EB 9 8 7 6 5 4 3 2 1

Preface to the Second Edition   xi
Preface to First Edition   xiii

I

Introduction — Molecular Structure and Reactivity

1. How to Write Lewis Structures and Calculate Formal Charges   2
   A. Determining the Number of Bonds   2
   B. Determining the Number of Rings and/or π Bonds (Degree of Unsaturation)   2
   C. Drawing the Lewis Structure   3
   D. Formal Charge   6
2. Representations of Organic Compounds   12
3. Geometry and Hybridization   14
4. Electronegativities and Dipoles   16
5. Resonance Structures   18
   A. Drawing Resonance Structures   18
   B. Rules for Resonance Structures   23
6. Aromaticity and Antiaromaticity   26
   A. Aromatic Carbocycles   26

    B. Aromatic Heterocycles   27
    C. Antiaromaticity   28
7. Tautomers and Equilibrium   29
8. Acidity and Basicity   32
9. Nucleophiles and Electrophiles   37
    A. Nucleophilicity   37
    B. Substrate   38
    C. Solvent   39

2

General Principles for Writing Reaction Mechanisms

1. Balancing Equations   64
2. Using Arrows to Show Moving Electrons   66
3. Mechanisms in Acidic and Basic Media   69
4. Electron-Rich Species: Bases or Nucleophiles?   76
5. Trimolecular Steps   78
6. Stability of Intermediates   79
7. Driving Forces for Reactions   82
    A. Leaving Groups   83
    B. Formation of a Small Stable Molecule   84
8. Structural Relationships between Starting Materials and Products   85
9. Solvent Effects   86
10. A Last Word   88

3

Reactions of Nucleophiles and Bases

1. Nucleophilic Substitution   106
    A. The S_N2 Reaction   106
    B. Nucleophilic Substitution at Aliphatic sp^2 Carbon
       (Carbonyl Groups)   112
    C. Nucleophilic Substitution at Aromatic Carbons   116
2. Eliminations at Saturated Carbon   120
    A. E2 Elimination   120
    B. Ei Elimination   122

3. Nucleophilic Addition to Carbonyl Compounds   123
    A. Addition of Organometallic Reagents   123
    B. Reaction of Nitrogen-Containing Nucleophiles with
       Aldehydes and Ketones   128
    C. Reactions of Carbon Nucleophiles with Carbonyl Compounds   130
4. Base-Promoted Rearrangements   141
    A. The Favorskii Rearrangement   141
    B. The Benzilic Acid Rearrangement   142
5. Additional Mechanisms in Basic Media   144

4

Reactions Involving Acids and Other Electrophiles

1. Stability of Carbocations   195
2. Formation of Carbocations   196
    A. Ionization   196
    B. Addition of an Electrophile to a π Bond   197
    C. Reaction of an Alkyl Halide with a Lewis Acid   199
3. The Fate of Carbocations   199
4. Rearrangement of Carbocations   200
    A. The Dienone–Phenol Rearrangement   204
    B. The Pinacol Rearrangement   206
5. Electrophilic Addition   208
    A. Regiospecificity   208
    B. Stereochemistry   209
6. Acid-Catalyzed Reactions of Carbonyl Compounds   213
    A. Hydrolysis of Carboxylic Acid Derivatives   213
    B. Hydrolysis and Formation of Acetals and Orthoesters   216
    C. 1,4-Addition   218
7. Electrophilic Aromatic Substitution   220
8. Carbenes   224
    A. Singlet and Triplet Carbenes   225
    B. Formation of Carbenes   226
    C. Reactions of Carbenes   227
9. Electrophilic Heteroatoms   231
    A. Electron-Deficient Nitrogen   232
    B. Rearrangements Involving Electrophilic Nitrogen   233
    C. Rearrangement Involving Electron-Deficient Oxygen   238

5

Radicals and Radical Anions

1. Introduction   283
2. Formation of Radicals   284
   A. Homolytic Bond Cleavage   284
   B. Hydrogen Abstraction from Organic Molecules   285
   C. Organic Radicals Derived from Functional Groups   286
3. Radical Chain Processes   287
4. Radical Inhibitors   290
5. Determining the Thermodynamic Feasibility of Radical Reactions   292
6. Addition of Radicals   294
   A. Intermolecular Radical Addition   294
   B. Intramolecular Radical Addition: Radical Cyclization Reactions   296
7. Fragmentation Reactions   299
   A. Loss of CO_2 299
   B. Loss of a Ketone   300
   C. Loss of N_2 300
   D. Loss of CO   300
8. Rearrangement of Radicals   303
9. The $S_{RN}1$ Reaction   307
10. The Birch Reduction   310
11. A Radical Mechanism for the Rearrangement of Some Anions   312

6

Pericyclic Reactions

1. Introduction   343
   A. Types of Pericyclic Reactions   343
   B. Theories of Pericyclic Reactions   344
2. Electrocyclic Reactions   346
   A. Selection Rules for Electrocyclic Reactions   346
   B. Stereochemistry of Electrocyclic Reactions (Conrotatory and
      Disrotatory Processes)   347
   C. Electrocyclic Reactions of Charged Species (Cyclopropyl Cations)   353
3. Cycloadditions   355
   A. Terminology of Cycloadditions   355
   B. Selection Rules for Cycloadditions   358
   C. Secondary Interactions   361
   D. Cycloadditions of Charged Species   362

4. Sigmatropic Rearrangements   366
   A. Terminology   366
   B. Selection Rules for Sigmatropic Rearrangements   368
5. The Ene Reaction   373
6. A Molecular Orbital View of Pericyclic Processes   379
   A. Orbitals   379
   B. Molecular Orbitals   380
   C. Generating and Analyzing π Molecular Orbitals   382
   D. HOMOs and LUMOs   387
   E. Correlation Diagrams   388
   F. Frontier Orbitals   392

7

Additional Problems   417

Appendix A: Lewis Structures of Common Functional Groups   453

**Appendix B: Symbols and Abbreviations Used in
                  Chemical Notation   455**

**Appendix C: Relative Acidities of Common Organic and
                  Inorganic Substances   457**

Index   465

In revising this text for the second edition, a major goal was to make the book more user-friendly for both graduate and undergraduate students. Introductory material has been fleshed out. Headings have been added to make it easier to locate topics. The structures have been redrawn throughout, with added emphasis on the stereochemical aspects of reaction mechanisms. Coverage of some topics such as solvent effects and neighboring group effects has been expanded, and Chapter 6 has been completely reorganized and extensively rewritten.

As in the previous edition, the focus of this book is on the *how* of writing organic mechanisms. For this reason and to keep the book compact and portable, the number of additional examples and problems has been minimized, and no attempt has been made to cover additional topics such as oxidation–reduction and organotransition metal reactions. The skills developed while working through the material in this book should equip the reader to deal with reactions whose mechanisms have been explored less thoroughly.

I am most grateful to the reviewers, who gave so generously of their time and experience in making suggestions for improving this book. Particular thanks go to series editor Jim Whitesell, who cast his eagle eye over the numerous structures and contributed to many stimulating discussions. Thanks also to John DiCesare and Hilton Weiss, and to John Murdzek, who meticulously annotated the entire manuscript both before and after revisions. Any comments regarding errors or suggestions for improvements in future editions will be most welcome.

Finally, my warmest thanks go to my husband, Dan, and to my children, Michael, Sarah, and Jeremy. Their loyal support, unflagging patience, and bizarre sense of humor bolster my spirits daily and shortened the long hours involved in preparing the manuscript.

Philippa Solomon

The ability to write feasible reaction mechanisms in organic chemistry depends on the extent of the individual's preparation. This book assumes the knowledge obtained in a one-year undergraduate course. A course based on this book is suitable for advanced undergraduates or beginning graduate students in chemistry. It can also be used as a supplementary text for a first-year course in organic chemistry.

Because detailed answers are given to all problems, the book also can be used as a tutorial and a review of many important organic reaction mechanisms and concepts. The answers are located conveniently at the end of each chapter. Examples of unlikely mechanistic steps have been drawn from my experience in teaching a course for beginning graduate students. As a result, the book clears up many aspects that are confusing to students. The most benefit will be obtained from the book if an intense effort is made to solve the problem before looking at the answer. It is often helpful to work on a problem in several different blocks of time.

The first chapter, a review of fundamental principles, reflects some of the deficiencies in knowledge often noted in students with the background cited above. The second chapter discusses some helpful techniques that can be utilized in considering possible mechanisms for reactions that may be found in the literature or during the course of laboratory research. The remaining chapters describe several of the common types of organic reactions and their mechanisms and propose mechanisms for a variety of reactions reported in the literature. The book does not cover all types of reactions. Nonetheless, anyone who works all the problems will gain insights that should facilitate the writing of reasonable mechanisms for many organic reactions.

Literature sources for most of the problems are provided. The papers cited do not always supply an answer to the problem but put the problem into a larger context. The answers to problems and examples often consider more than one possible mechanism. Pros and cons for each mechanism are provided. In order to emphasize the fact that frequently more than one reasonable pathway to a product may be written, in some cases experimental evidence supporting a particular mechanism is introduced only at the end of consideration of the problem. It is hoped that this approach will encourage users of this book to consider more than one mechanistic pathway.

I acknowledge with deep gratitude the help of all the students who have taken the course upon which this book is based. Special thanks to Drs. David Kronenthal, Tae-Woo Kwon, and John Freilich and Professor Hilton Weiss for reading the manuscript and making extremely helpful suggestions. Many thanks to Dr. James Holden for his editing of the entire manuscript and to my editor, Nancy Olsen, for her constant encouragement.

Audrey Miller

Introduction—Molecular Structure and Reactivity

Reaction mechanisms offer us insights into how molecules react, enable us to manipulate the course of known reactions, aid us in predicting the course of known reactions using new substrates, and help us to develop new reactions and reagents. In order to understand and write reaction mechanisms, it is essential to have a detailed knowledge of the structures of the molecules involved and to be able to notate these structures unambiguously. In this chapter, we present a review of fundamental principles relating to molecular structure and of ways to convey structural information. A crucial aspect of structure from the mechanistic viewpoint is the distribution of electrons, so this chapter outlines how to analyze and notate electron distributions. Mastering the material in this chapter will provide you with the tools you need to propose reasonable mechanisms and to convey these mechanisms clearly to others.

I. HOW TO WRITE LEWIS STRUCTURES AND CALCULATE FORMAL CHARGES

The ability to construct Lewis structures is fundamental to writing or understanding organic reaction mechanisms. It is particularly important because lone pairs of electrons frequently are crucial to the mechanism but often are omitted from structures appearing in the chemical literature.

There are two methods commonly used to show Lewis structures. One shows all electrons as dots. The other shows all bonds (two shared electrons) as lines and all unshared electrons as dots.

A. Determining the Number of Bonds

Hint I.I   To facilitate the drawing of Lewis structures, estimate the number of bonds. For a stable structure with an even number of electrons, the number of bonds is given by the equation:

(Electron Demand − Electron Supply) / 2 = Number of Bonds

The electron demand is two for each hydrogen and eight for all other atoms usually considered in organic chemistry. (The tendency of most atoms to acquire eight valence electrons is known as the octet rule.) For elements in group IIIA (e.g., B, Al, Ga), the electron demand is six. Other exceptions are noted, as they arise, in examples and problems.

For neutral molecules, the contribution of each atom to the electron supply is the number of valence electrons of the neutral atom. (This is the same as the group number of the element when the periodic table is divided into eight groups.) For ions, the electron supply is decreased by one for each positive charge of a cation and is increased by one for each negative charge of an anion.

Use the estimated number of bonds to draw that number of two-electron bonds in your structure. This may involve drawing a number of double and triple bonds (see the following section).

B. Determining the Number of Rings and / or π Bonds (Degree of Unsaturation)

The total number of rings and/or π bonds can be calculated from the molecular formula, bearing in mind that in an acyclic saturated hydrocarbon the number of hydrogens is $2n + 2$, where n is the number of carbon atoms.

Each time a ring or π bond is formed, there will be two fewer hydrogens needed to complete the structure.

On the basis of the molecular formula, the degree of unsaturation for a hydrocarbon is calculated as $(2m + 2 - n) / 2$, where m is the number of carbons and n is the number of hydrogens. The number calculated is the number of rings and / or π bonds. For molecules containing heteroatoms, the degree of unsaturation can be calculated as follows:

Nitrogen: For each nitrogen atom, subtract 1 from n.
Halogens: For each halogen atom, add 1 to n.
Oxygen: Use the formula for hydrocarbons.

This method cannot be used for molecules in which there are atoms like sulfur and phosphorus whose valence shell can expand beyond eight.

Example 1.1. *Calculate the number of rings and/or π bonds corresponding to each of the following molecular formulas.*

a. $C_2H_2Cl_2Br_2$

There are a total of four halogen atoms. Using the formula $(2m + 2 - n)/2$, we calculate the degree of unsaturation to be $[2(2) + 2 - (2 + 4)]/2 = 0$.

b. C_2H_3N

There is one nitrogen atom, so the degree of unsaturation is $[2(2) + 2 - (3 - 1)] = 2$.

C. Drawing the Lewis Structure

Start by drawing the skeleton of the molecule, using the correct number of rings or π bonds, then attach hydrogen atoms to satisfy the remaining valences. For organic molecules, the carbon skeleton frequently is given in an abbreviated form.

Once the atoms and bonds have been placed, add lone pairs of electrons to give each atom a total of eight valence electrons. When this process is complete, there should be two electrons for hydrogen, six for B, Al, or Ga, and eight for all other atoms. The total number of valence electrons for each element in the final representation of a molecule is obtained by counting each electron around the element as one electron, even if the electron is shared with another atom. (This should not be confused with counting electrons for charges or formal charges; see Section 1.D.) The number of valence electrons around each atom equals the electron demand. Thus, when the number of valence electrons around each element equals the electron demand, the number of bonds will be as calculated in Hint 1.1.

Atoms of higher atomic number can expand the valence shell to more than eight electrons. These atoms include sulfur, phosphorus, and the halogens (except fluorine).

Hint 1.3   When drawing Lewis structures, make use of the following common structural features.

1. Hydrogen is always on the periphery because it forms only one covalent bond.

2. Carbon, nitrogen, and oxygen exhibit characteristic bonding patterns. In the examples that follow, the R groups may be hydrogen, alkyl, or aryl groups, or any combination of these. These substituents do not change the bonding pattern depicted.

(a) Carbon in neutral molecules usually has four bonds. The four bonds may all be σ bonds, or they may be various combinations of σ and π bonds (i.e., double and triple bonds).

$$R-\overset{\overset{\displaystyle R}{|}}{\underset{\underset{\displaystyle R}{|}}{C}}-R \quad \text{or} \quad R\,\overset{\overset{\displaystyle R}{\cdot\cdot}}{\underset{\underset{\displaystyle R}{\cdot\cdot}}{:C:}}\,R$$

$$R-C\equiv C-R \quad \text{or} \quad R:C:::C:R$$

There are exceptions to the rule that carbon has four bonds. These include CO, isonitriles (RNC), and carbenes (neutral carbon species with six valence electrons; see Chapter 4).

(b) Carbon with a single positive or negative charge has three bonds.

$$\underset{R}{\overset{\overset{\displaystyle R}{|}}{\underset{}{C^{+}}}}{\diagdown}R \quad \text{or} \quad R\,\overset{\overset{\displaystyle R}{}}{\underset{}{\cdot\overset{\cdot\cdot}{C}\cdot^{+}}}\,R \quad \text{or} \quad {}^{+}CR_3$$

$$R-\overset{\overset{\displaystyle R}{|}}{\underset{\underset{\displaystyle R}{|}}{C}\!:^{-}} \quad \text{or} \quad R\,\overset{\overset{\displaystyle R}{\cdot\cdot}}{\underset{\underset{\displaystyle R}{}}{:C:^{-}}} \quad \text{or} \quad {}^{-}CR_3$$

(c) Neutral nitrogen, with the exception of nitrenes (see Chapter 4), has three bonds and a lone pair.

$$R-\overset{\overset{\displaystyle R}{|}}{\underset{\underset{\displaystyle R}{|}}{N}}\!: \quad \text{or} \quad R\,\overset{\overset{\displaystyle R}{\cdot\cdot}}{\underset{\underset{\displaystyle R}{}}{:\overset{\cdot\cdot}{N}:}} \quad \text{or} \quad NR_3$$

(d) Positively charged nitrogen has four bonds and a positive charge; exceptions are nitrenium ions (see Chapter 4).

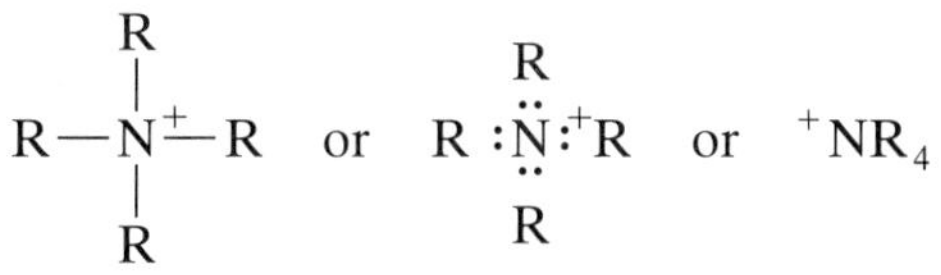

$$
R-\overset{\overset{\displaystyle R}{|}}{\underset{\underset{\displaystyle R}{|}}{N^{+}}}-R \quad \text{or} \quad R:\overset{\overset{\displaystyle R}{}}{\underset{\underset{\displaystyle R}{}}{\overset{..}{N}}}{}^{+}R \quad \text{or} \quad {}^{+}NR_4
$$

(e) Negatively charged nitrogen has two bonds and two lone pairs of electrons.

$$
R-\overset{..}{\underset{\underset{\displaystyle R}{|}}{N}}{}^{-} \quad \text{or} \quad R:\overset{..}{\underset{\underset{\displaystyle R}{}}{N}}{}^{-} \quad \text{or} \quad {}^{-}NR_2
$$

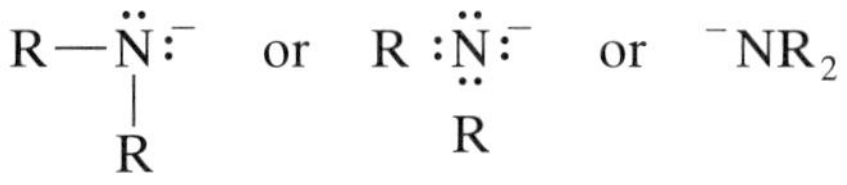

(f) Neutral oxygen has two bonds and two lone pairs of electrons.

$$
\underset{R\diagup \quad \diagdown R}{\overset{..}{\overset{\displaystyle O}{}}} \quad \text{or} \quad R:\overset{..}{\underset{\underset{\displaystyle R}{..}}{O}}: \quad \text{or} \quad R_2O
$$

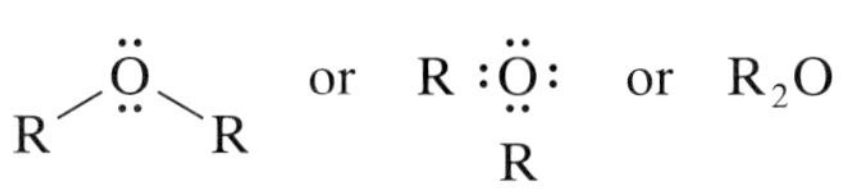

(g) Oxygen–oxygen bonds are uncommon; they are present only in peroxides, hydroperoxides, and diacyl peroxides (see Chapter 5). The formula, RCO_2R, implies the following structure:

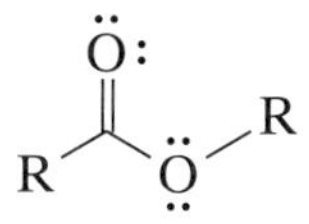

$$
\underset{R\diagup\quad\diagdown\overset{..}{\underset{..}{O}}\diagdown^{R}}{\overset{\overset{\displaystyle \overset{..}{O}:}{\|}}{C}}
$$

(h) Positive oxygen usually has three bonds and a lone pair of electrons; exceptions are the very unstable oxenium ions, which contain a single bond to oxygen and two lone pairs of electrons.

$$
\underset{R\diagup\quad\diagdown R}{\overset{+}{\overset{\overset{\displaystyle R}{|}}{O}}{}^{..}} \quad \text{or} \quad R:\overset{..}{\underset{\underset{\displaystyle R}{}}{O}}:{}^{+} \quad \text{or} \quad R_3O^{+}
$$

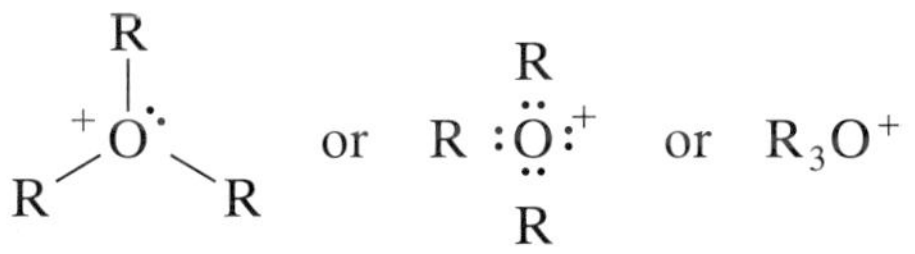

3. Sometimes a phosphorus or sulfur atom in a molecule is depicted with 10 electrons. Because phosphorus and sulfur have *d* orbitals, the outer shell can be expanded to accommodate more than eight electrons. If the shell, and therefore the demand, is expanded to 10 electrons, one more bond will be calculated by the equation used to calculate the number of bonds. See Example 1.5.

In the literature, a formula often is written to indicate the bonding skeleton for the molecule. This severely limits, often to just one, the number of possible structures that can be written.

Example 1.2. *The Lewis structure for acetaldehyde, CH_3CHO.*

	electron supply	electron demand
2C	8	16
4H	4	8
1O	6	8
	18	32

The estimated number of bonds is $(32 - 18)/2 = 7$.

The degree of unsaturation is determined by looking at the corresponding saturated hydrocarbon C_2H_6. Because the molecular formula for acetaldehyde is C_2H_6O and there are no nitrogen, phosphorus, or halogen atoms, the degree of unsaturation is $(6 - 4)/2 = 1$. There is either one double bond or one ring.

The notation CH_3CHO indicates that the molecule is a straight-chain compound with a methyl group, so we can write

$$CH_3 — C — O$$

We complete the structure by adding the remaining hydrogen atom and the remaining valence electrons to give

$$CH_3 — \overset{\displaystyle |}{\underset{\displaystyle H}{C}} = \overset{..}{\underset{..}{O}}$$

Note that if we had been given only the molecular formula C_2H_6O, a second structure could be drawn

$$\underset{H}{\overset{H}{>}} C \text{———} \overset{H}{\underset{H}{<}} C$$
$$\text{.O..}$$

A third possible structure differs from the first only in the position of the double bond and a hydrogen atom.

$$\underset{H}{\overset{H}{>}} C = C \overset{H}{\underset{.\overset{..}{O}{}^{.} — H}{<}}$$

This enol structure is unstable relative to acetaldehyde and is not isolable, although in solution small quantities exist in equilibrium with acetaldehyde.

D. Formal Charge

Even in neutral molecules, some of the atoms may have charges. Because the total charge of the molecule is zero, these charges are called formal charges to distinguish them from ionic charges.

Formal charges are important for two reasons. First, determining formal charges helps us pinpoint reactive sites within the molecule and can help us in choosing plausible mechanisms. Also, formal charges are helpful in determining the relative importance of resonance forms (see Section 5).

To calculate formal charges, use the completed Lewis structure and the following formula:

Hint 1.4

> Formal Charge = Number of Valence Shell Electrons
>
> − (Number of Unshared Electrons
>
> + Half the Number of Shared Electrons)

The formal charge is zero if the number of unshared electrons, plus the number of shared electrons divided by two, is equal to the number of valence shell electrons in the neutral atom (as ascertained from the group number in the periodic table). As the number of bonds formed by the atom increases, so does the formal charge. Thus, the formal charge of nitrogen in $(CH_3)_3 N$ is zero, but the formal charge on nitrogen in $(CH_3)_4 N^+$ is +1.

Note: *An atom always "owns" all unshared electrons.* This is true both when counting the number of electrons for determining formal charge and in determining the number of valence electrons. However, in determining *formal charge*, an atom "owns" *half of the bonding electrons*, whereas in determining the *number of valence electrons*, the atom "owns" *all the bonding electrons*.

Example 1.3. *Calculation of formal charge for the structures shown.*

(a)

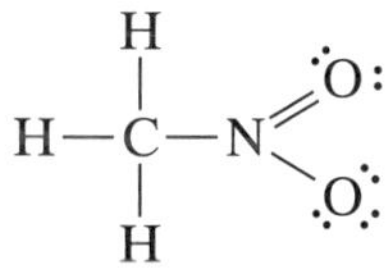

$$
\begin{array}{c}
\text{H} \\
| \\
\text{H—C—N} \\
| \\
\text{H}
\end{array}
\begin{array}{c}
\overset{\cdot\cdot}{\text{O}}: \\
\diagup \\
\diagdown \\
\overset{\cdot\cdot}{\underset{\cdot\cdot}{\text{O}}}:
\end{array}
$$

The formal charges are calculated as follows:

Hydrogen
    1 (no. of valence electrons) − 2/2 (2 bonding electrons divided by 2) = 0

Carbon
    4 (no. of valence electrons) − 8/2 (8 bonding electrons divided by 2) = 0

Nitrogen
    5 − 8/2 (8 bonding electrons) = +1

There are two different oxygen atoms:

Oxygen (double bonded)
   $6 - 4$ (unshared electrons) $- 4/2$ (4 bonding electrons) $= 0$

Oxygen (single bonded)
   $6 - 6$ (unshared electrons) $- 2/2$ (2 bonding electrons) $= -1$.

$$\text{(b)}\quad \begin{array}{ccccc} \text{H} & & \ddot{\text{:O:}} & & \text{H} \\ | & & | & & | \\ \text{H}-\text{C} & - & \text{S} & - & \text{C}-\text{H} \\ | & & | & & | \\ \text{H} & & \ddot{\text{:O:}} & & \text{H} \end{array}$$

The calculations for carbon and hydrogen are the same as those for part (a).

Formal charge for each oxygen:
   $6 - 6 - (2/2) = -1$

Formal charge for sulfur:
   $6 - 0 - (8/2) = +2$

Example 1.4. *Write possible Lewis structures for* C_2H_3N.

	electron supply	electron demand
3H	3	6
2C	8	16
1N	5	8
	16	30

The estimated number of bonds is $(30 - 16)/2 = 7$.

As calculated in Example 1.1, this molecular formula represents molecules that contain two rings and/or π bonds. However, because it requires a minimum of three atoms to make a ring, and since hydrogen cannot be part of a ring because each hydrogen forms only one bond, two rings are not possible. Thus, all structures with this formula will have either a ring and a π bond or two π bonds. Because no information is given on the order in which the carbons and nitrogen are bonded, all possible bonding arrangements must be considered.

Structures **1-1** through **1-9** depict some possibilities. The charges shown in the structures are formal charges. When charges are not shown, the formal charge is zero.

Structure **1-1** contains seven bonds using 14 of the 16 electrons of the electron supply. The remaining two electrons are supplied as a lone pair of electrons on the carbon, so that both carbons and the nitrogen have eight electrons around them. This structure is unusual because the right-hand carbon does not have four bonds to it. Nonetheless, isonitriles such as **1-1** (see Hint 1.3) are isolable. Structure **1-2** is a resonance form of **1-1**. (For a discussion of resonance forms, see Section 5.) Traditionally, **1-1** is written instead of **1-2**, because both carbons have an octet in **1-1**. Structures **1-3** and **1-4** represent resonance forms for another isomer. When all the atoms have an octet of electrons, a neutral structure like **1-3** is usually preferred to a charged form like **1-4** because the charge separation in **1-4** makes this a higher energy (and, therefore, less stable) species. Alternative forms with greater charge separation can be written for structures **1-5** to **1-9**. Because of the strain energy of three-membered rings and cumulated double bonds, **1-6** through **1-9** are expected to be quite unstable.

It is always a good idea to check your work by counting the number of electrons shown in the structure. The number of electrons you have drawn must be equal to the supply of electrons.

Example 1.5. *Write two possible Lewis structures for dimethyl sulfoxide,* *$(CH_3)_2SO$, and calculate formal charges for all atoms in each structure.*

	electron supply	electron demand
2C	8	16
6H	6	12
1S	6	8
1O	6	8
	26	44

According to Hint 1.1, the estimated number of bonds is $(44 - 26)/2 = 9$. Also, Hint 1.3 calculates 0 rings and/or π bonds. The way the formula is given indicates that both methyl groups are bonded to the sulfur, which is also bonded to oxygen. Drawing the skeleton gives the following:

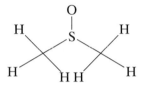

The nine bonds use up 18 electrons from the total supply of 26. Thus there are eight electrons (four lone pairs) to fill in. In order to have octets at sulfur and oxygen, three lone pairs are placed on oxygen and one lone pair on sulfur.

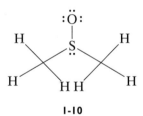

1-10

The formal charge on oxygen in **1-10** is −1. There are six unshared electrons and $2/2 = 1$ electron from the pair being shared. Thus, the number of electrons is seven, which is one more than the number of valence electrons for oxygen.

The formal charge on sulfur in **1-10** is +1. There are two unshared electrons and $6/2 = 3$ electrons from the pairs being shared. Thus, the number of electrons is five, which is one less than the number of valence electrons for sulfur.

All of the other atoms in **1-10** have a formal charge of 0.

There is another reasonable structure, **1-11**, for dimethyl sulfoxide, which corresponds to an expansion of the valence shell of sulfur to accommodate 10 electrons. Note that our calculation of electron demand counted eight electrons for sulfur. The 10-electron sulfur has an electron demand of 10 and leads to a total demand of 46 rather than 44 and the calculation of 10 bonds rather than 9 bonds. All atoms in this structure have zero formal charge.

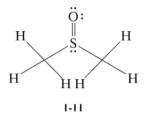

1-11

Hint 1.3 does not predict the π bond in this molecule, because the valence shell of sulfur has expanded beyond eight. Structures **1-10** and **1-11** correspond to different possible resonance forms for dimethyl sulfoxide (see Section 5), and each is a viable structure.

Why don't we usually write just one of these two possible structures for dimethyl sulfoxide, as we do for a carbonyl group? In the case of the carbonyl group, we represent the structure by a double bond between carbon and oxygen, as in structure **1-12**.

1-12   1-13

In structure **1-12**, both carbon and oxygen have an octet and neither carbon nor oxygen has a charge, whereas in structure **1-13**, carbon does not have an octet and both carbon and oxygen carry a charge. Taken together, these factors make structure **1-12** more stable and therefore more likely. Looking at the analogous structures for dimethyl sulfoxide, we see that in structure **1-10** both atoms have an octet and both are charged, whereas in structure **1-11**, sulfur has 10 valence electrons, but both sulfur and oxygen are neutral. Thus, neither **1-10** nor **1-11** is clearly favored, and the structure of dimethyl sulfoxide is best represented by a combination of structures **1-10** and **1-11**.

Note: No hydrogen atoms are shown in structures **1-12** and **1-13**. In representing organic molecules, it is assumed that the valence requirements of carbon are satisfied by hydrogen unless otherwise specified. Thus, in structures **1-12** and **1-13**, it is understood that there are six hydrogen atoms, three on each carbon.

When the electron supply is an odd number, the resulting unpaired electron will produce a radical; that is, the valence shell of one atom, other than hydrogen, will not be completed. This atom will have seven electrons instead of eight. Thus, if

Hint 1.5

you get a 1/2 when you calculate the number of bonds, that 1/2 represents a radical in the final structure.

PROBLEM 1.1   **Write Lewis structures for each of the following and show any formal charges.**

a. $CH_2=CHCHO$
b. $NO_2^+BF_4^-$
c. hexamethylphosphorous triamide, $[(CH_3)_2N]_3P$
d. $CH_3N(O)CH_3$
e. CH_3SOH (methylsulfenic acid)

Lewis structures for common functional groups are listed in Appendix A.

2. REPRESENTATIONS OF ORGANIC COMPOUNDS

As illustrated earlier, the bonds in organic structures are represented by lines. Often, some or all of the lone pairs of electrons are not represented in any way. The reader must fill them in when necessary. To organic chemists, the most important atoms that have lone pairs of electrons are those in groups VA, VIA, and VIIA of the periodic table: N, O, P, S, and the halogens. The lone pairs on these elements can be of critical concern when writing a reaction mechanism. Thus, you must remember that lone pairs may be present even if they are not shown in the structures as written. For example, the structure of anisole might be written with or without the lone pairs of electrons on oxygen:

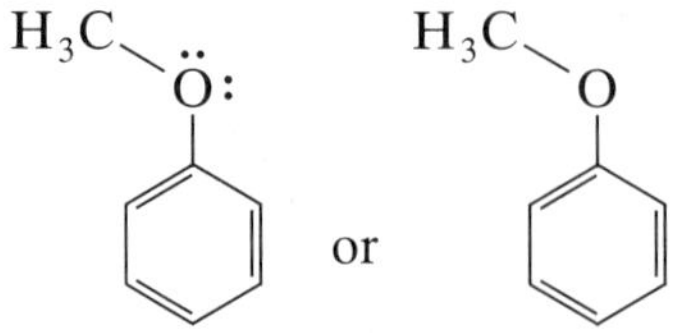

Other possible sources of confusion, as far as electron distribution is concerned, are ambiguities you may see in literature representations of cations and anions. The following illustrations show several representations of the resonance forms of the cation produced when anisole is protonated in the *para* position by concentrated sulfuric acid. There are three features to note in the first representation of the product, **1-14**: (i) Two lone pairs of electrons are shown on the oxygen. (ii) The positive charge shown on carbon means that the carbon has one less electron than neutral carbon. The number of electrons on carbon = (6 shared electrons)/2 = 3, whereas neutral carbon has four electrons. (iii) Both hydrogens are drawn in the *para* position to emphasize the fact that this carbon is now sp^3-hybridized. The second structure for the product, **1-15-1**, represents the overlap of one of the lone pairs of electrons on the oxygen with the rest of the π system. The electrons originally shown as a lone pair now are forming the second bond between oxygen and carbon. Representation **1-15-2**, the kind of structure commonly found in the literature, means exactly the same thing as **1-15-1**, but, *for simplicity, the lone pair on oxygen is not shown.*

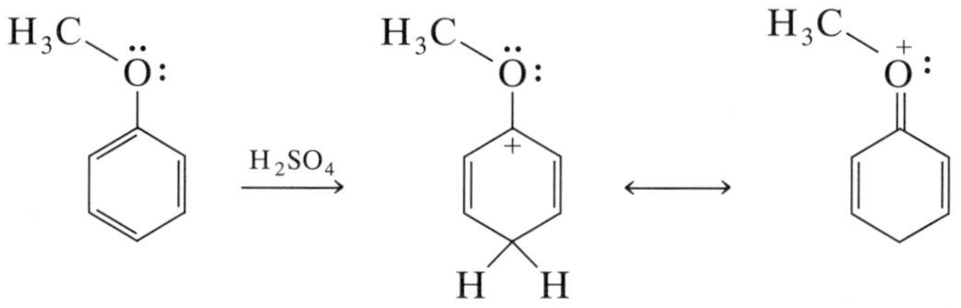

1-14   **1-15-1**

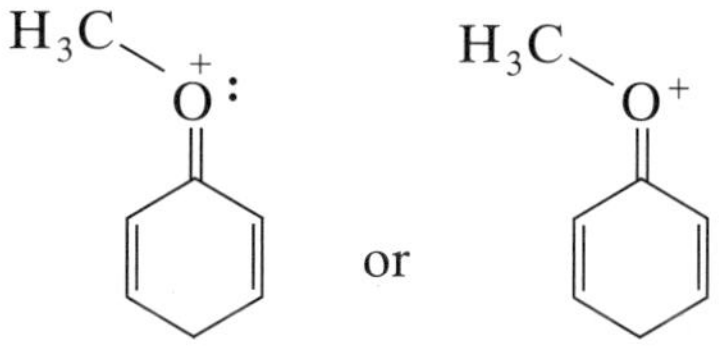

1-15-1   **1-15-2**

Similarly, there are several ways in which anions are represented. Sometimes a line represents a pair of electrons (as in bonds or lone pairs of electrons), sometimes a line represents a negative charge, and sometimes a line means both. The following structures represent the anion formed when a proton is removed from the oxygen of isopropyl alcohol.

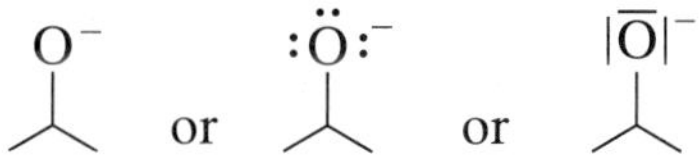

All three representations are equivalent, though the first two are the most commonly used.

A compilation of symbols used in chemical notation appears in Appendix B.

3. GEOMETRY AND HYBRIDIZATION

Particular geometries (spatial orientations of atoms in a molecule) can be related to particular bonding patterns in molecules. These bonding patterns led to the concept of hybridization, which was derived from a mathematical model of bonding. In that model, mathematical functions (wave functions) for the s and p orbitals in the outermost electron shell are combined in various ways (hybridized) to produce geometries close to those deduced from experiment.

The designations for hybrid orbitals in bonding atoms are derived from the designations of the atomic orbitals of the isolated atoms. For example, in a molecule with an sp^3 carbon atom, the carbon has four sp^3 hybrid orbitals, which are derived from the combination of the one s orbital and three p orbitals in the free carbon atom. The number of hybrid orbitals is always the same as the number of atomic orbitals used to form the hybrids. Thus, combination of one s and three p orbitals produces four sp^3 orbitals, one s and two p orbitals produce three sp^2 orbitals, and one s and one p orbital produce two sp orbitals.

We will be most concerned with the hybridization of the elements C, N, O, P, and S, because these are the atoms, besides hydrogen, that are encountered most commonly in organic compounds. If we exclude situations where P and S have expanded octets, it is relatively simple to predict the hybridization of any of these common atoms in a molecule. By counting X, the number of atoms, and E, the number of lone pairs surrounding the atoms C, N, O, P, and S, the hybridization and geometry about the central atom can be determined by applying the principle of valence shell electron pair repulsion to give the following:

1. If $X + E = 4$, the central atom will be sp^3-hybridized and the ideal geometry will have bond angles of 109.5°. *In exceptional cases, atoms with $X + E = 4$ may be sp^2-hybridized.* This occurs if sp^2 hybridization enables a lone pair to occupy a p orbital that overlaps a delocalized π electron system, as in the heteroatoms of structures **1-30** through **1-33** in Example 1.12.

2. If $X + E = 3$, the central atom will be sp^2-hybridized. There will be three hybrid orbitals and an unhybridized p orbital will remain. Again, the hybrid orbitals will be located as far apart as possible. This leads to an ideal geometry with 120° bond angles between the three coplanar hybrid orbitals and 90° between the hybrid orbitals and the remaining p orbital.

3. If $X + E = 2$, the central atom will be sp-hybridized and two unhybridized p orbitals will remain. The hybrid orbitals will be linear (180° bond angles), and the p orbitals will be perpendicular to the linear system and perpendicular to each other.

TABLE 1.1 Geometry and Hybridization in Carbon and Other Second Row Elements

Number of atoms + lone pairs $(X + E)$	Hybridization	Number of hybrid orbitals	Number of p orbitals	Geometry (bond angle)	Example^a
4	sp^3	4	0	Tetrahedral (109.5°)	
3	sp^2	3	1	Planar (120°)	
2	sp	2	2	Linear (180°)	

^{a}The geometry shown is predicted by VSEPR (valence shell electron pair repulsion) theory, in which orbitals containing valence electrons are directed so that the electrons are as far apart as possible. An asterisk indicates a hybridized atom.

The geometry and hybridization for compounds of second row elements are summarized in Table 1.1.

Example 1.6. *The hybridization and geometry of the carbon and oxygen atoms in 3-methyl-2-cyclohexen-1-one.*

The oxygen atom contains two lone pairs of electrons, so $X + E = 3$. Thus, oxygen is sp^2-hybridized. Two of the sp^2 orbitals are occupied by the lone pairs of electrons. The third sp^2 orbital overlaps with a sp^2-hybridized orbital at C-2 to form the C—O σ bond. The lone pairs and C-2 lie in a plane approximately 120° from one another. There is a *p* orbital perpendicular to this plane.

C-2 is *sp*²-hybridized. The three *sp*²-hybridized orbitals overlap with orbitals on O-1, C-3, and C-7 to form three σ bonds that lie in the same plane approximately 120° from each other. The *p* orbital, perpendicular to this plane, is parallel to the *p* orbital on O-1 so these *p* orbitals can overlap to produce the C—O π bond.

Carbons 3, 4, 5, and 8 are *sp*³-hybridized. (The presence of hydrogen atoms is assumed.) Bond angles are approximately 109.5°.

Carbons 6 and 7 are *sp*²-hybridized. They are doubly bonded by a σ bond, produced from hybrid orbitals, and a π bond produced from their *p* orbitals.

Because of the geometrical constraints imposed by the *sp*²-hybridized atoms, atoms 1, 2, 3, 5, 6, 7, and 8 all lie in the same plane.

PROBLEM 1.2   **Discuss the hybridization and geometry for each of the atoms in the following molecules or intermediates.**

   a. $CH_3—C≡N$
   b. $PhN=C=S$
   c. $(CH_3)_3P$

---

4. ELECTRONEGATIVITIES AND DIPOLES

Many organic reactions depend upon the interaction of a molecule that has a positive or fractional positive charge with a molecule that has a negative or fractional negative charge. In neutral organic molecules, the existence of a fractional charge can be inferred from the difference in electronegativity, if any, between the atoms at the ends of a bond. A useful scale of relative electronegativities was established by Linus Pauling. These values are given in Table 1.2, which also reflects the relative position of the elements in the periodic table. The larger the electronegativity value, the more electron-attracting the element. Thus, fluorine is the most electronegative element shown in the table.

Hint 1.6   Carbon, phosphorus, and iodine have about the same electronegativity. Within a row of the periodic table, electronegativity increases from left to right. Within a column of the periodic table, electronegativity increases from bottom to top.

TABLE 1.2   Relative Values for Electronegativities[a]

H	B	C	N	O	F
2.1	**2.0**	**2.5**	**3.0**	**3.5**	**4.0**
2.53	2.2	2.75	3.19	3.65	4.00
	Al	**Si**	**P**	**S**	**Cl**
	1.5	**1.8**	**2.1**	**2.5**	**3.0**
	1.71	2.14	2.5	3.96	3.48
					Br
					2.8
					3.22
					I
					2.5
					2.78

[a] The boldface values are those given by Linus Pauling in *The Nature of the Chemical Bond*, 3rd ed.; Cornell University Press; Ithaca, NY, 1960; p. 93. The second set of values is from Sanderson, R. T. *J. Am. Chem. Soc.* **1983**, *105*, 2259–2261; *J. Chem. Educ.* **1983**, *65*, 112.

From the relative electronegativities of the atoms, the relative fractional charges can be ascertained for bonds.

Example 1.7.   *Relative dipoles in some common bonds.*

$$\overset{\delta^+ \quad \delta^-}{C-O} \quad \overset{\delta^+ \quad \delta^-}{C=O} \quad \overset{\delta^+ \quad \delta^-}{C-Br} \quad \overset{\delta^+ \quad \delta^-}{C-N}$$

In all cases the more electronegative element has the fractional negative charge. There will be more fractional charge in the second structure than in the first, because the π electrons in the second structure are held less tightly by the atoms and thus are more mobile. The C—Br bond is expected to have a weaker dipole than the C—O single bond because bromine is not as electronegative as oxygen. You will notice that the situation is not so clear if we are comparing the polarity of the C—Br and C—N bonds. The Pauling scale would suggest that the C—N bond is more polar than the C—Br bond, whereas the Sanderson scale would predict the reverse. Thus, although attempts have been made to establish quantitative electronegativity scales, electronegativity is, at best, a qualitative guide to bond polarity.

PROBLEM 1.3   **Predict the direction of the dipole in the bonds highlighted in the following structures.**

    a. 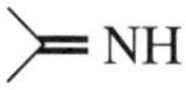 $\searrow\joinrel=NH$

    b. $Br-F$

    c. $CH_3-N(CH_3)_2$

    d. $CH_3-P(CH_3)_2$

---

5. RESONANCE STRUCTURES

When the distribution of valence electrons in a molecule cannot be represented adequately by a single Lewis structure, the structure can be approximated by a combination of Lewis structures that differ only in the placement of electrons. Lewis structures that differ only in the placement of electrons are called <u>resonance structures</u>. We use resonance structures to show the delocalization of electrons and to help predict the most likely electron distribution in a molecule.

A. Drawing Resonance Structures

A simple method for finding the resonance structures for a given compound or intermediate is to draw one of the resonance structures and then, by using arrows to show the movement of electrons, draw a new structure with a different electron distribution. This movement of electrons is formal only; that is, no such electron flow actually takes place in the molecule. The actual molecule is a hybrid of the resonance structures that incorporates some of the characteristics of each resonance structure. Thus, resonance structures themselves are not structures of actual molecules or intermediates but are a formality that helps to predict the electron distribution for the real structures. *Resonance structures, and only resonance structures, are separated by a double-headed arrow.*

Note: Chemists commonly use the following types of arrows:

- A double-headed arrow links two *resonance* structures

- Two half-headed arrows indicate an *equilibrium*

- A curved arrow indicates the movement of an *electron pair* in the direction of the arrowhead

- A curved half-headed arrow indicates the movement of a *single electron* in the direction of the arrowhead

A summary of symbols used in chemical notation appears in Appendix B.

Example 1.8. *Write the resonance structures for naphthalene.*

First, draw a structure, **1-16**, for naphthalene that shows alternating single and double bonds around the periphery. This is one of the resonance structures that contributes to the character of delocalized naphthalene, a resonance hybrid.

1-16   **1-17**

Each arrow drawn within **1-16** indicates movement of the π electron pair of a double bond to the location shown by the head of the arrow. This gives a new structure, **1-17**, which can then be manipulated in a similar manner to give a third structure, **1-18**.

1-17   **1-18**

Finally, when the forms have been figured out, they can be presented in the following manner:

How do you know that all possible resonance forms have been written? This is accomplished only by trial and error. If you keep pushing electrons around the naphthalene ring, you will continue to draw structures, but they will be identical to one of the three previously written.

What are some of the pitfalls of this method? If only a single electron pair in **1-17** is moved, **1-19** is obtained. However, this structure does not make sense. At the carbon labeled 1, there are five bonds to carbon; this is a carbon with 10 electrons. However, it is not possible to expand the valence shell of carbon. Similar rearrangement of other π bonds in either **1-16**, **1-17**, or **1-18** would lead to similarly nonsensical structures.

1-17   **1-19**

A second possibility would be to move the electrons of a double bond to just one of the terminal carbons; this leads to a structure like **1-20**. However, when more than one neutral resonance structure can be written, doubly charged resonance structures, like **1-20** and **1-21**, contribute an insignificant amount to the resonance hybrid and are usually not written.

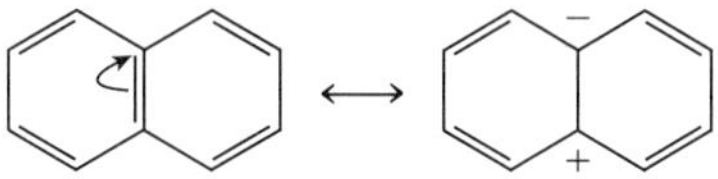

1-17   **1-20**

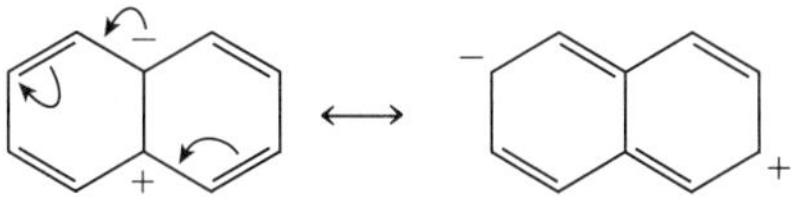

1-20   **1-21**

Example 1.9. *Write resonance forms for the intermediate in the nitration of anisole at the* **para** *position.*

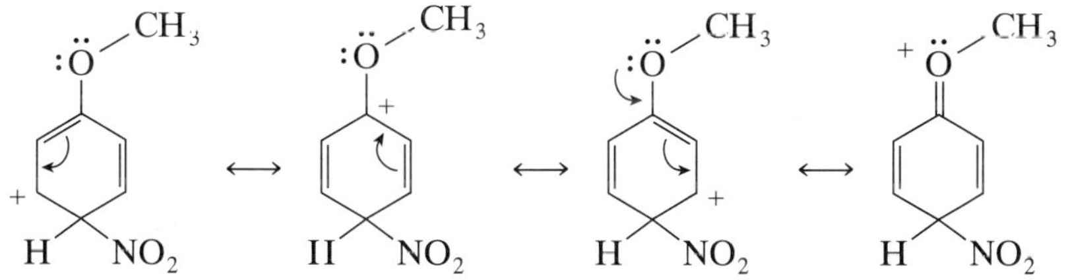

There are actually twice as many resonance forms as those shown because the nitro group is also capable of electron delocalization. Thus, for each resonance form written previously, two resonance forms can be substituted in which the nitro group's electron distribution has been written out as well:

Because the nitro group is attached to an sp^3-hybridized carbon, it is not conjugated with the electrons in the ring and is not important to their delocalization. Thus, if resonance forms were being written to rationalize the stability of the intermediate in the nitration of anisole, the detail in the nitro groups would not be important because it does not contribute to the stabilization of the carbocation intermediate.

Note: When an atom in a structure is shown with a negative charge, this is usually taken to imply the presence of an electron pair; often a pair of electrons and a negative sign are used interchangeably (see Section 2). This can sometimes be confusing. For example, the cyclooctatetraenyl anion (Problem 1.4e) can be depicted in several ways:

Notice that every representation shows two negative charges, so that we can be sure of the fact that this is a species with a double negative charge. In general, a negative charge sign drawn next to an atom indicates the presence of an electron pair associated with that atom. For some of the representations of the cyclooctatetraenyl anion, however, it is not clear how many electrons are in the π system (there is no ambiguity about the electrons in the σ bonds). In a situation like this, there is no hard and fast rule about how to count the electrons, based on the structural representation. To reach more solid ground, you need to know that cyclooctatetraene forms a relatively stable aromatic dianion with 10π electrons (see Section 6). Fortunately, these ambiguous situations are not common.

Draw resonance structures for each of the following.   **PROBLEM 1.4**

a. anthracene

b.

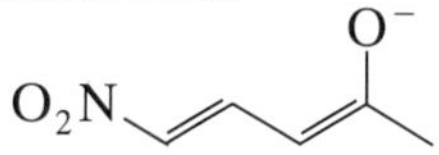

c. $PhCH_2^+$

d.

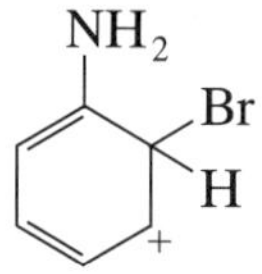

e.

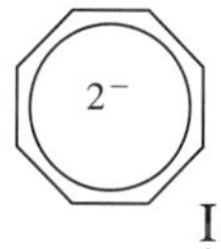

f.

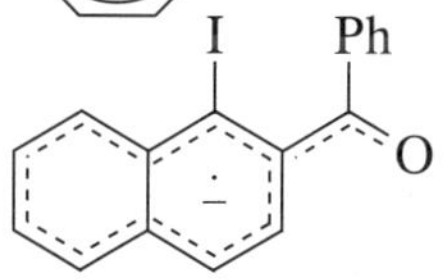

(This is the anion radical of 1-iodo-2-benzoylnaphthalene. The dashed lines indicate a delocalized π system. The symbol "Ph" stands for a phenyl group.)

Either *p*-dinitrobenzene or *m*-dinitrobenzene commonly is used as a radical trap in electron transfer reactions. The compound that forms the most stable radical anion is the better trap. Consider the radical anions formed when either of these starting materials adds an electron and predict which compound is commonly used.   **PROBLEM 1.5**

B. Rules for Resonance Structures

1. All of the electrons involved in delocalization are π electrons or, like lone pairs, they can readily be put into p orbitals.

2. Each of the electrons involved in delocalization must have some overlap with the other electrons. This means that if the orbitals are oriented at a 90° angle, there will be no overlap. The best overlap will occur when the orbitals are oriented at a 0° angle.

3. Each resonance structure must have the same number of π electrons. Count two for each π bond; only two electrons are counted for a triple bond because only one of the π bonds of a triple bond can overlap with the conjugated π system. Also, when a π system carries a charge, count two for an anion and zero for a positive charge.

4. The same number of electrons must be paired in each structure. Structures **1-22** and **1-23** are not resonance structures because they do not have the same number of paired electrons. In **1-22** there are two pairs of π electrons: a pair of electrons for the π bond and a pair of electrons for the anion. In **1-23** there is one pair of π electrons and two unpaired electrons (shown by the dots).

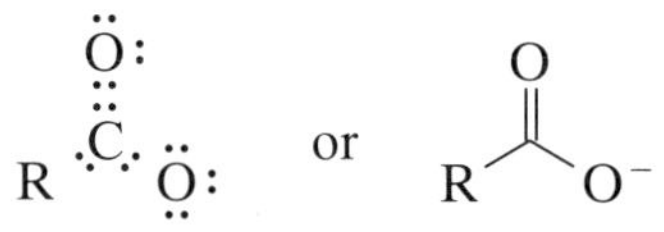

1-22

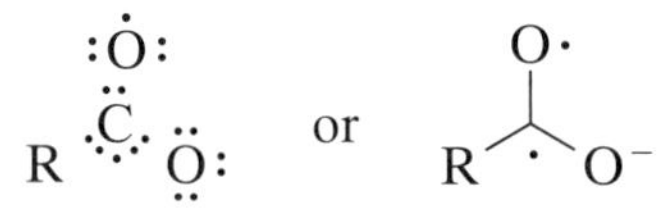

1-23

5. All resonance structures must have identical geometries. Otherwise they do not represent the same molecule. For example, the following structure (known as Dewar benzene) is not a resonance form of benzene because it is not planar and has two less π electrons. Because molecular geometry is linked to hybridization, it follows that hybridization also is unchanged for the atoms in resonance structures. (Note: If it is assumed that the central bond in this structure is a π bond, then it has the same number of electrons as benzene. However, in order for the p orbitals to overlap, the central carbon atoms would have to be much closer than they are in benzene, and this is yet another reason why Dewar benzene is an isolable compound rather than a resonance form of benzene.)

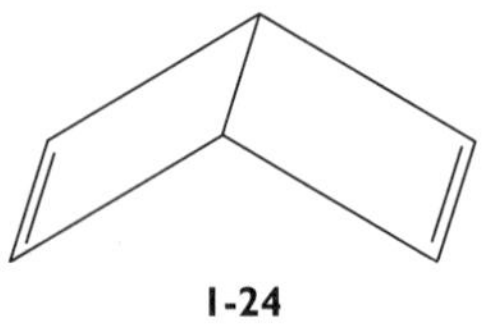

1-24

6. Resonance structures that depend on charge separation are of higher energy and do not contribute as significantly to the resonance hybrid as those structures that do not depend on charge separation.

much more important than

much more important than

7. Usually, resonance structures are more important when the negative charge is on the most electronegative atom and the positive charge is on the most electropositive atom.

contributes more than

In some cases, aromatic anions or cations are exceptions to this rule (see Section 6).

In the example that follows, **1-26** is less favorable than **1-25**, because the more electronegative atom in **1-26**, oxygen, is positive. In other words, although neither the positive carbon in **1-25** nor the positive oxygen in **1-26** has an octet, it is especially destabilizing when the much more electronegative oxygen bears the positive charge.

contributes more than

1-25   **1-26**

Electron stabilization is greatest when there are two or more structures of lowest energy.

The resonance hybrid is more stable than any of the contributing structures.

---

a. Do you think delocalization as shown by the following resonance structures is important? Explain why or why not.

PROBLEM 1.6

PROBLEM 1.6
continued

b. If the charges were negative instead of positive, would your answer be different? Explain.

PROBLEM 1.7   **Write Lewis structures for each of the following and show any formal charges. Also, draw all resonance forms for these species.**

a. CH_3NO_2
b. PhN_2^+
c. CH_3COCHN_2 (diazoacetone)
d. N_3CN
e. $CH_2{=}CHCH_2^- Li^+$

6. AROMATICITY AND ANTIAROMATICITY

A. Aromatic Carbocycles

Certain cyclic, completely conjugated, π systems show unusual stability. These systems are said to be aromatic. Hückel originally narrowly defined aromatic compounds as those completely conjugated, monocyclic carbon compounds that contain $(4n + 2)$ π electrons. In this designation, n can be 0, 1, 2, 3,..., so that systems that contain 2, 6, 10, 14, 18,... π electrons are aromatic. This criterion is known as Hückel's rule.

Example 1.10.   *Some aromatic compounds that strictly obey Hückel's rule.*

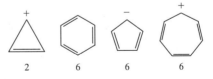

(Number of π electrons)

In these systems, each double bond contributes two electrons, each positive charge on carbon contributes none, and the negative charge (designation of an anion) contributes two electrons. If the first and last two structures did not have a charge on the singly bonded atom, they would not be aromatic because the π system could not be completely delocalized. That is, if the cyclopropenyl ring is depicted as uncharged, then there are two hydrogens on the carbon with no double bond. This carbon is then sp^3-hybridized and has no p orbital available to complete a delocalized system.

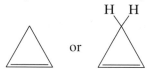

Hückel's rule has been expanded to cover fused polycyclic compounds because when these compounds have the requisite number of electrons, they also show unusual stability.

Example 1.11. *Some aromatic fused ring systems.*

　　　　1-27　　　　　　1-28　　　　　　　1-29

Structures **1-27** and **1-28**, which contain 10 conjugated π electrons, are examples of Hückel's rule with $n = 2$. Structure **1-29** obeys Hückel's rule with $n = 4$. In fused ring systems, only the electrons located on the periphery of the structure are counted when Hückel's rule is applied. (See Problem 1.8e for an example.)

B. Aromatic Heterocycles

Hückel's rule also can be extended to heterocycles, ring systems that incorporate noncarbon atoms. In heterocyclic compounds, a lone pair of electrons on the heterocyclic atom may be counted as part of the conjugated π system to attain the correct number for an aromatic system.

Example 1.12. *Some aromatic heterocycles.*

　　1-30　　1-31　　1-32　　　1-33　　　1-34

These examples illustrate how the lone pairs of electrons are considered in determining aromaticity. In each of the examples, the carbons and the heteroatoms are sp^2-hybridized, ensuring a planar system with a p orbital perpendicular to this plane at each position in the ring. In **1-30**, two electrons must be contributed by the nitrogen to give a total of 6π electrons. Thus, the lone pair of electrons would be in the p orbital on the nitrogen. In **1-31**, one of the lone pairs of electrons on the oxygen is also in a p orbital parallel with the rest of the π system in order to give a 6π electron aromatic system. Thus, the other lone pair on oxygen must be in an sp^2-hybridized orbital, which, by definition, is perpendicular to the conjugated π system and therefore cannot contribute to the number of electrons in the overlapping π system. The considerations concerning the sulfur in **1-32** are identical to those for oxygen in **1-31**, so this is a 6π electron system. Compound **1-33** is similar to **1-30** with overlap of the additional fused six-membered ring, making this an aromatic 10π electron system. In sharp contrast to the other examples, a totally conjugated 6π electron system is formed in **1-34** with the contribution of only one electron from the nitrogen. The lone pair of electrons will then be in an sp^2-hybridized orbital perpendicular to the aromatic 6π electron system. Thus, these two electrons are not part of the delocalized system. In conclusion, *the heteroatom is hybridized in such a way that one or two of its electrons can become part of an aromatic system.*

C. Antiaromaticity

Are conjugated systems that contain $4n$ (4, 8, 12, 16, . . .) π electrons also aromatic? These systems actually are destabilized by delocalization and are said to be antiaromatic.

Example 1.13. *Some antiaromatic systems.*

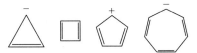

The first three examples contain 4π electrons, whereas the last one contains 8π electrons. All are highly unstable species. On the other hand, cyclooctatetraene, **1-35**, an 8π electron system, is much more stable than any of the preceding compounds. This is because the π electrons in cyclooctatetraene are not delocalized significantly: the eight-membered ring is bent into a tublike structure and adjacent π bonds are not parallel.

I-35

Classify each of the following compounds as aromatic, antiaromatic,   PROBLEM 1.8
or nonaromatic.

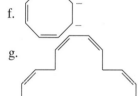

a.

b. CH_2^-

c.   NH N N

d.

e.

f. $^-$ $^-$

g.

---

7. TAUTOMERS AND EQUILIBRIUM

Tautomers are isomers that differ in the arrangement of single and double bonds and a small atom, usually hydrogen. Under appropriate reaction conditions, such isomers can equilibrate by a simple mechanism.

Equilibrium exists when there are equal rates for both the forward and reverse processes of a reaction. Equilibrium usually is designated by half-headed arrows shown for both the forward and reverse reactions. If it is known that one side of the equilibrium is favored, this may be indicated by a longer arrow pointing to the side that is favored.

Example 1.14.   *An acid–base equilibrium.*

$$CH_3CO_2H + H_2O \rightleftharpoons CH_3CO_2^- + H_3O^+$$

Example 1.15.   *Tautomeric equilibria of ketones.*

The keto and enol forms of aldehydes and ketones represent a common example of tautomerism. The tautomers interconvert by an equilibrium process that involves the transfer of a hydrogen atom from oxygen to carbon and back again.

Hint 1.7

To avoid confusing resonance structures and tautomers, use the following criteria:

1. *Tautomers* are readily converted *isomers*. As such they differ in the placement of a double bond and a hydrogen atom. The equilibration between the isomers is shown with a pair of half-headed arrows.

2. *Resonance structures* represent *different π bonding patterns*, not different chemical species. Different resonance structures are indicated by a double-headed arrow between them.

3. All resonance structures for a given species have identical σ bonding patterns (with a few unusual exceptions) and identical geometries. In tautomers, the σ bonding pattern differs.

Example 1.16. *Tautomerism versus resonance.*

Compounds **1-36** and **1-37** are tautomers; they are isomers and are in equilibrium with each other:

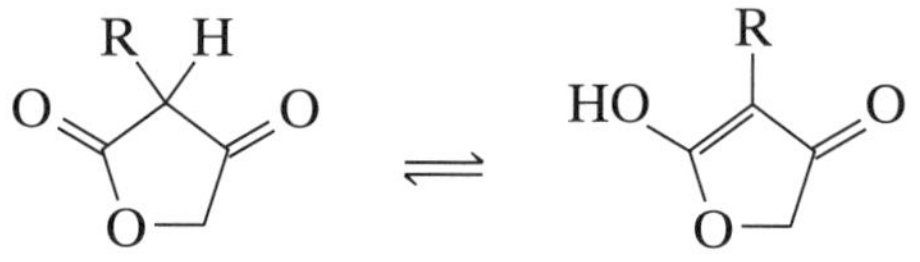

1-36	1-37

On the other hand, **1-38**, **1-39**, and **1-40** are resonance forms. The hybrid of these structures can be formed from **1-36** or **1-37** by removing the acidic proton.

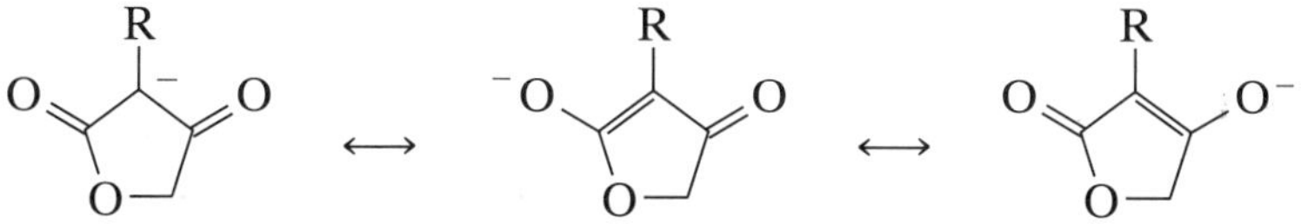

1-38	1-39	1-40

Note that **1-38**, **1-39**, and **1-40** have the same atoms attached at all positions, whereas the tautomers **1-36** and **1-37** differ in the position of a proton.

Write tautomeric structures for each of the following compounds. The number of tautomers you should write, in addition to the original structure, is shown in parentheses.   **PROBLEM 1.9**

a. 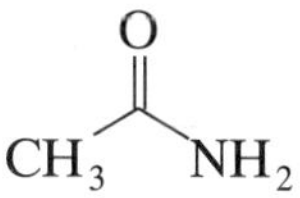 (2)

b. CH_3CHO (1)

c. $CH_3CH{=}CHCHO$ (1)

d. 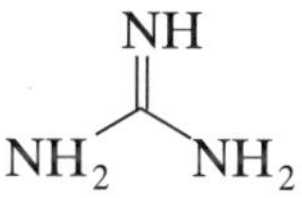 (2)

e. 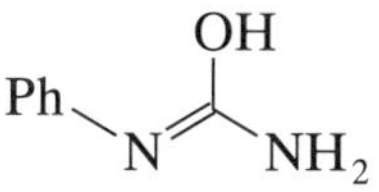 (5)

PROBLEM 1.10   **For each of the following sets of structures, indicate whether they are tautomers, resonance forms, or the same molecule.**

a.

b.

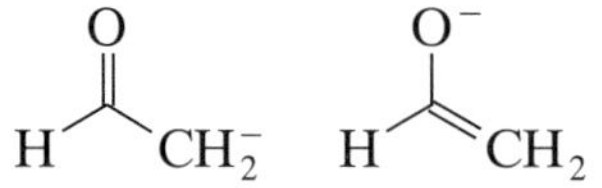

c.

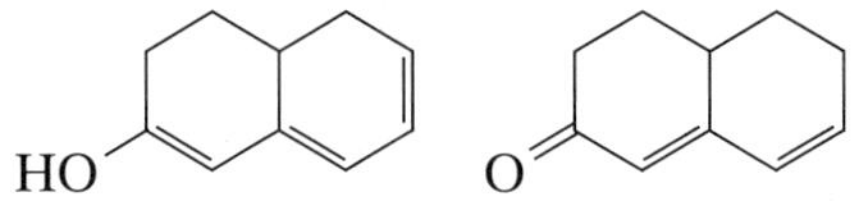

PROBLEM 1.11   **In a published paper, two structures were presented in the following manner and referred to as resonance forms. Are the structures shown actually resonance forms? If not, what are they and how can you correct the picture?**

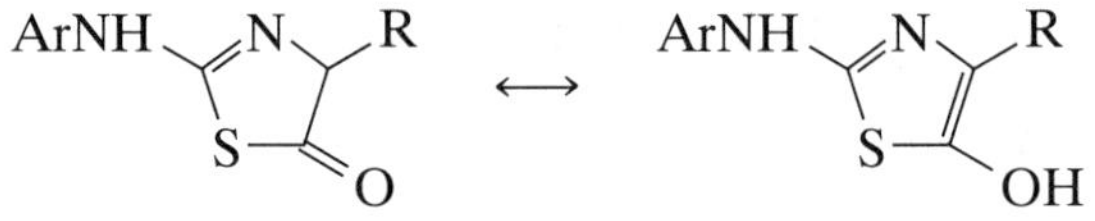

8. ACIDITY AND BASICITY

A Bronsted acid is a proton donor. A Bronsted base is a proton acceptor.

$$CH_3CO_2H + CH_3NH_2 \rightleftharpoons CH_3CO_2^- + CH_3\overset{+}{N}H_3$$
$$\text{acid} \qquad \text{base} \qquad \text{conjugate base} \quad \text{conjugate acid}$$

If this equation were reversed, the definitions would be similar:

$$CH_3\overset{+}{N}H_3 + CH_3CO_2^- \rightleftharpoons CH_3NH_2 + CH_3CO_2H$$
$$\text{acid} \qquad \text{base} \qquad \text{conjugate base} \quad \text{conjugate acid}$$

In each equation, the acids are the proton donors and the bases are proton acceptors.

There is an inverse relationship between the acidity of an acid and the basicity of its conjugate base. That is, the more acidic the acid, the weaker the basicity of the conjugate base and vice versa. For example, if the acid is very weak, like methane, the conjugate base, the methyl carbanion, is a very strong base. On the other hand, if the acid is very strong, like sulfuric acid, the conjugate base, the HSO_4^- ion, is a very weak base. Because of this reciprocal relationship between acidity and basicity, most references to acidity and basicity use a single scale of pK_a values, and relative basicities are obtained from the relative acidities of the conjugate acids.

Table 1.3 shows the approximate pK_a values for common functional groups. The lower the pK_a value, the more acidic the protonated species. Any conjugate acid in the table will protonate a species lying below it and will be protonated by acids listed above it. Thus, a halogen acid (pK_a of -10 to -8) will protonate any species listed in this table, whereas a carboxylic acid (pK_a of $4-5$) would be expected to protonate aliphatic amines (pK_a values of $9-11$), but not primary and secondary aromatic amines (pK_a values of -5 and 1, respectively).

Appendix C contains a more detailed list of pK_a values for a variety of acids. Especially at very high pK_a values, the numbers may be inaccurate because various approximations have to be made in measuring such values. This is often the reason why the literature contains different pK_a values for the same acid.

It is helpful to think of the pK_a as the pH of a solution in which the acid is 50% ionized.

Hint 1.8

Weak acids with a high pK_a require a high pH (strong alkali) to lose their protons. Acids with a pK_a greater than 15.7 cannot be deprotonated in aqueous solution; those with a pK_a less than 15.7 will be soluble (to some extent) in 5% NaOH solution. (The value 15.7 is used because we are considering the pK_a of water. The pH scale is based on the ion product of water, $[H^+][OH^-]$, which equals 10^{-14}. The pK_a value is based on an H_2O concentration of 55 mol/liter.) Bases whose conjugate acids have a pK_a less than 0 (i.e., negative) will not be protonated in aqueous solution because water, being a stronger base, is protonated preferentially. Bases whose conjugate acids have a positive pK_a are basic enough to dissolve (to some extent) in 5% HCl.

For each of the following pairs, indicate which is the strongest base. For b and d use resonance structures to rationalize the relative basicities. **PROBLEM 1.12**

a. $H\bar{C}{=}CH_2$, $\bar{C}{\equiv}CH$

b. $\bar{C}H_2CON(Et)_2$, $[(CH_3)_2CH]_2\bar{N}$

c. CH_3O^-, $(CH_3)_3CO^-$

d. *p*-nitrophenolate, *m*-nitrophenolate

TABLE I.3   Typical Acidities of Common Organic and Inorganic Substances

Group	Conjugate acid[a]	Conjugate base[a]	Typical pK_a
Halogen acids	HX (X=I, Br, Cl)	X⁻	−10 to −8
Nitrile	$R-C\equiv \overset{+}{N}-H$	$RC\equiv N$	−10
Aldehyde, ketone	$R-\overset{\overset{+}{O}-H}{\underset{\|}{C}}-R'$	$R-\overset{O}{\underset{\|}{C}}-R'$	−7
Thiol, sulfide	$R-\overset{H}{\underset{+}{S}}-R'$	$R-S-R'$	−7 to −6
Phenol, aromatic ether	Ph—$\overset{H}{\underset{+}{O}}$—R′	Ph—O—R	−7 to −6
Ester, acid	$R-\overset{\overset{+}{O}-H}{\underset{\|}{C}}-OR'$	$R-\overset{\overset{+}{O}-H}{\underset{\|}{C}}-OR'$	−7 to −6
Sulfonic acid[b,c]	$R-\overset{O}{\underset{O}{S}}-O-H$	$R-\overset{O}{\underset{O}{S}}-O^-$	−5 to −2
Alcohol, ether	$R-\overset{H}{\underset{+}{O}}-R'$	$R-O-R'$	−3 to −2
H_2O	$H_2\overset{+}{O}-H$	H_2O	−1.7
Amide	$R-\overset{\overset{+}{O}-H}{\underset{\|}{C}}-NR_2$	$R-\overset{O}{\underset{\|}{C}}-NR_2$	−1
Carboxylic acid[b,c]	$R-\overset{O}{\underset{\|}{C}}-O-H$	$R-\overset{O}{\underset{\|}{C}}-O^-$	3–5
Aromatic amine	Ph—$\overset{H}{\underset{+}{N}}$—R′₂	Ph—NR′₂ [d]	4–5
Pyridine	Py—$\overset{\pm}{N}$—H	Py—N [d]	5
Alkylamine	$R-\overset{H}{\underset{+}{N}}-R'_2$	$R-N-R'_2$ [d]	9–11
Phenol[c]	Ph—O—H	Ph—O⁻	9–11
Thiol[c]	R—S—H	R—S⁻	9–11

(continues)

TABLE 1.3 (continued)

Group	Conjugate acid[a]	Conjugate base[a]	Typical pK_a
Sulfonamide[c]	R—S—N—H (with O double bonds above and below S, H on N)	R—S—N—H (with O double bonds, negative charge on N)	10
Nitro[c]	R—CH—NO$_2$ (H on CH)	R—CH—NO$_2$ (negative on CH)	10
Amide	R—C—N—R (O double bond, H on N)	R—C—N—R (O double bond, negative on N)	15–17
Alcohol	R—O—H	R—O$^-$	15–19
Aldehyde, ketone	R—C—C—R' (H and R' on first C, O double bond on second C)	R'—C—C—R' (negative and R on first C, O double bond)	17–20
Ester	R—C—C—OR (H and R' on first C, O double bond)	R—C—C—OR (negative and R' on first C, O double bond)	20–25
Nitrile	R—C—C≡N (H and R' on C)	R—C—C≡N (negative and R' on C)	25
Alkyne	R—C≡C—H	R—C≡C$^-$	25
Amine	R$_2$N—H	R$_2$N$^-$	35–40
Alkane	R$_3$C—H	R$_3$C$^-$	50–60

[a]Abbreviations: R = alkyl; R' = alkyl or H.
[b]Dissolves in 5% NaHCO$_3$ solution.
[c]Dissolves in 5% NaOH solution.
[d]Dissolves in 5% HCl solution.

For each of the following compounds, indicate which proton is more   **PROBLEM 1.13**
likely to be removed when the compound is treated with base and
rationalize your answer. Assume that equilibria are involved in each
case.

a. $CH_3COCH_2COCH_3$

b. $H_2NCH_2CH_2OH$

c.

d. HN

Example 1.17. *Calculating the equilibrium constant for an acid–base reaction in order to predict whether the reaction is likely to proceed.*

$$Br^- + EtOH \rightleftharpoons HBr + EtO^-$$

For this reaction, the equilibrium constant is

$$K_a = \frac{[HBr][EtO^-]}{[Br^-][EtOH]}$$

Using the values listed in Appendix C, this equilibrium constant can be calculated by an appropriate combination of the equilibrium constant for the ionization of ethanol and the equilibrium constant for the ionization of HBr. The equilibrium constant for the ionization of ethanol is

$$10^{-15.9} = \frac{[EtO^-][H^+]}{[EtOH]}$$

The equilibrium constant for the ionization of HBr is

$$10^9 = \frac{[Br^-][H^+]}{[HBr]}$$

If the equilibrium constant for ethanol ionization is divided by that for HBr ionization, the equilibrium constant for the reaction of bromide ion with ethanol is obtained:

$$K_a = \frac{10^{-15.9}}{10^9} = \frac{[HBr][EtO^-][H^+]}{[H^+][Br^-][EtOH]} = \frac{[HBr][EtO^-]}{[Br^-][EtOH]}$$

K_a equals $10^{-24.9}$. Thus, it can be concluded that bromide ion does not react with ethanol.

PROBLEM 1.14

Using the values listed in Appendix C, calculate the equilibrium constants for each of the following reactions and predict which direction, forward or reverse, is favored.

a. $CH_3CO_2Et + EtO^- \rightleftharpoons {}^-CH_2CO_2Et + EtOH$

b.

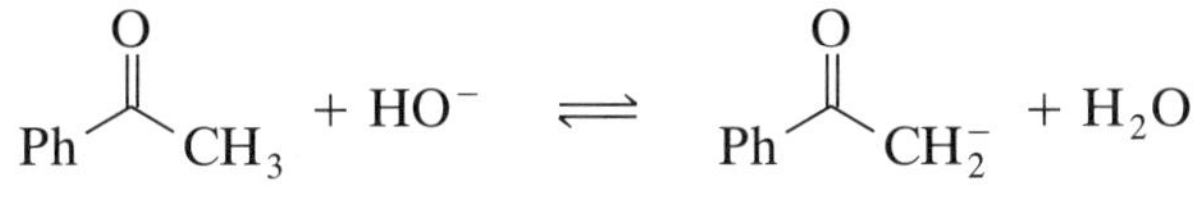

$+ HO^- \rightleftharpoons$

$+ H_2O$

c. $Ph_3CH + {}^-N(i\text{-}Pr)_2 \rightleftharpoons Ph_3C^- + HN(i\text{-}Pr)_2$

Note: i-Pr = -CH(CH$_3$)$_2$.

9. NUCLEOPHILES AND ELECTROPHILES

Nucleophiles are reactive species that seek an electron-poor center. They have an atom with a negative or partial negative charge, and this atom is referred to as the nucleophilic atom. Reacting species that have an electron-poor center are called electrophiles. These electron-poor centers usually have a positive or partial positive charge, but electron-deficient species can also be neutral (radicals and carbenes, see Chapter 5). Table 1.4 lists common nucleophiles and Table 1.5 common electrophiles.

Reactivity in reactions involving nucleophiles depends on several factors, including the nature of the nucleophile, the substrate, and the solvent.

A. Nucleophilicity

Nucleophilicity measures the ability of a nucleophile to react at an electron-deficient center. It should not be confused with basicity, although often there are parallels between the two. Whereas nucleophilicity considers the *reactivity* (i.e., the rate of reaction) of an electron-rich species *at an electron-deficient center* (usually carbon), basicity is a measure of the *position of equilibrium* in reaction with a *proton*.

Table 1.6 shows nucleophiles ranked by one measure of nucleophilicity. These nucleophilicities are based on the relative reactivities of the nucleophile and water with methyl bromide at 25°C. The nucleophilicity n is calculated according to the Swain–Scott equation:

$$\log \frac{k}{k_o} = sn$$

TABLE 1.4   Common Nucleophiles[a]

F^-, Cl^-, Br^-, I^-

$ROH, RO^-, RO-O^-$

$$[R-\overset{\overset{O}{\|}}{C}\overset{*}{=}O]^-, \quad [Ar-\overset{\overset{O}{\|}}{C}\overset{*}{=}O]^-$$

$RSH, RS^-, ArSH, ArS^-$

$RSR', [S-S]^{2-}$ (disulfide)

$RNH_2, RR'NH, RR'R''N, ArNH_2,$ etc.

H_2NNH_2 (hydrazine)

$[N=N=N]^-$ (azide)

$[N=C=O]^-$ (isocyanate)

N^- (phthalimidate)

R_3P (phosphine)

$(RO)_3P$ (phosphite)

$LiAlH_4, NaBH_4, LiEt_3BH$

$RMgX, ArMgX$

$RCuLi$

$RC{\equiv}C^-$

[a]An asterisk indicates a nucleophilic atom.

where k is the rate constant for reactions with the nucleophile, k_o is the rate constant for reaction when water is the nucleophile, $s = 1.00$ (for methyl bromide as substrate), and n is the relative nucleophilicity. The larger the n value, the greater the nucleophilicity. Thus, Table 1.6 shows that the thiosulfate ion ($S_2O_3^{2-}$, $n = 6.4$) is more nucleophilic than iodide (I^-, $n = 5.0$).

B. Substrate

The structure of the substrate influences the rate of reaction with a nucleophile, and this effect is reflected in the s values defined in the previous

TABLE 1.5 Common Electrophiles[a]

$\overset{*}{Z}nCl_2$, $\overset{*}{A}lCl_3$, $\overset{*}{B}F_3$

$\overset{*}{P}Br_3$

$\overset{*}{S}OCl_2$

$-\overset{|}{\underset{|}{\overset{*}{C}}}-X$ (X = Cl, Br, I)

$-\overset{|}{\underset{|}{\overset{*}{C}}}-O-SO_2R$ (R = p-tolyl, CF_3, CH_3)

$-\overset{X}{\underset{|}{\overset{|}{\overset{*}{C}}}}-CO_2H$, $-\overset{X}{\underset{|}{\overset{|}{C}}}-CO_2R$, $-\overset{X}{\underset{|}{\overset{|}{C}}}-\overset{O}{\overset{||}{C}}-R$

$-\overset{O}{\overset{||}{C}}-OR$, $-\overset{O}{\overset{||}{C}}R$

$\overset{*}{C}H_2N_2^{b}$ (diazomethane)

$H-\overset{*}{O}-\overset{*}{O}-H$ (hydrogen peroxide)

$\underset{*}{\overset{}{C}}\overset{O}{\diagup\diagdown}\underset{*}{C}$ (epoxide)

$\overset{*}{N}=O^{b}$ (generated from HNO_2)

---

[a]An asterisk indicates an electrophilic atom.
[b]To react as an electrophile, CH_2N_2 (diazomethane) must first be protonated to form the methyl diazonium cation:

$$H_2\overset{-}{C}=\overset{+}{N}=N \xrightarrow{H^+} H_3C-\overset{*}{N}\equiv N.$$

section. For example, methyl bromide and chloroacetate both have s values of 1.00, so they react at the same rate. On the other hand, iodoacetate, with an s value of 1.33, reacts faster than methyl bromide, whereas benzyl chloride, with an s value of 0.87, reacts more slowly.

C. Solvent

Nucleophilicity is often solvent-dependent, but the relationship is a complex one and depends on a number of different factors. Ritchie and co-workers have measured solvent-dependent relative nucleophilicities, N_+, in various

TABLE 1.6   Nucleophilicities toward Carbon[a]

Nucleophile	n	Nucleophile	n
$S_2O_3^{2-}$	6.4	pyridine	3.6
SH^-	5.1	Br^-	3.5
CN^-	5.1	PhO^-	3.5
SO_3^{2-}	5.1	$CH_3CO_2^-$	2.7
I^-	5.0	Cl^-	2.7
$PhNH_2$	4.5	$HOCH_2CO_2^-$	2.5
SCN^-	4.4	SO_4^{2-}	2.5
OH^-	4.2	$ClCH_2CO_2^-$	2.2
$(NH_2)_2CS$	4.1	F^-	2.0
N_3^-	4.0	NO_3^-	1.0
HCO_3^-	3.8	H_2O	0.0
$H_2PO_4^-$	3.9		

[a] From Wells, P. R. *Chem. Rev.* **1963**, *63*, 171–219.

solvents, using the equation

$$\log \frac{k_n}{k_{H_2O}} = N_+$$

where k_n is the rate constant for reaction of a cation with a nucleophile in a given solvent, and k_{H_2O} is the rate constant for reaction of the same cation with water in water. Some N_+ values are given in Table 1.7. Note that nucleophilicity is greater in dipolar aprotic solvents like dimethyl sulfoxide and dimethylformamide than in protic solvents like water or alcohols. For this reason, dimethyl sulfoxide is often used as a solvent for carrying out nucleophilic substitutions.

Sometimes relative nucleophilicities change in going from a protic to an aprotic solvent. For example, the relative nucleophilicities of the halide ions in water are $I^- > Br^- > Cl^-$, whereas in dimethylformamide, the nucleophilicities are reversed, i.e., $Cl^- > Br^- > I^-$.

TABLE 1.7   Relative Nucleophilicities in Common Solvents[a]

Nucleophile (solvent)	N_+	Nucleophile (solvent)	N_+
H_2O (H_2O)	0.0	PhS^- (CH_3OH)	10.51
CH_3OH (CH_3OH)	1.18	PhS^- [$(CH_3)_2SO$]	12.83
CN^- (H_2O)	3.67	N_3^- (H_2O)	7.6
CN^- (CH_3OH)	5.94	N_3^- (CH_3OH)	8.85
CN^- [$(CH_3)_2SO$]	8.60	N_3^- [$(CH_3)_2SO$]	10.07
CN^- [$(CH_3)_2NCHO$]	9.33		

[a] From Ritchie, C. D. *J. Am. Chem. Soc.* **1975**, *97*, 1170–1179.

In each of the following reactions, label the electrophilic or nucleophilic **PROBLEM 1.15**
center in each reactant. In a, b, and c, different parts of acetophenone
are behaving as either a nucleophile or an electrophile.

a. 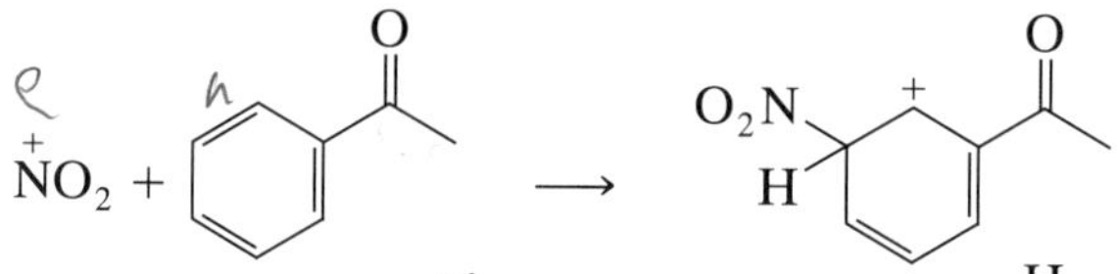 $\overset{+}{N}O_2$ +

b. 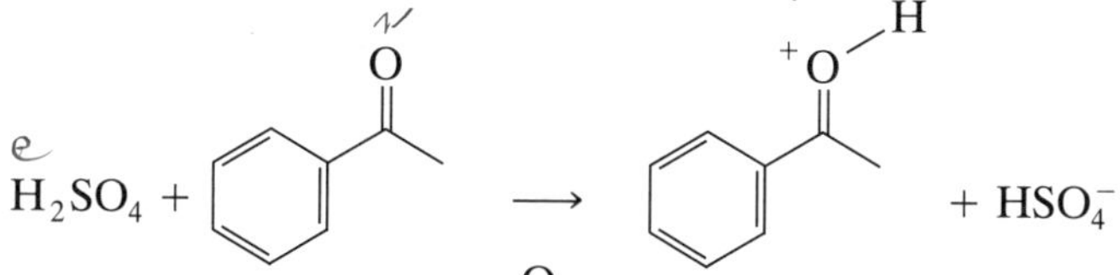 H_2SO_4 +

c. 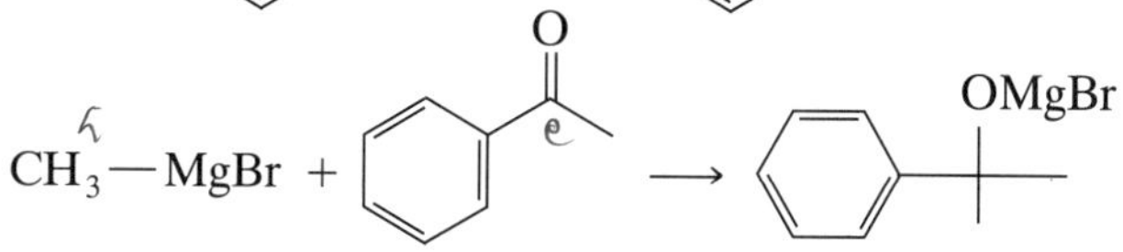 CH_3-MgBr +

d.

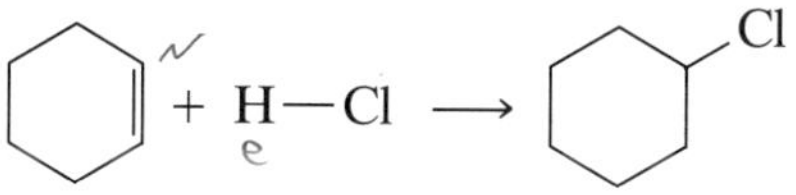

ANSWERS TO PROBLEMS

For all parts of this problem, the overall carbon skeleton is given. There-   Problem 1.1
fore, a good approach is to draw the skeleton of the molecule with single
bonds and fill in extra bonds, if necessary, to complete the octet of atoms
other than hydrogen.

a. Electron supply – $(3 \times 4)(C) + (1 \times 6)(O) + (4 \times 1)(H) = 22$. Electron
demand $= (3 \times 8)(C) + (1 \times 8)(O) + (4 \times 2)(H) = 40$. Estimate of bonds
$= (40 - 22)/2 = 9$. This leaves two bonds left over after all atoms are
joined by single bonds; thus, there is a double bond (as shown) between
the CH_2 and CH groups and a double bond between the second CH and
the oxygen to give the following skeleton:

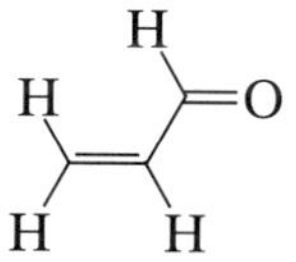

Calculation of the number of rings and/or π bonds also shows that two π
bonds are present. The molecular formula is C_3H_4O. The number of
hydrogens for a saturated hydrocarbon is $(2 \times 3) + 2 = 8$. There are
$(8 - 4)/2 = 2$ rings and/or π bonds.

Eighteen electrons are used in making the nine bonds in the molecule. There are four electrons left (from the original supply of 22); these can be used to complete the octet on oxygen by giving it two lone pairs of electrons.

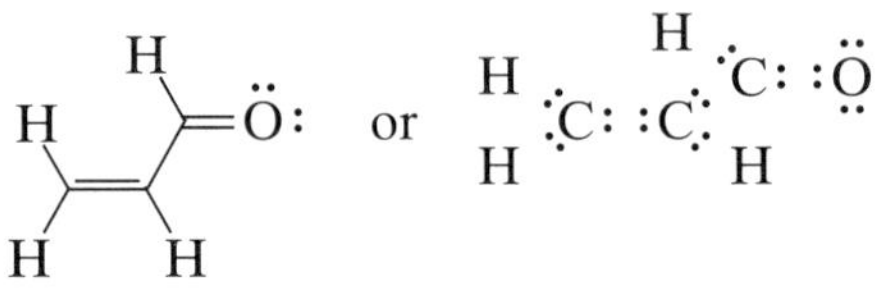

b. The way the charges are written in the structure indicates that the NO_2^+ and BF_4^- are separate entities. For NO_2^+, electron demand is $3 \times 8 = 24$. Electron supply is $(1 \times 5)(N) + (2 \times 6)(O) - 1$ (positive charge) $= 16$. Estimate of bonds $= (24 - 16)/2 = 4$. A reasonable structure can be drawn with two double bonds. The remaining eight electrons (electron supply $-$ electrons used for bonds) are used to complete the octets of the two oxygens to give the structure shown.

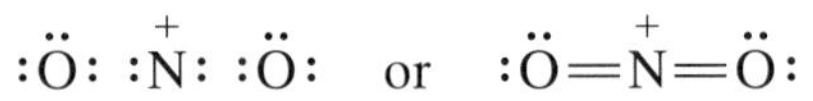

$$:\overset{..}{\underset{..}{O}}: :\overset{+}{N}: :\overset{..}{\underset{..}{O}}: \quad \text{or} \quad :\overset{..}{O}=\overset{+}{N}=\overset{..}{O}:$$

An alternative structure is much less stable. It has two adjacent positively charged atoms, both of which are electronegative.

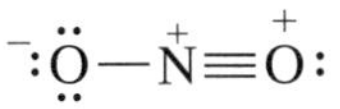

$$^-:\overset{..}{\underset{..}{O}}-\overset{+}{N}\equiv\overset{+}{O}:$$

By using an electron demand of six for B, we calculate that the electron demand for the atoms in BF_4^- is $(4 \times 8)(F) + 6(B) = 38$. The number of bonds predicted is then $(38 - 32)/2 = 3$. Because there are five atoms, clearly we need at least four bonds to join them together. In a neutral molecule, the electron demand for boron is six. In this negative ion, a reasonable move would be to assign an octet of electrons to B so that we can form four bonds and join all the atoms. These bonds use up eight electrons, and we can use the remaining 24 electrons to complete the octets around the four F atoms.

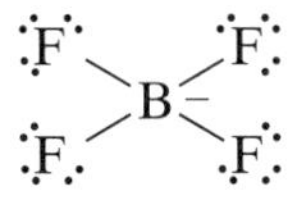

Although the rules presented in Chapter 1 for obtaining Lewis structures work most of the time, there are situations in which these approximations are not applicable. Boron is an element that displays unusual bonding properties in a number of its compounds.

c. Electron supply $= (6 \times 4)(C) + (18 \times 1)(H) + (3 \times 5)(N) + (1 \times 5)(P) = 62$. Electron demand $= (10 \times 8) + (18 \times 2) = 116$. The number of bonds predicted is $(116 - 62)/2 = 27$. After the bonds are placed in the molecule, there are 8 electrons left $[62 - 2(27)]$, which are used to complete the octets of phosphorus and nitrogen. All charges are zero

d. Electron supply = $(2 \times 4)(C) + (6 \times 1)(H) + (1 \times 5)(N) + (1 \times 6)(O) =$ 25. Electron demand = $(4 \times 8) + (6 \times 2) = 44$. Estimate of bonds = $(44 - 25) = 9.5$. This means that there must be nine bonds and an odd electron somewhere in the molecule. There are two likely structures (**1-41** and **1-42**) that agree with the carbon skeleton given:

1-41   1-42

Because of the chemistry of the nitroxide functional group, the odd electron is usually written on oxygen, as in **1-41**, even though this means that the most electronegative element in the molecule is the one that does not have an octet. None of the atoms in **1-41** have a formal charge. In **1-42** there is a +1 charge on nitrogen and a −1 charge on oxygen. If you did not have an odd electron in your structure, either you had too many electrons, as in structure **1-43**, or you had too few electrons.

1-43

Structure **1-43** has an overall negative charge, which is placed on the oxygen. Because this compound is not neutral, it would not be stable in the absence of a counterion. If you did not put enough electrons in the structure, it will be positively charged. You should always count the number of electrons in your finished Lewis structure to make sure that you have exactly the number calculated for the electron supply.

Unlike the sulfur in dimethyl sulfoxide (Example 1.5), nitrogen and oxygen do not have *d* orbitals. Thus, it is not possible for this compound to have a double bond between nitrogen and oxygen, because this will leave either nitrogen or oxygen with more than eight electrons.

e. Electron supply = $(1 \times 4)(C) + (4 \times 1)(H) + (2 \times 6)(S$ and $O) = 20$. Electron demand = $24(C, S, O) + 8(H) = 32$. Estimate of bonds = $(32 - 20)/2 = 6$. The molecular formula, CH_4OS, suggests there are no rings or π bonds when sulfur has an octet.

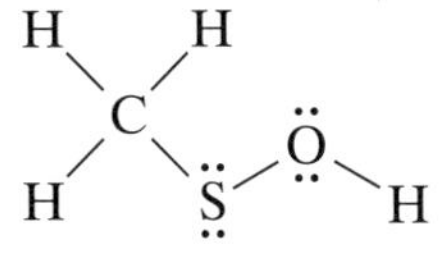

1-44

After insertion of the bonds, the eight electrons left over are used to complete the octets of sulfur and oxygen by filling in two lone pairs on each atom. There are no charges on any of the atoms in **1-44**.

The way the structure is given in the problem indicates that the hydrogen is on oxygen. Another possible structure, which used to be written for sulfenic acids, puts the acidic hydrogen on sulfur rather than oxygen:

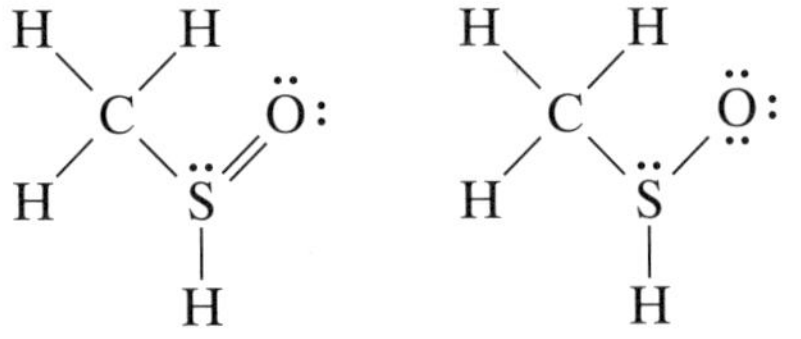

1-45   **1-46**

In **1-45** there are seven bonds because the sulfur has expanded its valence shell to accommodate 10 electrons. This structure has no formal charges. In **1-46** the sulfur has a charge of +1 and the oxygen has a charge of −1.

It is always a good idea to write a complete Lewis structure before deciding on hybridization so that lone pairs are not missed.

Problem 1.2
continued

a. The CH_3 carbon is sp^3-hybridized. The carbon is attached to four distinct groups, three hydrogens and one carbon, so $X = 4$. The other carbon and the nitrogen are both sp-hybridized because $X + E = 2$. The sp-hybridized carbon is attached to the CH_3 carbon and to the nitrogen. The nitrogen is attached to carbon and also has a lone pair. The C—C—N skeleton is linear because of the central sp-hybridized atom. The H—C—H bond angles of the methyl group are approximately 109°, the tetrahedral bond angle.

b. All of the carbons in the phenyl ring are sp^2-hybridized. This ring is also conjugated with the external π system, so the entire molecule is planar. The nitrogen and sulfur are sp^2-hybridized. Nitrogen is bonded to the phenyl group and the external carbon and also contains a lone pair, so $X + E = 3$; the sulfur is attached to carbon and has two lone pairs, so $X + E = 3$. The external carbon is sp-hybridized. It is attached to only two atoms, the nitrogen and the sulfur, so $X + E = 2$. The N=C=S group is linear. Note, however, that the Ph—N=C bond angle is 120°.

c. For phosphorus $X + E = 4$ (three ligands and one lone pair); thus, the carbons and the phosphorus all are sp^3-hybridized with approximate tetrahedral bond angles.

Problem 1.3

a. $\overset{\delta^+}{\diagup}\!\!=\!\!\overset{\delta^-}{NH}$ b. $\overset{\delta^+}{Br}\!-\!\overset{\delta^-}{F}$ c. $\overset{\delta^+}{H_3C}\!-\!\overset{\delta^-}{N(CH_3)_2}$

d. $\overset{\delta^-}{H_3C}\!-\!\overset{\delta^+}{P(CH_3)_2}$

Problem 1.4

a.

I-47

Is the resonance form **1-48** different from those written above? The answer is no.

I-48

This represents exactly the same electron distribution as **1-47**. Although the double bond between the two right-hand rings is written to the right of the σ bond in **1-47** and to the left in **1-48**, the double bond is still between the same two carbons. Thus, **1-47** and **1-48** are identical.

Would the following redistribution of electrons lead to a correct resonance form? The answer is no.

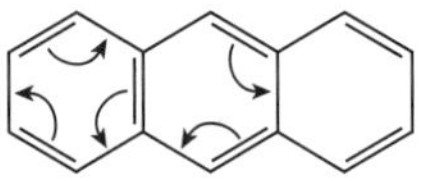

If a double bond is written at every single bond to which an arrow is drawn, we obtain structure **1-49**:

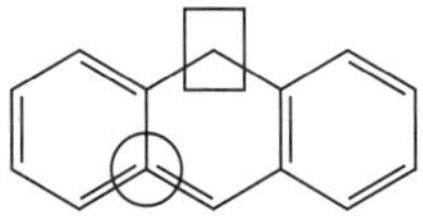

1-49

In **1-49** the circled carbon has 10 electrons, which is impossible for carbon. Also, the boxed carbon does not have enough electrons. (This carbon is attached only to one hydrogen, not two.) Thus, this is an impossible structure.

b. The nitro group is conjugated with the π system. Hence, there are two forms for the nitro group whenever the second negative charge is located on an atom external to the nitro group.

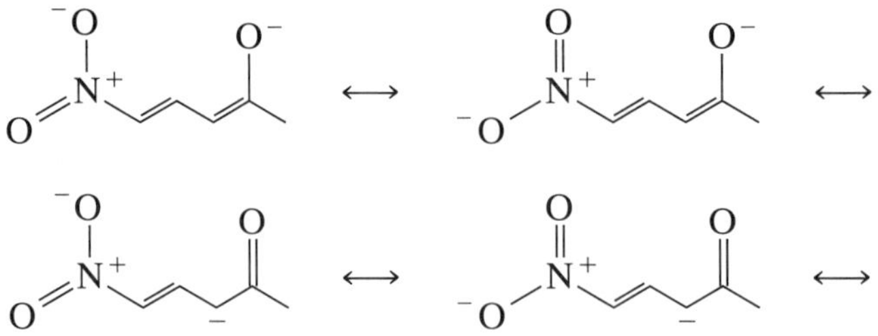

Problem 1.4
continued

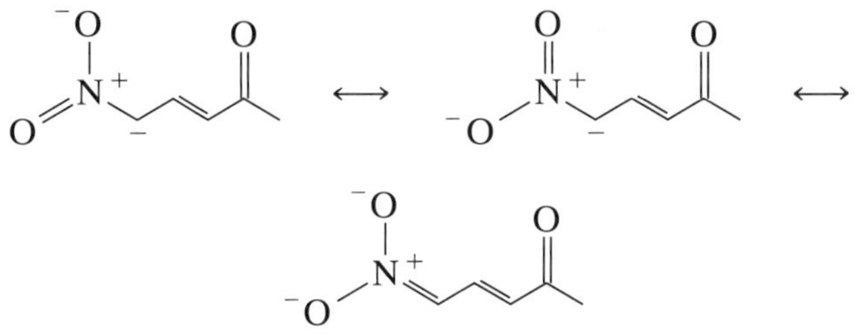

A structure like **1-50** would be incorrect because it represents a different chemical species; in fact, this compound has one less hydrogen at the terminal carbon and an additional negative charge. Remember that all resonance forms must have identical geometries.

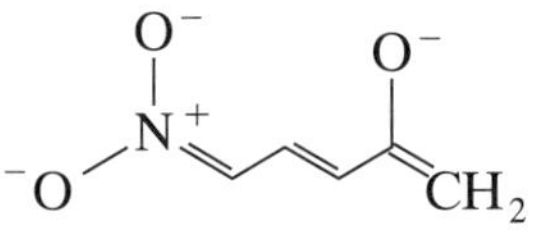

I-50

c.

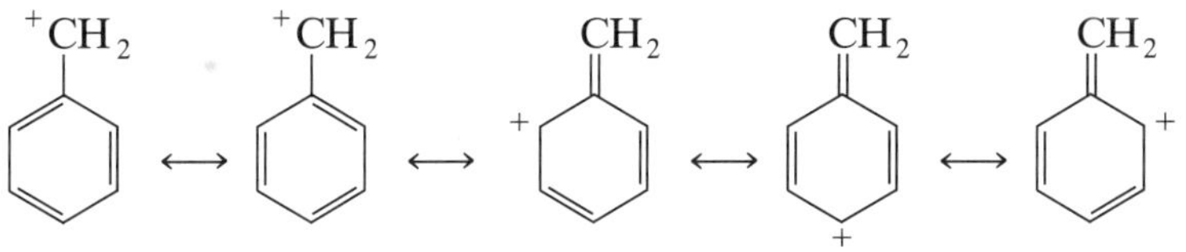

d.

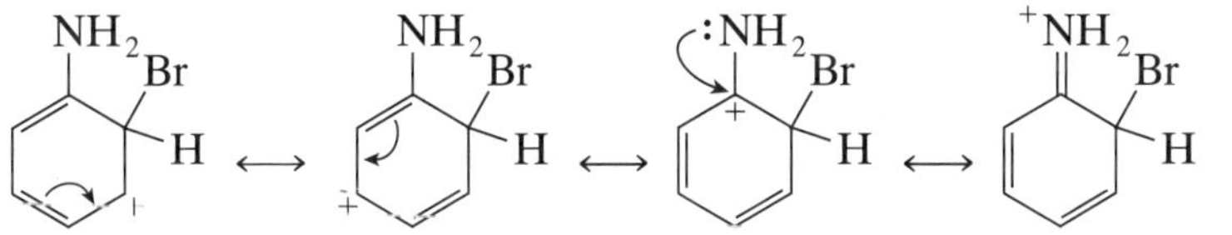

e. A circle drawn inside a ring means that there is a totally conjugated π system (single and double bonds alternate around the ring). The dinegative charge means that this conjugated system has two extra electrons, that is, two electrons have been added to the neutral, totally conjugated system (cyclooctatetraene). To draw the first structure, start with the parent compound. From it we can derive the highly unstable resonance structure with a negative and positive charge. This structure makes very little contribution to the actual structure of cyclooctatetraene and is shown purely for the purpose of "electron bookkeeping." We now add two electrons to form the dianion.

Problem 1.4
continued

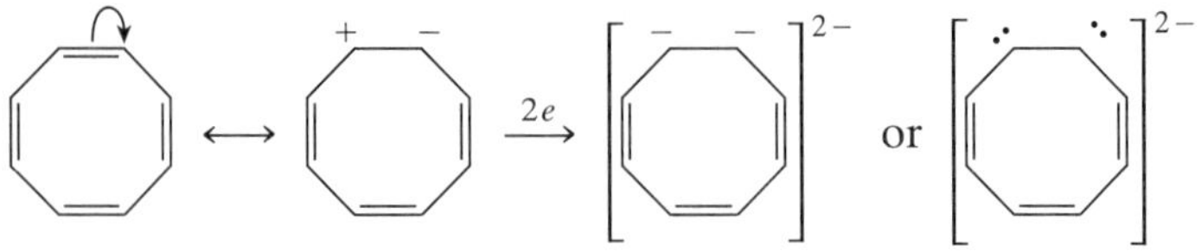

The resonance forms drawn here are a few of the many possible forms that can be drawn:

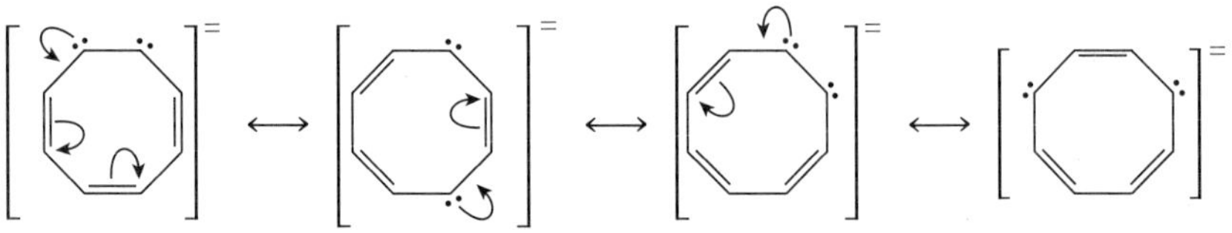

f. To figure out the first structure, from which the others can be derived by "electron pushing," examine the structure one step at a time. The dotted lines indicate that the ring system of the parent compound consists of alternating single and double bonds. From the parent compound, the radical anion can be derived by the addition of an electron (e.g., from a reducing agent such as sodium). Draw one of the resonance structures for the conjugated π system of the parent compound and then add one electron to form the radical anion. (Note the use of a half-headed arrow to show the movement of the single electron.) Use of a stepwise approach helps to make sure that you have the correct number of double bonds, charges, and electrons.

There are many resonance forms possible. Just a few are shown here. Notice that the aromatic Ph group also presents opportunities for drawing additional resonance structures.

Problem 1.4
continued

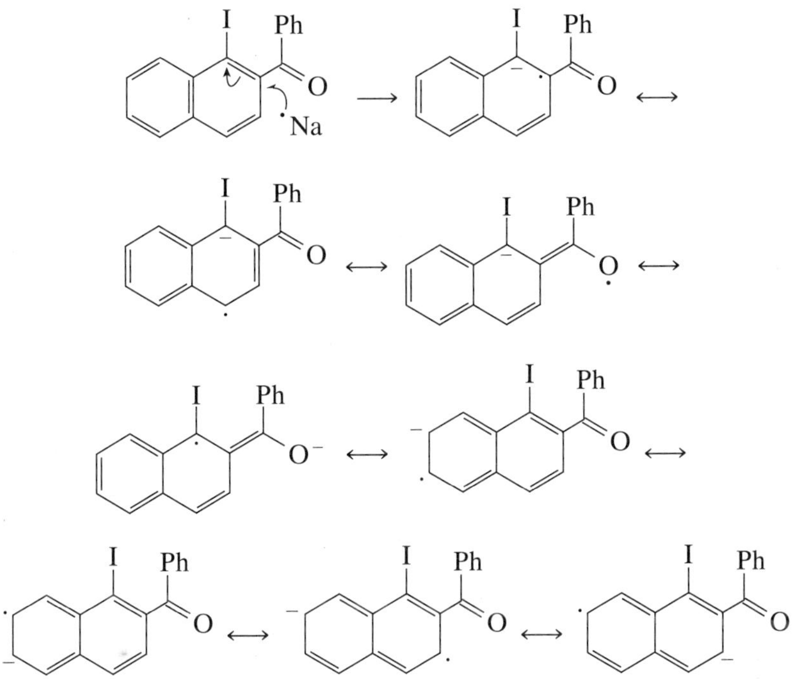

Once again, start by drawing the structure of the conjugated parent compound and then add an electron to form the radical anion. Only a few of the possible resonance forms are drawn. Nonetheless, it can be seen that the anion and radical can be delocalized onto both nitro groups simultaneously for the *p*-dinitrobenzene, and this leads to more possible resonance forms. Because there is more delocalization in the intermediate from the *para* compound, it should be easier to transfer an electron to *p*-dinitrobenzene, and hence, it should be a better radical trap.

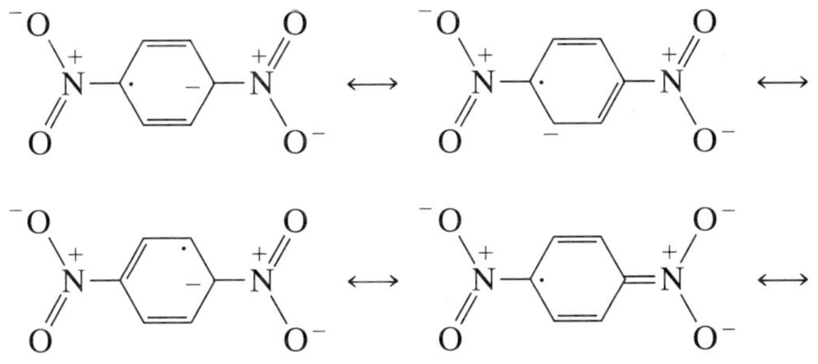

Problem 1.5

Problem 1.5
continued

Problem 1.6

a. The resonance forms with a positive charge on nitrogen or oxygen do not contribute significantly to stabilization of the positive charge, because the nitrogen and oxygen do not have an octet of electrons when they bear the positive charge. The fact that oxygen or nitrogen does not have an octet is a critical point. When oxygen and nitrogen do have an octet, they can contribute significantly to a resonance hybrid in which they are positively charged. For example, the 1-methoxyethyl carbocation is stabilized significantly by the adjacent oxygen:

b. When nitrogen and oxygen are negatively charged, they have an octet of electrons. A negative charge on these atoms, both more electronegative than carbon, is significantly stabilizing. Thus, this delocalization would be very important.

Problem 1.7

a. Electron supply = $(1 \times 4)(C) + (3 \times 1)(H) + (1 \times 5)(N) + (2 \times 6)(O) = 24$. Electron demand = $(4 \times 8) + (3 \times 2) = 38$. Estimate of bonds = $(38 - 24)/2 = 7$. By using Hint 1.2, one ring or one π bond is expected. Two possible skeleton structures are **1-51** and **1-52**.

1-51　　　　　　1-52

The combination of a highly strained three-membered ring and a weak O—O bond makes **1-52** unlikely. When you see the grouping NO_2 in a molecule, you can assume that it is a nitro group. Filling in the five lone pairs of electrons gives **1-53** or the resonance structure **1-54**.

Problem 1.7
continued

1-53   **1-54**

In a neutral molecule, the sum of the formal charges must equal 0. Thus, if oxygen is given a formal charge of −1, some other atom in the molecule must have a formal charge of +1. Calculation shows that this atom is the nitrogen.

Structure **1-55** is incorrect because the nitrogen has ten electrons and nitrogen does not have orbitals available to expand its valence shell.

1-55

b. Electron supply = $(6 \times 4)(C) + (5 \times 1)(H) + (2 \times 5)(N) - 1 = 38$. Electron demand = $(8 \times 8) + (5 \times 2) = 74$. Estimate of bonds = $(74 - 38)/2 = 18$. This cation does not have an odd electron. One electron was removed from the electron supply because the species is positively charged. The resonance structures that contribute most to the resonance hybrid are **1-56** and **1-57**.

1-56   **1-57**

Resonance forms **1-58** and **1-59** contribute much less than **1-56** and **1-57** to the resonance hybrid because the positively charged nitrogen in these forms does not have an octet of electrons.

1-58   1-59

c. Electron supply = $(3 \times 4)(C) + (4 \times 1)(H) + (2 \times 5)(N) + (1 \times 6)(O) =$ 32. Electron demand = $(6 \times 8) + (4 \times 2) = 56$. Estimate of total bonds = $(56 - 32)/2 = 12$. Several possible structures are shown. All are resonance forms.

A resonance structure that will not contribute as significantly to the hybrid is **1-60**.

1-60

In this structure, the positively charged nitrogen does not have an octet of electrons, so that it will be less stable than the previous structures in which positively charged nitrogen does have an octet. Notice also that the number of bonds in **1-60** is one less than that estimated.

You might have interpreted the molecule's skeleton to be that shown in resonance structures **1-61** through **1-63**. However, none of these structures is expected to be stable. Structure **1-61** has a large number of charged atoms, **1-62** has an uncharged carbon with only six valence electrons (a carbene, see Chapter 4), and **1-63** has a carbene carbon and a nitrogen that has only six electrons. The first answers drawn for this problem are much more likely.

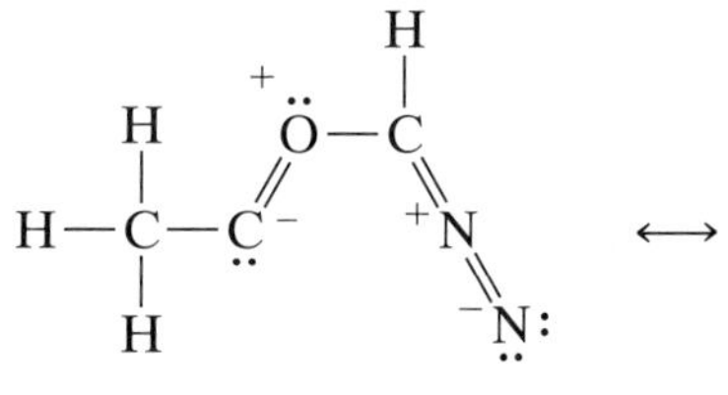

Problem 1.7
continued

1-61

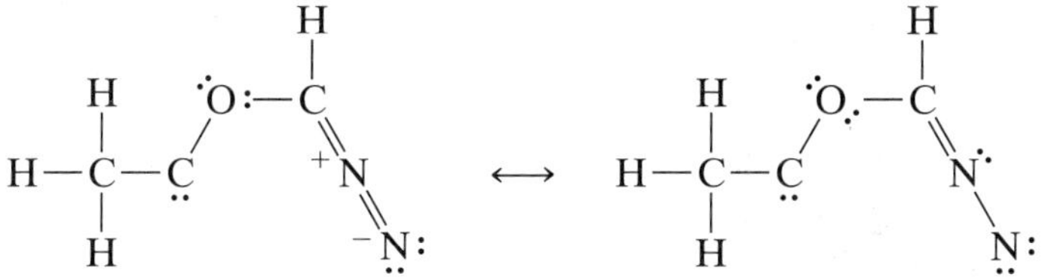

1-62 **1-63**

d. Electron supply = $(1 \times 4)(C) + (4 \times 5)(N) = 24$. Electron demand = $5 \times 8 = 40$. Estimate of bonds = $(40 - 24)/2 = 8$. Because they are resonance structures, all of the following are correct:

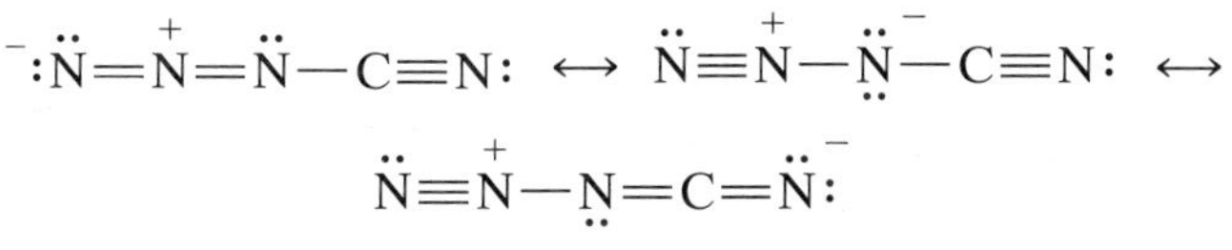

1-64

In a neutral molecule, the sum of the formal charges must equal 0. Thus, if one nitrogen is given a formal of +1, some other atom in the molecule must have a formal charge of −1.

Another acceptable structure is **1-65**.

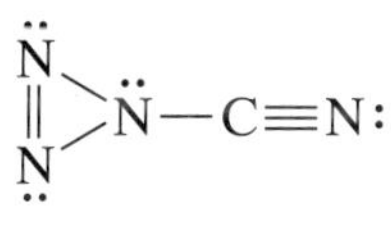

1-65 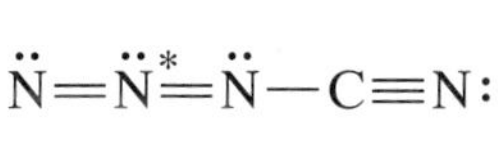 **1-66**

On the other hand, **1-66** is unacceptable. The starred nitrogen has 10 electrons, which is not possible for any element in period 2 of the periodic table. Also, the terminal nitrogen next to this nitrogen has only six electrons instead of an octet. Other structures, **1-67** and **1-68**, in which one

Problem 1.7
continued

of the nitrogens does not have an octet, are less stable than the resonance structure **1-64**.

$$:N{=}\ddot{N}{-}\ddot{N}{-}C{\equiv}N: \qquad :\ddot{N}{-}\ddot{N}{=}\ddot{N}{-}C{\equiv}N:$$

1-67   **1-68**

Structures **1-69** and **1-70**, possible alternatives to a linear nitrogen array for the azide group, are not acceptable. In **1-69** there is a nitrogen with 10 electrons, and in **1-70** the nitrogen at the top does not have an octet of electrons. In addition, the electron-deficient nitrogen is located next to a positively charged nitrogen, an extremely unstable electron arrangement.

1-69   **1-70**

e. Electron supply for the allyl anion $= (3 \times 4)(C) + (5 \times 1)(H) + 1 = 18$. Electron demand $= (3 \times 8) + (5 \times 2) = 34$. Estimate of bonds $= (34 - 18)/2 = 8$.

The lithium cation sits equidistant from the two concentrations of negative charge at either end of the linear π system.

Problem 1.8

a. Thiophene is aromatic. The sulfur is sp^2-hybridized. One of the lone pairs of electrons is in an sp^2-hybridized orbital, and the other lone pair is in a p orbital parallel to the p orbitals on each of the carbons. Thus, this is a 6π electron system.

b. The pentadienyl anion is nonaromatic. Although the structure contains 6π electrons that are conjugated, it is not cyclic.

c. This heterocycle is antiaromatic or nonaromatic. If the NH nitrogen is sp^2-hybridized, the lone pair of electrons on the nitrogen can be in a p

orbital that overlaps with the other six p electrons in the ring. This would be an antiaromatic system. However, because of the destabilization that delocalization would cause, it is unlikely that good overlap of the nitrogen lone pair electrons with the π system would occur.

Problem 1.8
continued

d. Cycloheptatriene is nonaromatic. The carbon at the top of the structure is sp^3-hybridized. Thus, the delocalization of the 6π electron system is interrupted by this carbon. For aromaticity, not only must the compound have the correct number of electrons, but those electrons must be completely delocalized.

e. Pyrene is aromatic. This compound contains 16π electrons, so, strictly speaking, it does not obey Hückel's rule and you might have expected this compound to be antiaromatic. However, if a compound has fused six-membered rings, all of which are totally conjugated, it exhibits the properties we associate with aromaticity and is usually considered to be aromatic. We can rationalize this aromaticity by looking at the alternative resonance structure that follows.

In this structure the π bonds on the periphery form a fully conjugated loop. We can consider that the molecule has two noninteracting π systems: the 14π electron system of the periphery, which is aromatic according to Hückel's rule, and the highlighted 2π electron system of the central double bond.

f. The cyclooctatetraenyl dianion is aromatic. This is a completely conjugated, cyclic, 10π electron system. The ring is planar. However, the situation is not so simple. It has been calculated that the energy reduction, due to aromaticity, is not large enough to compensate for the electron–electron repulsion energy. Apparently, such anionic compounds exist primarily because the metallic counterions are strongly solvated by the reaction solvents.

g. This macrocyclic alkyne is aromatic. The completely conjugated system contains 14π electrons. Only two of the π electrons of the triple bond are counted, because the other two π electrons are perpendicular to the first two and cannot conjugate with the rest of the π system.

Problem 1.9

a.

OH
CH₂ ⟍C⟋ NH₂

OH
CH₃ ⟍C⟋ NH

b. CH₂=CHOH

c. CH₂=CH—CH=CHOH

d.

NH₂
HN ⟍C⟋ NH₂

NH₂
H₂N ⟍C⟋ NH

e.

O
Ph ⟍NH ⟍C⟋ NH₂

OH
Ph ⟍NH ⟍C⟋ NH

O   NH₂
⟍C⟋
N
H
H

O   NH₂
⟍C⟋
H   H
N

O   NH₂
⟍C⟋
N
H
H

There are two more possible structures in which there is isomerism about the imine nitrogen:

N   O
⟍C⟋
H   NH₂
H

H   H
N   O
⟍C⟋
NH₂

Problem 1.10

a. The first structure is the same compound as the last; the second is the same as the fourth. The first, second, and third structures meet the basic requirement for tautomers: interconversion involves only movement of a double bond and one hydrogen atom. However, most chemists would not call them tautomers because the allylic proton that moves is not very acidic. Its pK_a can be estimated from the pK_a of the allylic proton of 1-propene: 47.1–48.0 (from Appendix C). Thus, the intermediate anion necessary for the interconversion of these isomers would be formed only under extremely basic conditions. When the equilibration is this difficult to effect, the different isomers usually are not called tautomers. On the other hand, if another double bond were added to the structure, the compound would be very acidic (from Appendix C the related cyclopentadiene has a

$pK_a = 18.1$ in DMSO, which is more acidic than the α protons in acetone, $pK_a = 19.2$). Thus, most chemists would call compounds **1-71** and **1-72** tautomers.

Problem 1.10
continued

I-71   **I-72**

The structures in the problem are not reasonable resonance forms, because conversion from one structure to another involves movement of a hydrogen atom and rehybridization of two carbon atoms. Therefore, if the double bond were moved in either direction to the next ring location, it would leave one carbon with only three bonds (carbon 5 in the right-hand structure following) and place five bonds on another (carbon 2 in the right-hand structure following). These other structures would have extremely high energies and contribute nothing to a resonance hybrid.

b. These two structures are resonance forms. They represent the simplest enolate ion that can exist.

c. These compounds are tautomers. They differ in the placement of a double bond and a proton, *and* they could readily be in equilibrium with each other.

The two structures are tautomers because they differ in the placement of a double bond and a proton, and they could readily be in equilibrum with each other. If the resonance arrow is changed to $\rightleftharpoons$, the picture will be correct.

Problem 1.11

a. $CH_2{=}CH^-$ is more basic than $HC{\equiv}C^-$. That is, according to Appendix C, ethene is less acidic than ethyne, so the conjugate base of ethene is more basic than the conjugate base of ethyne.

Problem 1.12

b. $[(CH_3)_2CH]_2N^-$ is more basic than $^-CH_2CON(Et)_2$. The diisopropylamide anion cannot be stabilized by resonance, because the N is bonded only to sp^3-hybridized carbons. On the other hand, $^-CH_2CON(Et)_2$ is

Problem 1.12
continued

stabilized by resonance. The negative charge is delocalized on both carbon and oxygen:

This anion is more stable than the diisopropylamide ion and, thus, diisopropylamide anion is the more basic.

c. $(CH_3)_3CO^-$ is more basic than CH_3O^-. The *t*-butoxide anion is not as well-solvated as methoxide and, thus, is more basic. Appendix C also indicates that methanol is more acidic than *t*-butyl alcohol, so the relative basicity of the conjugate bases is the reverse.

d. The *m*-nitrophenolate anion is less stable than the *p*-nitrophenolate anion and, therefore, more basic. The resonance forms for *m*-nitrophenolate are as follows:

The resonance forms for *p*-nitrophenolate are as follows:

The *p*-nitrophenolate anion is more stable because the negative charge can be delocalized onto the nitro group. Because it is more stable, it will be less basic. Note that in the drawings, the other Kekule form of the ring and the other form of the nitro group were omitted where appropriate. Because these forms also stabilize the starting materials, they do not explain a difference in energy between the starting phenol and product phenolate. It is the size of this energy difference that determines the acidity of the starting material.

Problem 1.12
continued

a. The carbanion produced by removal of a proton on the CH_2 (methylene) group is the one stabilized most effectively by resonance because it is conjugated with two carbonyl groups.

Problem 1.13

Removal of a proton from a methyl group will give a carbanion that is stabilized by conjugation with only one carbonyl group.

b. The proton will be removed from the oxygen, because this is the most electronegative element in the molecule and can stabilize the negative charge most effectively.

c. Removal of a proton from the γ position gives the most stabilized anion because more delocalization is possible.

Removal of the α proton on the sp^3-hybridized carbon gives a less stabilized anion because the charge is located on only two centers, not three.

Problem 1.13
continued

d. The proton will be removed from the exocyclic amino nitrogen to form the resonance-stabilized anion, **1-73**.

1-73

The neutral molecule has three different types of acidic hydrogens attached to nitrogen atoms. Two of these are attached to amino nitrogens and one is attached to an imine group. To compare the acidity of the various hydrogen atoms, we need to compare the stability of the different anions formed by loss of the different protons. Ordinarily, by analogy to the acidity of carbon acids ($HC\equiv CH > H_2C=CH_2 > CH_3-CH_3$), we might expect the imine to be more acidic than either of the amines. However, the exocyclic amine proton actually is the most acidic because the anion **1-73**, formed by loss of the exocyclic amine proton, can be represented by two resonance forms of equal energy in which the charge is distributed over two nitrogen atoms. Resonance stabilization is not available to either anion **1-74**, formed by loss of the imine proton, or anion **1-75**, derived by loss of the endocyclic amine proton.

1-74   **1-75**

Other factors being equal, an anion with lone pairs in an sp^2 orbital is less basic than an anion with lone pairs in an sp^3 orbital. In general, the more *s* character in the orbital forming the bond to hydrogen, the more acidic the proton. A good way to remember this is to note the high acidity of acetylene ($HC\equiv CH$, $pK_a = 28.8$) compared with ethylene ($H_2C=CH_2$, $pK_a = 44$) and ethane (CH_3CH_3, $pK_a \approx 50$).

Problem 1.14   a.

$$K_a = 10^{-30.5}/10^{-15.9} = 10^{-14.6}$$

This value tells us that there is very little ester anion present at equilibrium.

b.

$$K_a = 10^{-24.7}/10^{-15.7} = 10^{-9}$$

This value is only an approximation, because the two acidities are not measured in the same solvent. Nonetheless, the value indicates that equilibrium favors starting material to a large extent.

Problem 1.14
continued

c.

$$K_a = 10^{-30.6}/10^{-35.7} = 10^{5.1}$$

In this case the reaction goes substantially to the right. By using the alternative pK_a of 39 for the amine, the answer would be $10^{8.4}$.

a. $^+NO_2$ is the electrophile. Writing a Lewis structure for this species $(\:\ddot{O}\!:\:\overset{+}{\ddot{N}}\!:\:\ddot{O}\!:)$ indicates that the nitrogen is positive and will be the atom that reacts with the nucleophile. In acetophenone, the π electrons of the ring act as the nucleophile. In keeping track of electrons, it is helpful to think of the nucleophile as the electron pair of the π bond to the carbon where the nitrogen becomes attached:

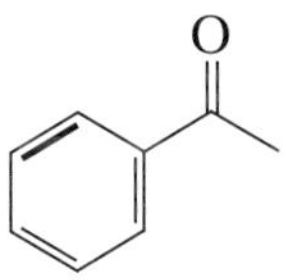

Problem 1.15

b. The electrophile is the proton of sulfuric acid that is transferred to the oxygen of acetophenone. Thus, the oxygen of acetophenone acts as the nucleophile in this reaction. The product shown would be a direct result of a lone pair of electrons on the oxygen acting as a nucleophile. One could also show the π electrons of the $C{=}O$ group acting as the nucleophile. This would give the following structure:

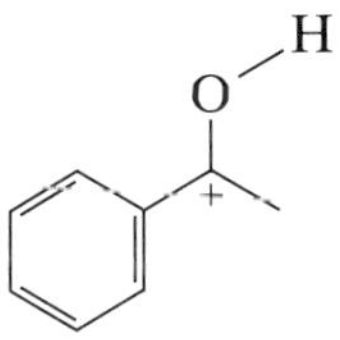

This structure and the structure drawn in the problem are the same, because they are both resonance forms that contribute to the same resonance hybrid.

   This reaction could also be described as an acid–base reaction, where the acid is the proton of sulfuric acid and the base is the oxygen of the carbonyl group.

Problem 1.15
continued

c. This bond of the Grignard reagent is a highly polarized covalent bond, so that the carbon bears a negative charge. This nucleophilic carbon of the Grignard reagent becomes attached to the electrophilic carbon of the carbonyl group in acetophenone.

d. The electron pair, constituting the π bond in cyclohexene, is the nucleophile, and a proton from HCl is the electrophile.

2

General Principles for Writing Reaction Mechanisms

In writing a reaction mechanism, we give a step-by-step account of the bond (electron) reorganizations that take place in the course of a reaction. These mechanisms do not have any objective existence; they are merely our attempt to represent what is going on in a reaction. Although experiments can suggest that some mechanisms are reasonable and others are not, for many reactions there is no evidence regarding the mechanism, and we are free to write whatever mechanism we choose, subject only to the constraint that we conform to generally accepted mechanistic patterns.

The purpose of this book is to help you figure out a number of pathways for a new reaction by showing you some of the steps that often take place under a particular set of reaction conditions. This chapter is devoted to some general principles, derived from the results of many experiments by organic chemists, that can be applied to writing organic mechanisms. Subsequent chapters will develop the ideas further under more specific reaction conditions.

It often is difficult to predict what will actually happen in the course of a reaction. If you were planning to run a reaction that had never been done before, you would plan the experiment on the basis of previously run reactions that look similar. You would assume that the steps of bond reorganization that take place in the new reaction are analogous to those in the reactions previously run. However, you might find that one or more steps in your reaction scheme give unanticipated results. In other words, although a number of general ideas about the course of reactions have been developed on the basis of experiments, it is sometimes difficult to choose which ideas apply to a particular reaction. Working through the problems in this book will help you develop the ability to make some of those choices. Nonetheless, often you will conclude that there is more than one possible pathway for a reaction.

I. BALANCING EQUATIONS

Hint 2.1

It can be assumed, unless otherwise stated, that when an organic reaction is written, the products shown have undergone any required aqueous workup, which may involve acid or base, to give a neutral organic molecule (unless salts are shown as the product). In other words, when an equation for a reaction is written in the literature or on an exam, an aqueous workup usually is assumed and intermediates, salts, etc. are not shown.

From the viewpoint of organic chemistry, an equation is usually considered to be balanced if it accounts for all the carbon atoms and is balanced with respect to charges and electrons. Ordinarily, no attempt is made to account for the changes in the inorganic species involved in the reaction.

Hint 2.2

Check that equations are balanced. First, balance all atoms on both sides of the equation, and then balance the charges. Be aware that when equations are written in the organic literature, they are frequently not balanced.

Example 2.1.   *Balancing atoms.*

In this equation, the carbon atoms balance but the hydrogen and oxygen atoms do not. The equation is balanced by adding a molecule of water to the right-hand side.

The equation is balanced with regard to charge: the positive charge on the left balances the positive charge on the right.

Example 2.2.   *Balancing charges.*

In this equation, the carbon, hydrogen, nitrogen, and oxygen atoms balance. At first glance, the charges also appear to balance because there is a single net negative charge on each side of the equation. However, the right-hand side of the equation contains an incorrect Lewis structure in which there is an electron-deficient carbon and the formal charge on nitrogen is omitted. The equation is balanced correctly by adding a negative charge on carbon and a positive charge on nitrogen.

In the following steps, supply the missing charges and lone pairs. **PROBLEM 2.1**
Assume that no molecules with unpaired electrons are produced.

b.

c.

2. USING ARROWS TO SHOW MOVING ELECTRONS

In writing mechanisms, bond-making and bond-breaking processes are shown by curved arrows. The arrows are a convenient tool for thinking about and illustrating what the actual electron redistribution for a reaction may be.

Hint 2.3

The arrows that are used to show the redistribution of electron density are drawn from a position of high electron density to a position that is electron-deficient. Thus, arrows are drawn leading away from negative charges or lone pairs and toward positive charges or the positive end of a dipole. In other words, they are drawn leading away from nucleophiles and toward electrophiles. Furthermore, it is only in *unusual* reaction mechanisms that two arrows will lead either away from or toward the same atom.

Example 2.3.   *Using arrows to show redistribution of electron density.*

The following equations show the electron flow for the transformations in Examples 2.1 and 2.2.

For an understanding of why the neutral oxygen in the first equation reacts with a proton of the hydronium ion rather than the positively charged oxygen, see Hint 2.9.

When you first start drawing reaction mechanisms, rewrite any intermediate structure before you try to manipulate it further. This avoids confusing the arrows associated with electron flow for one step with the arrows associated with electron flow for a subsequent step. As you gain experience, you will not need to do this. It will also be helpful to write the Lewis structure for at least the reacting atom and to write lone pairs on atoms such as nitrogen, oxygen, halogen, phosphorus, and sulfur.

Hint 2.4

For the following reactions, supply the missing charges and then use curved arrows to show the bond breaking and bond making for each step.

PROBLEM 2.2

a.

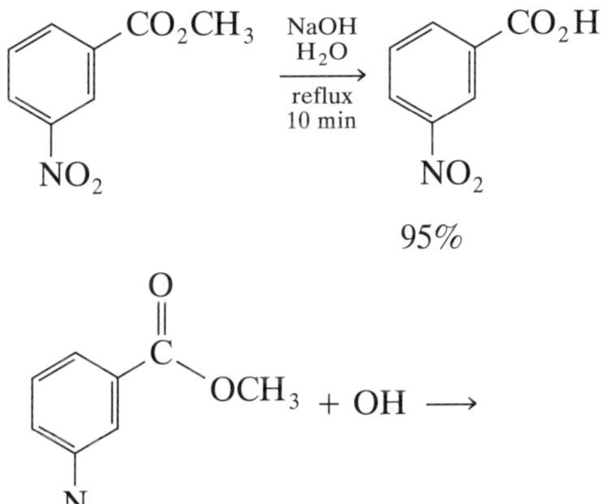

$$95\%$$

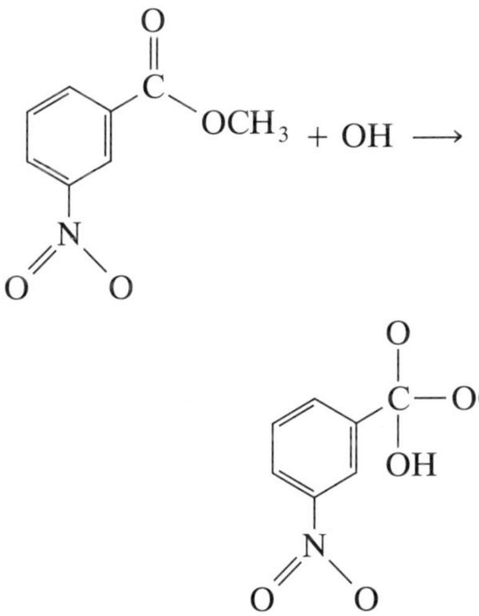

PROBLEM 2.2
continued

b.

PROBLEM 2.2
continued

c.

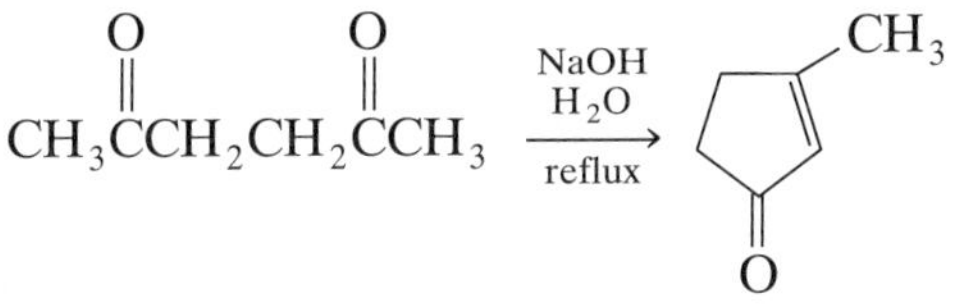

42%

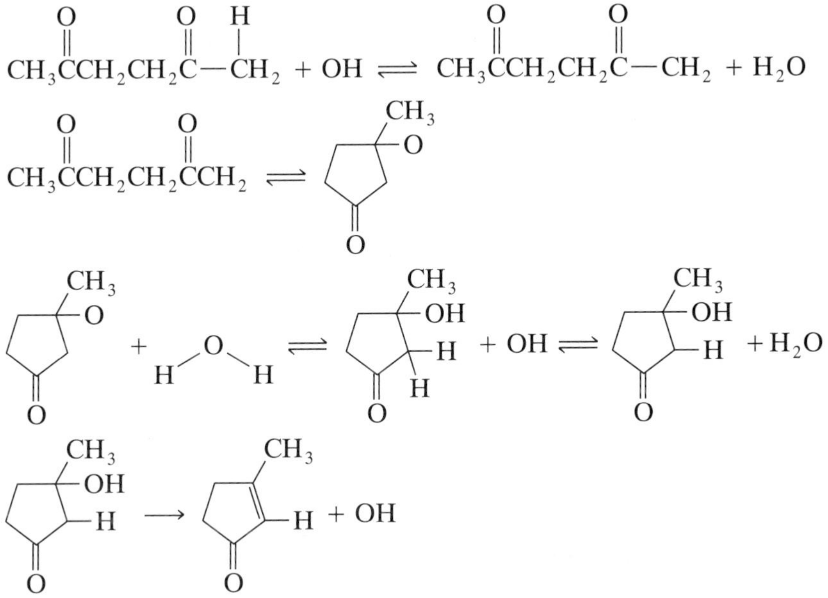

3. MECHANISMS IN ACIDIC AND BASIC MEDIA

If a reaction is run in a strongly basic medium, any *positively* charged species must be weak acids. If a reaction is run in a strongly acidic medium, any *negatively* charged species must be weak bases. In a weak acid or base (like water), both strong acids and strong bases may be written as part of the mechanism.

Hint 2.5

We can look at what this means in some specific situations. Using the pK_a values listed in Appendix C, we find that in a strongly basic solution like 5% aqueous sodium hydroxide (pH ~ 14), the only protonated species would be those with a $pK_a > 12$ (e.g., guanidine, pK_a 13.4). In strongly acidic solutions like 5% aqueous hydrochloric acid (pH ~ 0), the only anions present would be those whose conjugate acid had a $pK_a < -2$ (e.g., $PhSO_3H$, $pK_a - 2.9$). In a solution closer to neutrality (e.g., 5% $NaHCO_3$, pH = 8.5), we would find positively charged guanidine and aliphatic amino groups (pK_a = 13.4 and 10.7, respectively), as well as neutral aromatic amines ($pK_a \sim 4$), and anions such as acetate and 2,4-dinitrophenolate (pK_a = 4.7 and 4.1, respectively).

Example 2.4.   *Writing a mechanism in a strong base.*

The following mechanism for the hydrolysis of methyl acetate in a strong base is consistent with the experimental data for the reaction.

As suggested in Hint 2.5, all of the charged species in this mechanism are negatively charged because the reaction occurs in strong base. Thus, the following steps would be incorrect for a reaction in base because they involve the formation of ROH_2^+ (the intermediate) and H_3O^+, both of which are strong acids.

Another way of looking at this is to realize that, in aqueous base, hydroxide ion has a significant concentration. Because hydroxide is a much better nucleophile than water, it will act as the nucleophile in the first step of the reaction.

Example 2.5. *Writing a mechanism in strong acid.*

The following step is consistent with the facts known about the esterification of acetic acid with methanol in strong acid.

Because the fastest reaction for a strong base, CH_3O^-, in acid is protonation, the concentration of CH_3O^- would be negligible and the following mechanistic step would be highly improbable for esterification in acid:

Example 2.6. *Writing a mechanism in a weak base or weak acid.*

When a reaction occurs in the presence of a weak acid or weak base, the intermediates do not necessarily carry a net positive or negative charge. For example, the following mechanism often is written for the hydrolysis of acetyl chloride in water. (Most molecules of acetyl chloride probably are protonated on oxygen before reaction with a nucleophile, because acid is produced as the reaction proceeds.)

In the first step, the weak base water acts as a nucleophile. In the second step, the weak base chloride ion is shown removing a proton. This second step also could have been written with water acting as the base. Notice that in this example most of the lone pairs have been omitted from the Lewis structures. Reactions in the chemical literature often are written in this way.

Example 2.7.   *Strong acids and bases as intermediates in the tautomerization of enols in water (neutral conditions).*

The first step is usually described as a proton transfer from the enol to a molecule of water. However, when arrows are used to show the flow of electrons, the arrow must proceed from the nucleophile to the electrophile.

This step produces a strong acid, hydronium ion, and a strong base, the enolate anion **2-1**. This anion is a resonance hybrid of structures **2-1** and **2-2**.

The hybrid can remove a proton from the hydronium ion to give the ketone form of the tautomers. Although not strictly correct (because resonance structures do not exist), such reactions commonly are depicted as arising from the resonance structure that bears the charge on the atom that is adding the proton.

2-2

Because enolization under neutral conditions produces both a strong acid and a strong base, the reaction is very slow. Addition of a very small amount of either a strong acid or a strong base dramatically increases the rate of enolization.

Approach writing the mechanism in a logical fashion. For example, if the reagent is a strong base, look for acidic protons in the substrate, then look for a reasonable reaction for the anion produced. If the anion formed by deprotonation has a suitable leaving group, its loss would lead to overall elimination. If the anion formed is a good nucleophile, look for a suitable electrophilic center at which the nucleophile can react. (For further detail, see Chapter 3.)

Hint 2.6

For mechanisms in acid, follow a similar approach and look for basic atoms in the substrate. Protonate a basic atom and consider what reactions would be expected from the resulting cation.

When writing mechanisms in acid and base, keep in mind that protons are removed by bases. Even very weak bases like HSO_4^-, the conjugate base of sulfuric acid, can remove protons. Protons do not just leave a substrate as H^+ because the bare proton is very unstable! Nonetheless the designation $-H^+$ is often used when a proton is removed from a molecule. (Whether this designation

Hint 2.7

is acceptable is up to individual taste. If you are using this book in a course, you will have to find out what is acceptable to your instructor.) A corollary is that when protons are added to a substrate, they originate from an acid, that is, protons are not added to substrates as freely floating (unsolvated) protons.

PROBLEM 2.3   **In each of the following reactions, the first step in the mechanism is removal of a proton. In each case, put the proton most likely to be removed in a box. The pK_a values listed in Appendix C may help you decide which proton is most acidic.**

a.

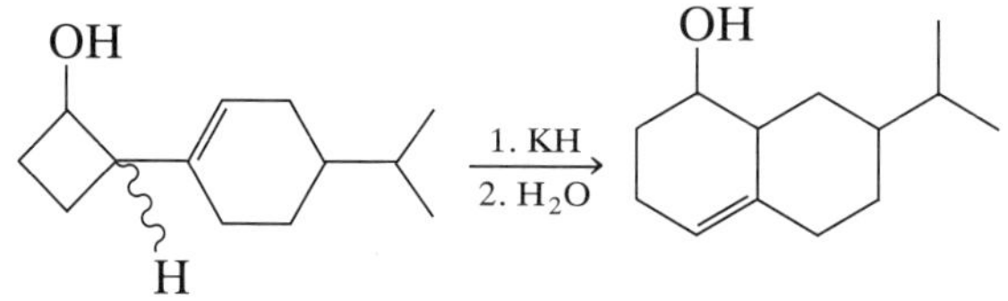

The wavy bond line means that the sterochemistry is unspecified.

Cohen, T.; Bhupathy, M. *J. Am. Chem. Soc.* **1983**, *105*, 520–525.

b.

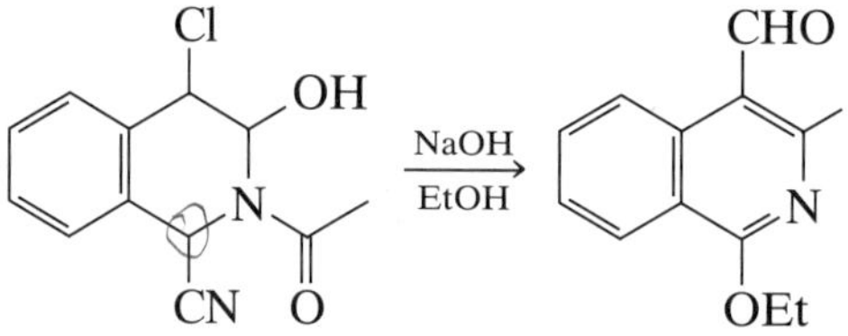

Kirby, G. W.; McGuigan, H.; Mackinnon, J. W. M.; Mallinson, R. R. *J. Chem. Soc., Perkin Trans. 1* **1985**, 405–408; Kirby, G. W.; Mackinnon, J. W. M.; Elliott, S.; Uff, B. C. *J. Chem. Soc., Perkin Trans. 1* **1979**, 1298–1302.

c.

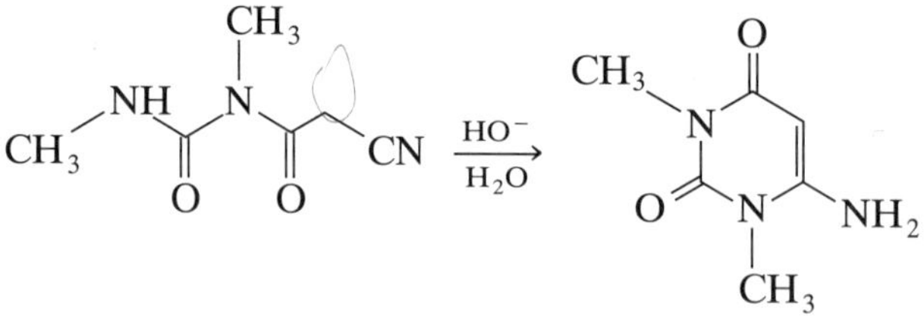

Bernier, J. L.; Henichart, J. P.; Warin, V.; Trentesaux, C.; Jardillier, J. C. *J. Med. Chem.* **1985**, *28*, 497–502.

In this example, removal of the most acidic proton does not lead to the product. Which is the most acidic proton and which is the one that must be removed in order to give the product? (Hint 2.14 may be helpful in relating the atoms of the starting material and the product.)

PROBLEM 2.3
continued

d.

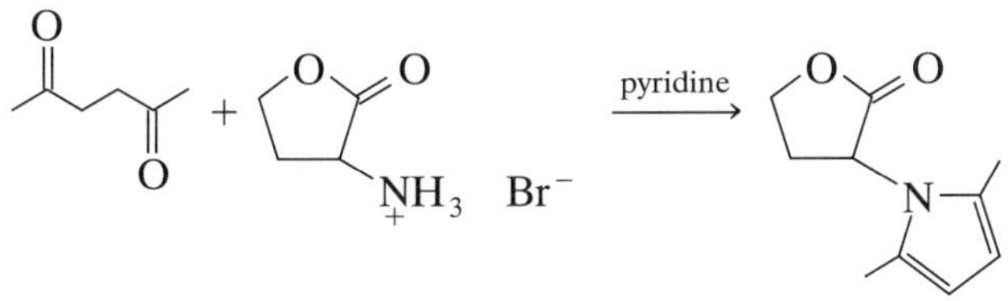

Jaffe, K.; Cornwell, M.; Walker, S.; Lynn, D. G. *Abstracts of Papers*, 190th National Meeting of American Chemical Society, Chicago; American Chemical Society: Washington DC, 1985; ORGN 267.

For each of the following reactions, the first step in the mechanism is protonation. In each case, put the atom most likely to be protonated in a box.

PROBLEM 2.4

a.

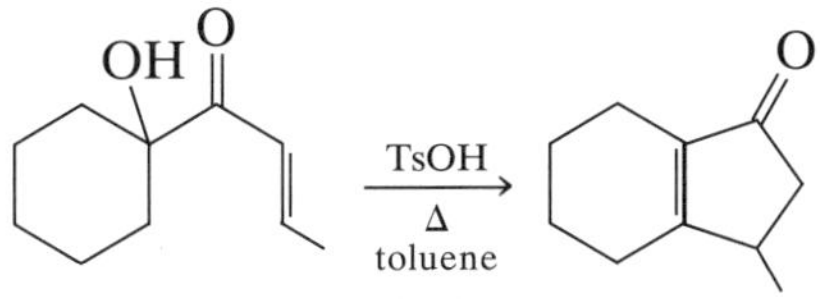

$TsOH = p$-toluenesulfonic acid

Jacobson, R. M.; Lahm, G. P. *J. Org. Chem.* **1979**, *44*, 462–464.

b.

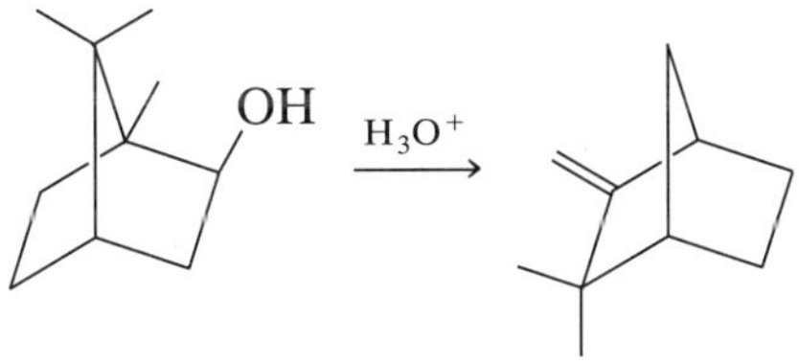

c.

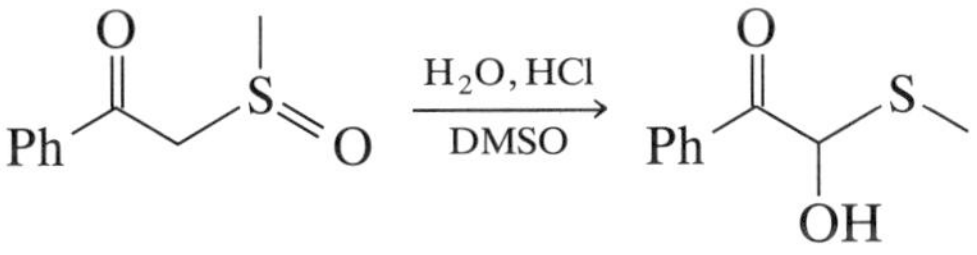

76–81%

House, H. O. *Modern Synthetic Reactions*, 2nd ed.; Benjamin: Menlo Park, CA, 1972; p. 726.

Hint 2.8   When a mechanism involves the removal of a proton, removal of the most acidic proton does not always lead to the product. An example is Problem 2.3.c, in which removal of the most acidic proton by base does not lead to the product. (A mechanism for this reaction is proposed in the answer to Problem 2.3.c.) Similarly, when a mechanism involves protonation, it is not always protonation of the most basic atom that leads to product. Such reactions are called unproductive steps. When equilibria are involved, they are called unproductive equilibria.

4. ELECTRON-RICH SPECIES: BASES OR NUCLEOPHILES?

Hint 2.9   A Lewis base, that is, a species with a lone pair of electrons, can function either as a base, abstracting a proton, or as a nucleophile, reacting with a positively charged atom (usually carbon). Which of these processes occurs depends on a number of factors, including the structure of the Lewis base, the structure of the substrate, the specific combination of base and substrate, and the solvent.

Example 2.8. *Abstraction of an acidic proton in preference to nucleophilic addition.*

Consider the reaction of methylmagnesium bromide with 2,4-pentanedione. This substrate contains carbonyl groups that might undergo nucleophilic reaction with the Grignard reagent. However, it also contains very acidic protons (see Appendix C), one of which reacts considerably faster with the Grignard reagent than the carbonyl groups. Thus, the reaction of methylmagnesium bromide with 2,4-pentanedione leads to methane and, after aqueous workup, starting ketone.

Example 2.9.   *Nucleophilic substitution in preference to proton abstraction.*

$$Ph-\overset{\overset{\displaystyle O}{\|}}{C}-OCH_2CH_2Cl \xrightarrow[\text{87°C, 24 h}]{\overset{\text{NaI}}{\text{Methyl ethyl ketone}}} Ph-\overset{\overset{\displaystyle O}{\|}}{C}OCH_2CH_2I + NaCl$$

$$80\%$$

$$Ph-\overset{\overset{\displaystyle O}{\|}}{C}-OCH_2CH_2 \overset{\frown}{-}Cl \longrightarrow Ph\overset{\overset{\displaystyle O}{\|}}{C}-OCH_2CH_2-I + Cl^-$$

$$I^-$$

Ford-Moore, A. H. *Organic Syntheses, Coll. Vol 4*, **1964**, 84.

With iodine ion (I^-), a good nucleophile that is a weak base, substitution is the predominant reaction.

Example 2.10.   *Competition between substitution and proton abstraction.*

$$CH_3-\underset{\underset{\displaystyle Br}{|}}{CH}-CH_3 \xrightarrow[\text{CH}_3\text{CH}_2\text{OH}]{\text{CH}_3\text{CH}_2\text{O}^-\text{Na}^+} CH_3-\underset{\underset{\displaystyle OCH_2CH_3}{|}}{CH}-CH_3 + CH_3-CH{=}CH_2$$

$$\sim 50\% \qquad \sim 50\%$$

$$CH_3CH_2O^-$$

$$CH_3-\underset{\underset{\displaystyle Br}{|}}{CH}-CH_3 \longrightarrow CH_3-\underset{\underset{\displaystyle OCH_2CH_3}{|}}{CH}-CH_3 + Br^-$$

$$CH_3CH_2O^- \quad H$$

$$CH_3-\underset{\underset{\displaystyle Br}{|}}{CH}{-}CH_2 \longrightarrow CH_3-CH{=}CH_2 + CH_3CH_2OH + Br^-$$

In the reaction of isopropyl bromide with sodium ethoxide in ethanol, the dual reactivity of sodium ethoxide is apparent. Ethoxide ion can act as a nucleophile, displacing bromide ion from carbon to produce isopropyl ethyl ether, or it can remove a proton, with simultaneous loss of bromide ion, to produce propene.

If you know the product of a reaction, usually it is not too difficult to determine whether an electron-rich reagent is acting as a base or as a nucleophile. Predicting the course of a reaction can be a more difficult task. However, as you work through a number of examples and problems, you will start to develop a feel for this as well.

5. TRIMOLECULAR STEPS

Trimolecular steps are rare because of the large decrease in entropy associated with three molecules simultaneously assuming the proper orientation for reaction.

Hint 2.10   Avoid formulating mechanisms involving trimolecular steps. Instead, try to break a trimolecular step into two or more bimolecular steps.

Example 2.11. *Breaking a trimolecular step into several bimolecular steps.*

When mechanisms for the following reaction are considered,

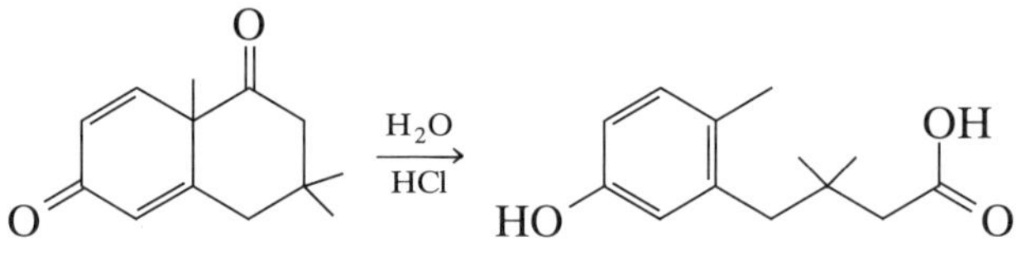

one of the steps could be written as a trimolecular reaction:

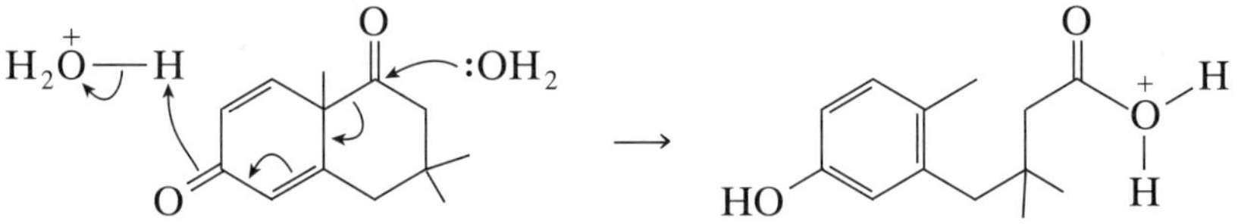

2-3

Note that, in the reaction that produces intermediate **2-3**, only one lone pair of electrons is shown on the water molecules. This follows the common practice of selectively omitting lone pairs from Lewis structures and showing only the lone pairs actually taking part in the reaction. Intermediate **2-3** can lose a proton to water to give the product.

However, we can also write another mechanism, which avoids the trimolecular step:

See Problem 4.11.a for an alternative mechanism for this reaction.

6. STABILITY OF INTERMEDIATES

Any intermediates written for a reaction mechanism must have reasonable stability. For example, second row elements (e.g., carbon, nitrogen, and oxygen) should not be written with more than eight valence electrons, although third row elements like sulfur and phosphorus can, and do, expand their valence shells to accommodate 10 (occasionally more) electrons. In addition, positively charged carbon, nitrogen, and oxygen species with only six valence electrons are generally formed with difficulty. Although carbocations (six electrons) are high-energy intermediates that are encountered in many reactions, the corresponding positively charged nitrogen and oxygen species with six electrons are rare, especially for oxygen. (For an example of an electron-deficient nitrogen species, the nitrenium ion, see Chapter 4.)

Hint 2.11

Nucleophilic reaction cannot occur at a positively charged oxygen or nitrogen that has a filled valence shell. Only eight electrons can be accommodated by elements in the second period of the periodic table. However, third period elements, like sulfur and phosphorus, can (and do) expand their valence shells to accommodate 10 (occasionally more) electrons.

Example 2.12. *Nitrogen cannot accommodate more than eight electrons in the valence shell.*

The following step is inappropriate because, in the product, nitrogen has expanded its valence shell to 10 electrons.

To avoid this situation, the π electrons in the double bond could move to the adjacent carbon, giving an internal salt called an ylide. Although they are rather unstable, ylides are intermediates in some well-known reactions.

If a neutral nucleophile reacted with nitrogen in a similar manner, there would be three charges on the product. The two positive charges on adjacent atoms would make this a very unstable intermediate.

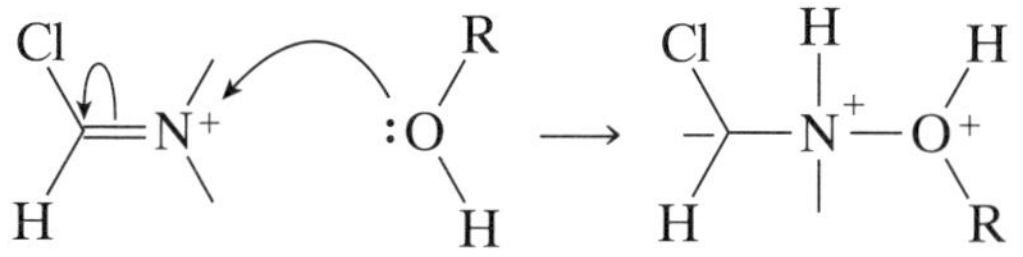

With the preceding reagents, a more appropriate reaction would be

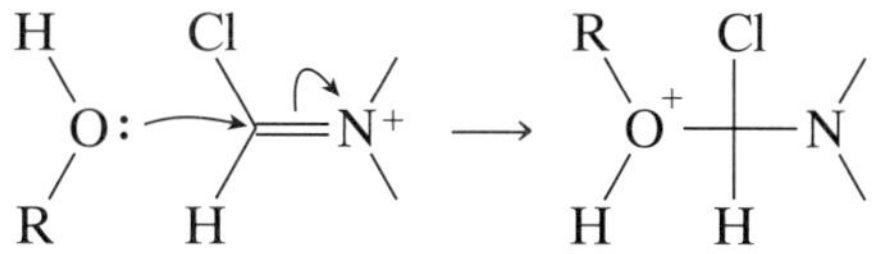

In writing a mechanism, avoid intermediates containing positively charged nitrogen or oxygen ions with less than eight electrons. These species are rare and have high energy because of the high electronegativities of oxygen and nitrogen.

Hint 2.12

Example 2.13. *How to avoid writing mechanisms with electron-deficient, positively charged oxygen and nitrogen species.*

Take a look at the following mechanistic steps:

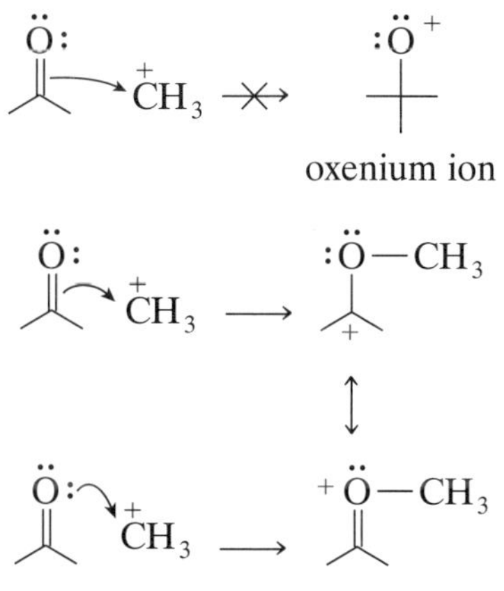

oxenium ion

If the electrophile bonds to the carbon, the process generates an oxenium ion, a highly unstable species. If the electrophile bonds to the oxygen, the process generates a resonance-stabilized carbocation. Note that if we depict the bond being formed by the bonding electron pair between carbon and oxygen, we obtain the product with the electrophile bonded either to carbon or to oxygen. If we use the lone pair of electrons on oxygen, we obtain the product in which the electrophile is bonded to oxygen. It does not matter which pair of electrons we use as long as we draw correct Lewis structures and obtain intermediates that have reasonable stability. (For another example of this, see the answer to Problem 2.5.b.) Our choice of mechanism is based not on which electrons we choose to "push," but on the stability of the intermediate formed.

The situation with nitrogen is analogous. The nitrenium ion is highly unstable, and the carbonium ion is resonance-stabilized.

nitrenium ion

The preceding reaction steps present another difficulty, namely, attack by the methyl cation. Primary carbocations that lack stabilizing groups are highly unstable, and the methyl cation is the least stable of the carbocations. In fact, even in "superacid" (FSO_3H-SbF_5), no primary carbocation is stable enough to be detected. Consequently, a mechanism that invokes such a species should be looked upon with suspicion.

7. DRIVING FORCES FOR REACTIONS

Hint 2.13   A viable reaction should have some energetic driving force. Examples include formation of a stable inorganic compound, formation of a stable double bond or aromatic system, formation of a stable carbocation, anion, or radical from a less stable one, and formation of a stable small molecule (see Hint 2.14).

A reaction may be driven by a decrease in enthalpy, an increase in entropy, or a combination of the two. Reactions driven by entropy often involve forming more product molecules from fewer starting molecules. Reactions that form more stable bonds are primarily enthalpy-driven. When writing a mechanism, constantly ask the following questions: Why would this reaction go this way? What is favorable about this particular step?

A. Leaving Groups

When a reaction step involves a nucleophilic substitution, the nature of the leaving group often is a key factor in determining whether the reaction will occur. In general, *leaving group ability is inversely related to base strength*. Thus, H_2O is a much better leaving group than OH^-, and I^- is a better leaving group than F^-. A list of common leaving groups appears in Table 3.1.

If the reaction involves a poor leaving group, then a very good nucleophile will be necessary to induce the reaction to occur, as the next example illustrates.

Example 2.14. *A rationale for the involvement of different leaving groups in the acid- and base-promoted hydrolysis of amides.*

Ammonia is the leaving group in the acid-promoted hydrolysis of amides. Amide ion, $^-NH_2$, is the leaving group in the base-promoted hydrolysis. The difference can be explained by the driving force of the intramolecular nucleophile relative to the ability of amide ion or ammonia to act as a leaving group. In acid, the intramolecular nucleophile is the oxygen of one of the hydroxyl groups of the tetrahedral intermediate.

In base, the intramolecular nucleophile is the oxyanion of the tetrahedral intermediate:

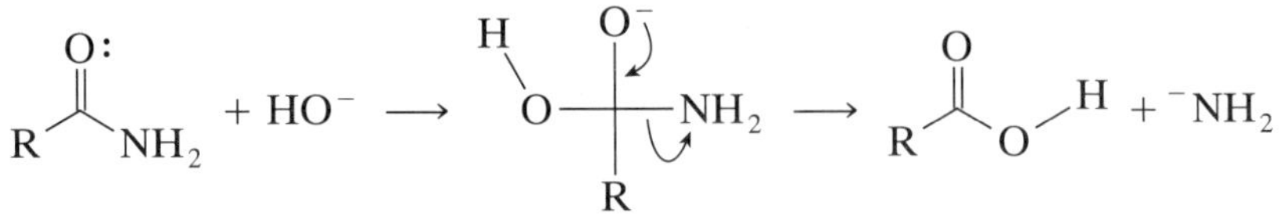

The hydroxyl group is not a strong enough nucleophile, even intramolecularly, to drive the loss of an amide ion, so that under acidic conditions, the nitrogen must be protonated in order to form a sufficiently good leaving group. The leaving group then becomes ammonia, a better leaving group than amide anion. In base, the oxyanion formed by reaction of the original amide with a hydroxyl ion is a strong enough nucleophile to drive the loss of an amide ion.

B. Formation of a Small Stable Molecule

Formation of a small stable molecule can be a significant driving force for a reaction because this involves a decrease in enthalpy and an increase in entropy.

Hint 2.14   A frequent driving force for a reaction is formation of the following small stable molecules: nitrogen, carbon monoxide, carbon dioxide, water, and sulfur dioxide.

Example 2.15. *Loss of carbon monoxide in the thermal reaction of tetraphenylcyclopentadienone with maleic anhydride.*

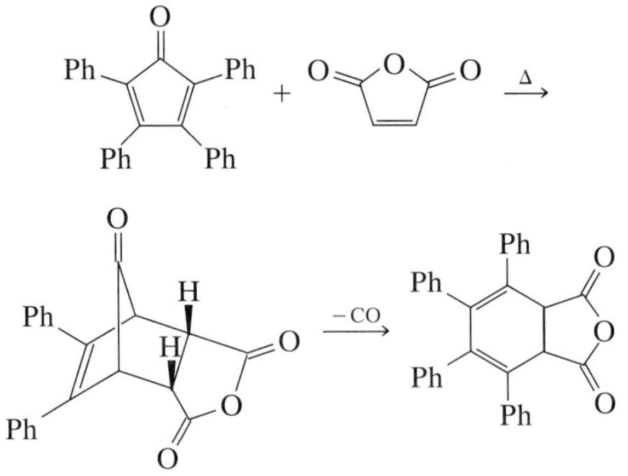

8. STRUCTURAL RELATIONSHIPS BETWEEN STARTING MATERIALS AND PRODUCTS

Numbering of the atoms in the starting material and the product can help you determine the relationship between the atoms in the starting material and those in the product.

Hint 2.15

Number the atoms of the starting material in any logical order. Next, by looking for common sequences of atoms and bonding patterns, identify atoms of the product that correspond with atoms of the starting material and assign to the atoms of the product the corresponding number of the atoms from the starting material. Then, *using the smallest possible number of bond changes*, fill in the rest of the numbers.

Example 2.16. *Using a numbering scheme when writing a mechanism.*

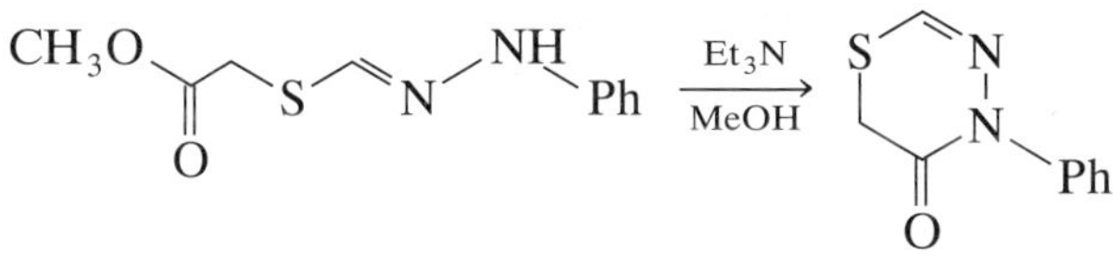

Numbering of the atoms in the starting material and product makes it clear that nitrogen-1 becomes attached to carbon-6.

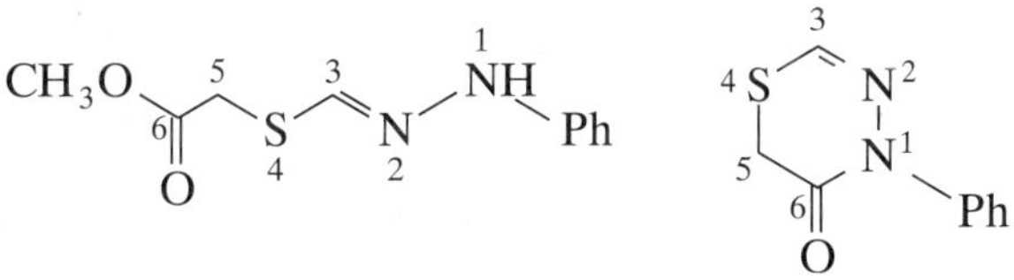

With this connection established, we can write the mechanism as follows:

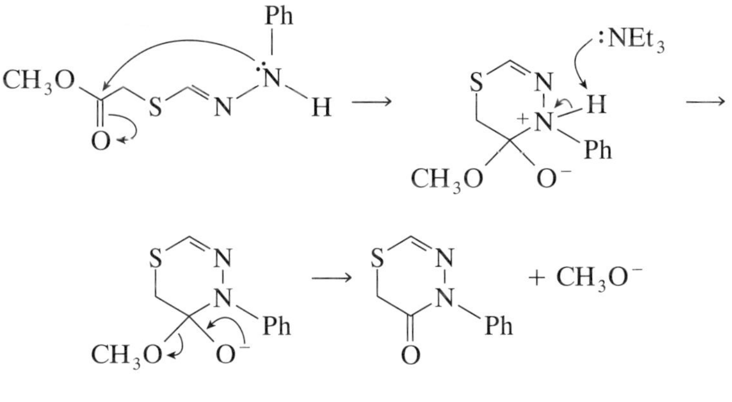

The other products, methoxide ion and triethylammonium ion, would equilibrate to give the weakest acid and weakest base (see Appendix C).

$$CH_3O^- + H\overset{+}{N}Et_3 \rightleftharpoons CH_3OH + NEt_3$$

Example 2.17. *Using a numbering scheme to decide which bonds have been formed and which broken.*

First, consecutively number the atoms in the starting material. In this example, the atoms in the product can be numbered by paying close attention to the location of the phenyl groups and nitrogens:

Without having to write any mechanistic steps, the numbering scheme allows us to decide that the bond between C-5 and O-1 breaks and that a new bond forms between O-1 and C-6. This numbering scheme gives the least possible rearrangement of the atoms when going from starting material to product. This information is invaluable when writing a mechanism for this reaction.

    This example is derived from Alberola, A.; Gonzalez, A. M.; Laguna, M. A.; Pulido, F. J. *J. Org. Chem.* **1984,** *49,* 3423–3424; a mechanism is suggested in this paper.

9. SOLVENT EFFECTS

    Usually, the primary function of a solvent is to provide a medium in which reactants and products can come into contact with one another and interact. Accordingly, solubility dictates the choice of solvent for many organic reactions. However, the nature of the solvent can influence the mechanism of a reaction, and sometimes the choice of solvent dictates the pathway by which a reaction proceeds. In terms of effect on the mechanism, interactions of polar solvents with polar reagents are the most important. Accordingly, solvents can be divided into three groups:

1. Protic solvents, e.g., water, alcohols, and acids.
2. Polar aprotic solvents, e.g., dimethylformamide (DMF), dimethyl sulfoxide (DMSO), acetonitrile (CH_3CN), acetone, sulfur dioxide, and hexamethylphosphoramide (HMPA).
3. Nonpolar solvents, e.g., chloroform, tetrahydrofuran (THF), ethyl ether, benzene, carbon tetrachloride.

Interactions between a polar solvent and a charged species are stabilizing. Protic solvents can stabilize both anionic and cationic species, whereas polar aprotic solvents stabilize only cationic species. Thus, protic solvents favor reactions in which charge separation occurs in the transition state, the high-energy point in the reaction pathway. In nucleophilic substitution reactions, the pathway where two charged species are formed (i.e., S_N1 reaction) is favored in protic solvents, whereas the pathway with a less polar transition state (i.e., S_N2 reaction) is favored in nonprotic solvents.

Example 2.18. *The influence of solvent on basicity.*

Chloride ion generally is a moderate nucleophile and a weak base. However, in the following dehydrohalogenation reaction, chloride ion functions as a base, removing a proton to bring about elimination of the elements of HCl.

$$LiCl, HCN(CH_3)_2$$
$$100°C$$
$$45\%$$

Warnhoff, E. W.; Martin, D. G.; Johnson, W. S. *Org. Synth. Coll. Vol. 4*, **1963**, 162.

In this reaction, chloride functions as a base because the reaction is carried out in the polar aprotic solvent DMF. In polar aprotic solvents, cations are stabilized by solvent interaction, but anions do not interact with the solvent. The "bare" chloride anion functions as a base because it is not stabilized by solvent interaction.

For another example of the effect of solvent on reaction mechanism, see Example 4.14.

Hint 2.16   Use the combination of reagents and solvents specified as a guide to the mechanism. Ionic reagents and polar solvents point to ionic mechanisms. The absence of ionic reagents and use of a nonpolar solvent may suggest a nonionic mechanism.

10. A LAST WORD

The fourteenth century English philosopher William of Occam introduced the principle known as Occam's razor. A paraphrase of this principle which can be applied to writing organic reaction mechanisms is expressed in Hint 2.17.

Hint 2.17   When more than one mechanistic scheme is possible, the simplest is usually the best.

PROBLEM 2.5   **For each of the following transformations, number all relevant non-hydrogen atoms in the starting materials, and number the same atoms in the product.**

a.

(See Problem 3.19.b for further exploration of this reaction.)

b.

(See Example 4.10 for further exploration of this reaction.)

In each of the following problems, an overall reaction is given, followed by a mechanism. For each mechanism shown, identify inappropriate steps, give the number of any applicable hint, and explain its relationship to the problem. Then write a more reasonable mechanism for each reaction.

PROBLEM 2.6

a.

2-4

2-4

b.

2-5

2-5

c.

2-6

2-6

Kirby, A. J.; Martin, J. J. *J. Chem. Soc., Perkin Trans. II* **1983**, 1627–1632.

PROBLEM 2.6
continued

d.

2-7

Kunieda, N.; Fujiwara, Y.; Suzuki, A.; Kinoshita, M. *Phosphorus Sulfur* **1983**, *16*, 223–232.

ANSWERS TO PROBLEMS

Problem 2.1

a.

b.

c.

Problem 2.2

a.

b.

Problem 2.2
continued

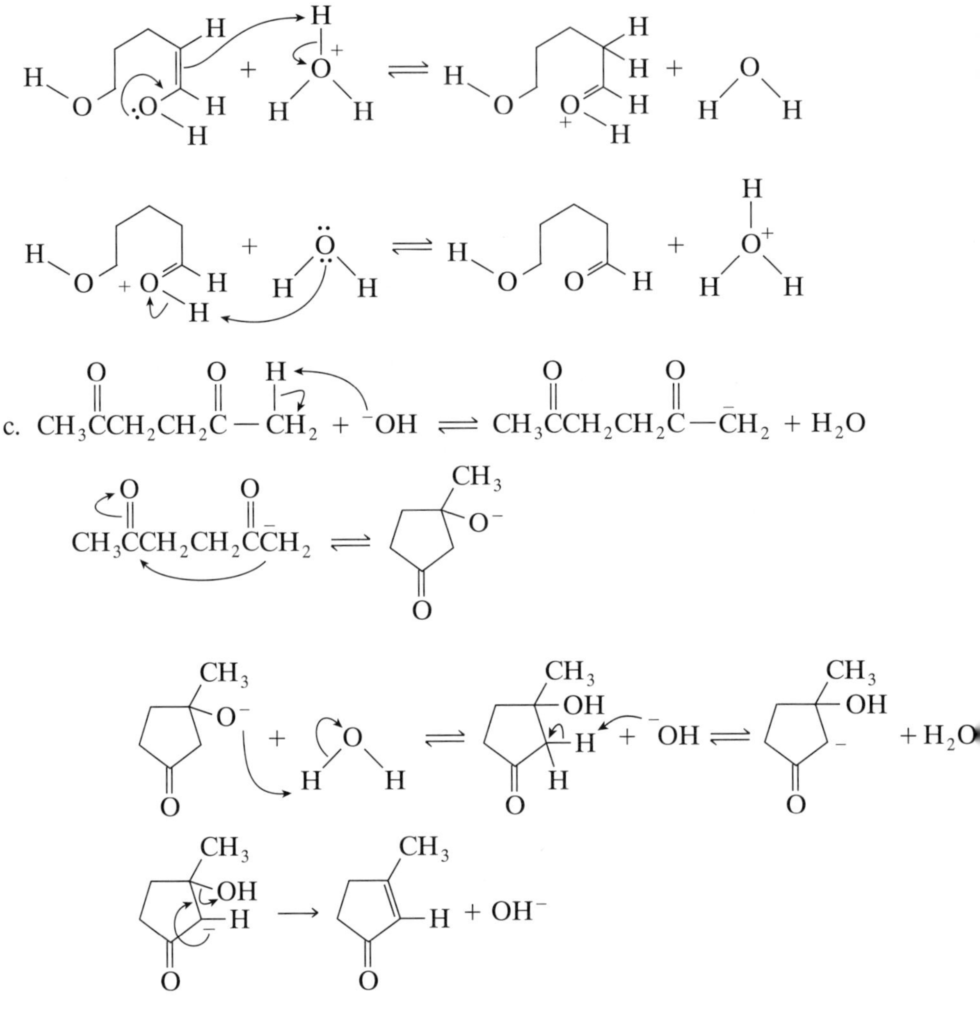

c.

$$CH_3\overset{O}{\overset{\|}{C}}CH_2CH_2\overset{O}{\overset{\|}{C}}-CH_2 + {}^-OH \rightleftharpoons CH_3\overset{O}{\overset{\|}{C}}CH_2CH_2\overset{O}{\overset{\|}{C}}-\bar{C}H_2 + H_2O$$

Problem 2.3

a.

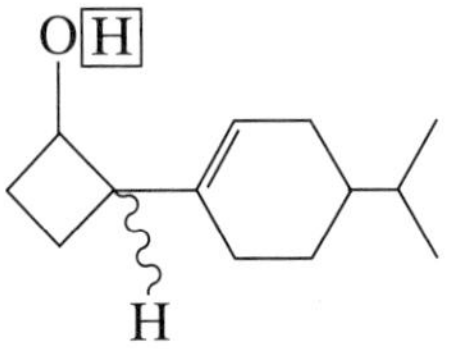

Note: There is some question about the mechanism for this rearrangement. See the paper cited for details.

Problem 2.3
continued

b.

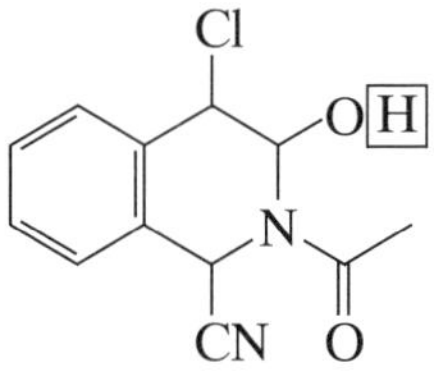

c. removal leads to product

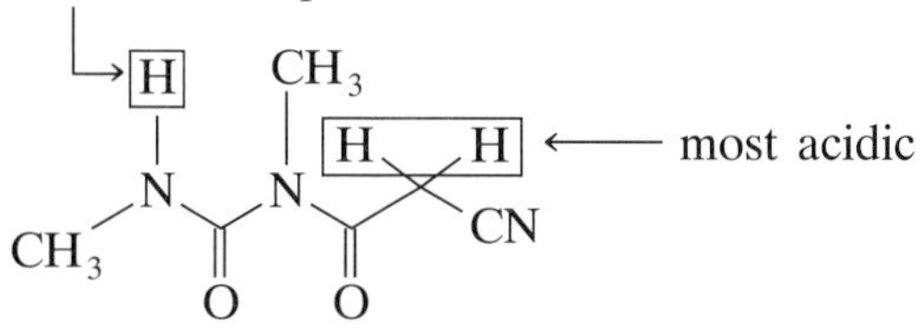

— most acidic

On the basis of acidity, the protons in this molecule can be divided into three groups: the methyl hydrogens, the amide hydrogen, and the methylene hydrogens. Of these, the methyl hydrogens are the least acidic, and the methylene hydrogens are the most acidic. We can estimate the acidity of the protons by examining the stabilities of the anions that result when different hydrogens are removed from the molecule. Removal of a proton from either methyl group gives an anion stabilized only by the inductive effects of nitrogen; no resonance forms are possible. Removal of hydrogen from the amide nitrogen gives an anion stabilized by two resonance forms, one with negative character on nitrogen and one with negative character on oxygen:

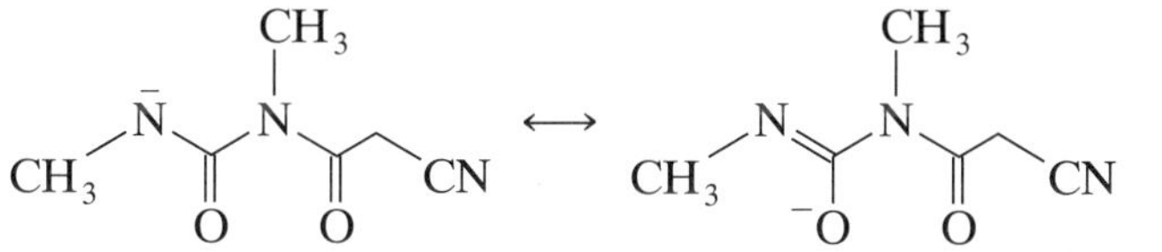

Finally, removal of hydrogen from the methylene group gives an anion with three resonance forms in which the negative charge is on nitrogen, oxygen, or carbon:

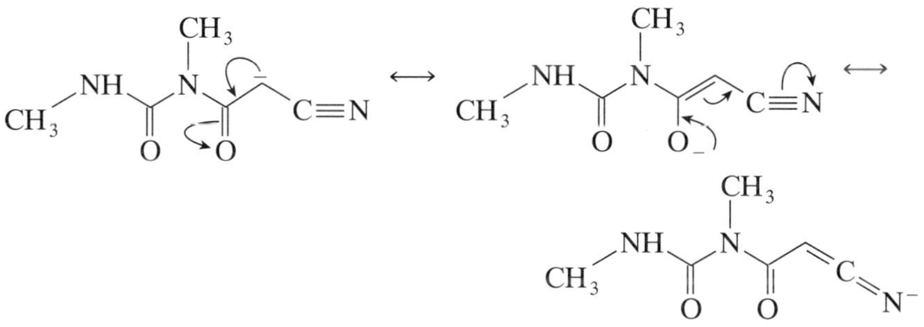

Thus, the anion formed by loss of a proton from the methylene group should be more stable.

We also can compare the acidities of the various protons by using the pK_a values listed in Appendix C.

Acid	Conjugate base	pK_a
$\underset{\substack{\parallel \\ \text{CH}_3\text{CNH}_2}}{\text{O}}$	$\underset{\substack{\parallel \\ \text{CH}_3\text{CNH}^-}}{\text{O}}$	25.5
$\underset{\substack{\parallel \\ \text{NCCH}_2\text{COCH}_3}}{\text{O}}$	$\underset{\substack{\parallel \\ \text{NCCHCOCH}_3}}{\text{O}}$	12.8
$\underset{\substack{\parallel \\ \text{CH}_3\text{COCH}_2\text{CH}_3}}{\text{O}}$	$\underset{\substack{\parallel \\ \text{CH}_2\text{COCH}_2\text{CH}_3}}{\text{O}}$	30
$\underset{\substack{\parallel \\ \text{CH}_3\text{CN(CH}_2\text{CH}_3)_2}}{\text{O}}$	$\underset{\substack{\parallel \\ \text{CH}_2\text{CN(CH}_2\text{CH}_3)_2}}{\text{O}}$	34.5

Using these values, there are several ways to compare the acidity of the protons in the boxes. The pK_a of the amide protons in acetamide is 25.5. The pK_a of the amide proton in this compound should be somewhat greater, because the second amido nitrogen will reduce the resonance stabilization of the anion formed when the amide proton is removed. The precedent for the other boxed protons is the pK_a of the methylene protons of methyl cyanoacetate, 12.8. Because the carbonyl group in the given compound is an amide rather than an ester, the pK_a of its protons will be somewhat higher, but not nearly as high as the value for the amide proton.

We can arrive at the same conclusion in a different way using an alternative analysis based on the pK_a values in Appendix C. The pK_a of the protons on the α carbon of ethyl acetate is 30, whereas the pK_a of those of methyl cyanoacetate is 12.8. Therefore, a cyano group lowers the pK_a of the α carbon protons by $30 - 12.8 = 17$ units. The pK_a of the α carbon protons of an N,N-disubstituted amide, N,N-diethylacetamide, is 34.5. If we assume that the cyano group enhances the acidity of the methylene protons in the given compound by the same amount, their pK_a would be $34.5 - 17 = 16.5$. Thus, the estimated pK_a of the methylene protons is at least 9 pK units lower than that of the amide proton, a factor of one billion.

The mechanism for the reaction could be written as follows. First, the proton is removed by base:

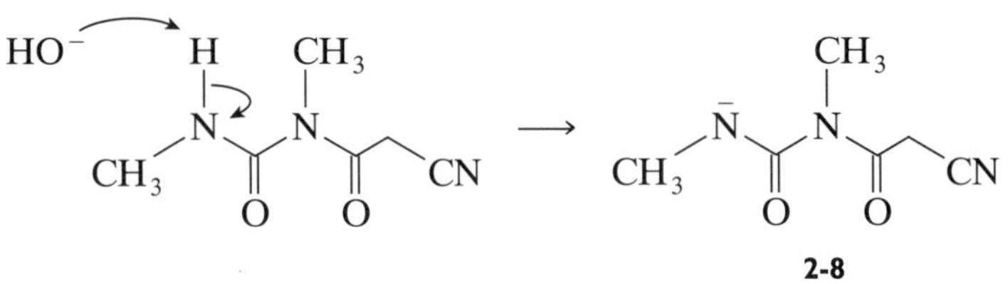

2-8

Problem 2.3
continued

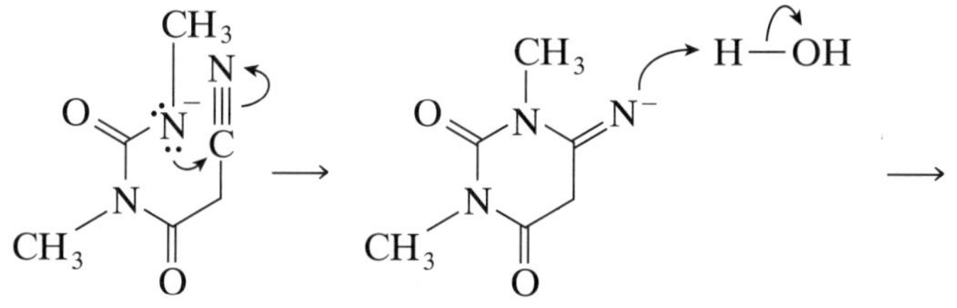

2-8

The resulting anion, **2-8**, can be written in a conformation that makes it clear how the subsequent cyclization takes place.
Now the intermediate tautomerizes to give the final product.

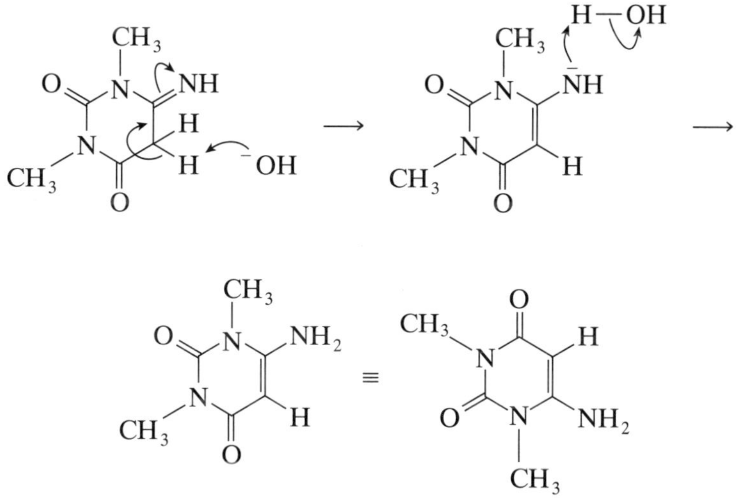

d.

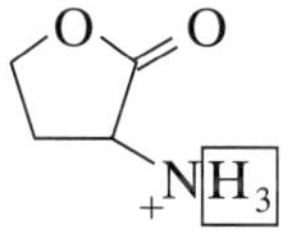

Problem 2.4

a.

Table 1.3 and Appendix C indicate that carbonyl oxygens are more basic than the oxygen of an alcohol. The protonated carbonyl is stabilized by resonance, whereas the protonated alcohol is not. In this case, the protonated carbonyl is further stabilized by delocalizations of charge onto the C=C double bond.

In the paper cited (Jacobsen and Lahm, 1979), the mechanism proposed suggests that the starting material is dehydrated before cyclization occurs. This means that protonation of the less basic group leads to the product. Satisfy yourself that this is reasonable by writing possible reaction mechanisms that would result from protonation of the hydroxyl and carbonyl hydrogens.

b.

c.

Appendix C is of help here. The pK_a of protonated acetophenone is -4.3, whereas the pK_a of protonated dimethyl sulfoxide is -1.5. Thus, dimethyl sulfoxide is almost 10^3 times more basic than acetophenone. These compounds are excellent models for the two functional groups in the compound given.

If hard data like those in Appendix C are unavailable, it is very difficult to decide which oxygen is more basic, by simply considering the functional groups and the structure of the molecule. On one hand, because sulfur is slightly more electronegative than carbon (see Table 1.2), one might predict that the carbonyl oxygen is more basic than the sulfoxide oxygen. Furthermore, protonation of the carbonyl group leads to a cation that is stabilized by resonance delocalization onto the aromatic ring:

Problem 2.4
continued

On the other hand, the π bond between C and O should be stronger than the π overlap between S and O because C and O both use $2p$ orbitals to form π bonds, whereas a π bond between S and O utilizes the less effective $3p$–$2p$ overlap. This should make the oxygen of the sulfoxide functional group more negative. Additionally, sulfur stabilizes charges more effectively than do carbon and oxygen because of its higher polarizability.

In the absence of hard data, qualitative arguments suggest that both oxygens are fairly basic. In a situation like this, it is usually best to adopt a trial-and-error approach, first writing mechanisms that start with the protonation of one functional group and then writing mechanisms that start with the protonation of the other functional group. Remember, a mechanism does not always proceed through protonation of the most basic atom.

a. Start numbering the product at the carboethoxy group, which has not been altered by the reaction. Once you notice that the keto group of the ketoester reactant is also unchanged, the remaining atoms easily fall into place. For the isocyanate, N-6 is obvious, and C-7 and O-8 follow.

Problem 2.5

b. The first thing to notice is that the position of the C—O bond, relative to the methyl group, has not changed. Thus, an initial attempt to number the product would leave those atoms in the same relative positions as in the starting material. This gives the following numbers:

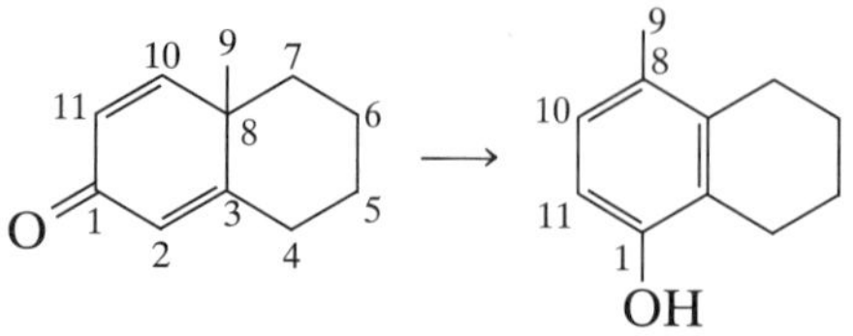

A strong possibility is that the left-hand ring of the product contains the same carbons that this ring contained in the starting material. By minimizing the changes in bonding, the following results:

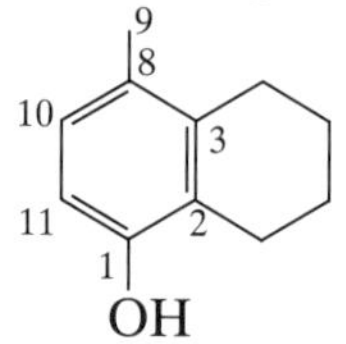

This leaves two possible numbering schemes for the right-hand ring:

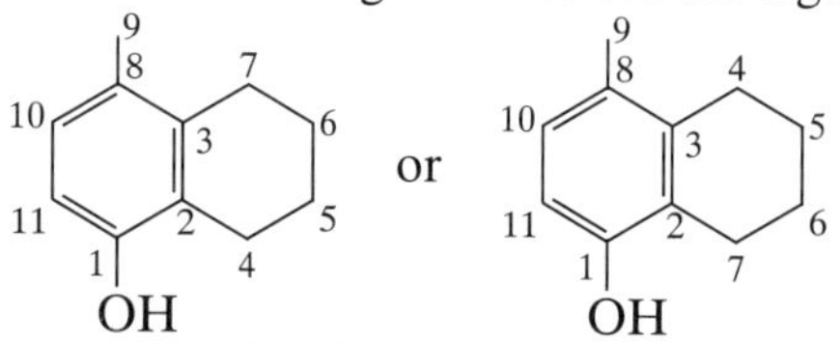

The structure on the right involves less rearrangement; however, if we examine the symmetrical intermediate involved in this reaction (see Example 4.10), we see that, in fact, the two numbering schemes are equivalent in terms of bond reorganization.

Problem 2.6   a. Hint 2.5. This reaction is taking place in a strongly acidic medium. Therefore, strong bases like ⁻OH will be in such low concentration that they cannot be effective reagents in the reaction. A better mechanism would be to protonate the oxygen of one of the hydroxyl groups to convert it into a better (and neutral) leaving group, the water molecule.

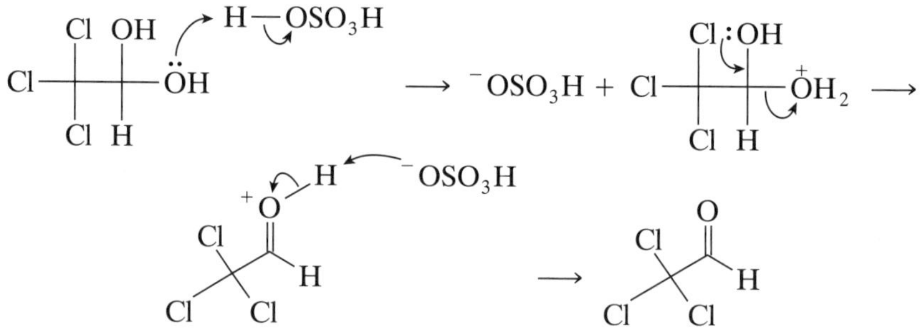

Problem 2.6
continued

b. Hint 2.8. Break up the trimolecular step. First, protonate the carbonyl group, converting it to a better electrophile. Follow this with nucleophilic reaction with the carbonyl group by the oxygen of another molecule of starting material.

The nucleophilic addition of the carbonyl compound to another proto-nated molecule can be written either as shown or by using the π bond as the nucleophile instead of a lone pair on oxygen. This second kind of addition leads to the second resonance form shown instead of the first:

The two representations for this step are equivalent.

The oxygen of one starting aldehyde molecule must be protonated before nucleophilic addition of the carbonyl oxygen of the second. Other-wise, a very basic anion is formed in a very acidic medium (the same problem discussed in part a).

Problem 2.6
continued

c. There are two major problems with the mechanism shown. Data from Appendix C indicate that the substituted phenol would not be ionized significantly in DMSO. Protonated dimethyl sulfoxide has a pK_a of −1.5, whereas *m*-nitrophenol has a pK_a of 8.3. Also, the intermediate carbanion, resulting from the addition, is very unstable; it is not stabilized by resonance in any way. The following mechanism avoids these problems by forming an intermediate cation, **2-9**, which is stabilized by resonance, and by leaving removal of the proton to the last step.

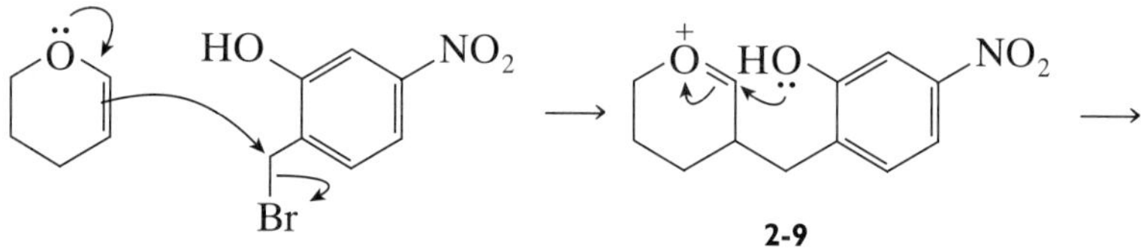

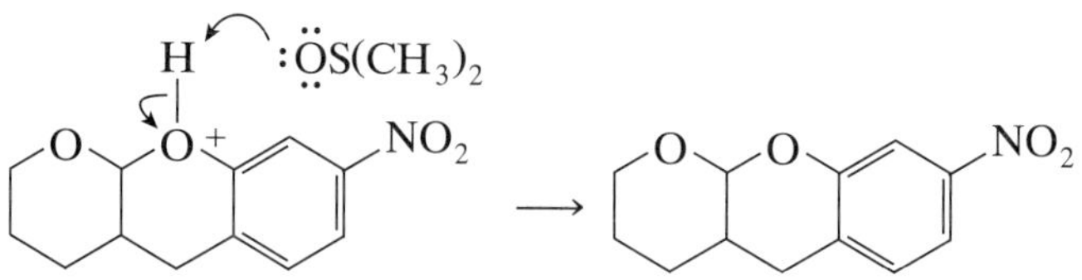

Intermediate **2-9** could also be written as the other resonance form, **2-10**, which contributes to the stabilization of the positive charge:

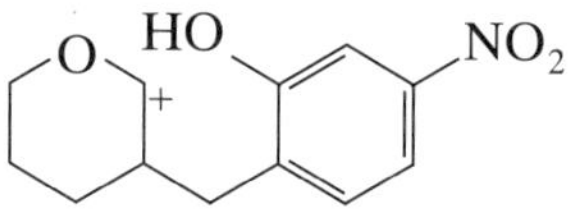

2-10

Often it is useful to draw resonance forms for proposed intermediates because their existence is an indication of stability, which represents a driving force for the reaction. Removal of a proton from the OH group of the resonance hybrid of **2-9** and **2-10** is unlikely because bromide ion is an even weaker base than DMSO (the pK_a of HBr is −9). Therefore, it is expected that ring closure, by nucleophilic reaction with the cation, takes place before removal of the proton.

d. Hint 2.5. The given mechanism has two steps that are unlikely in strong acid: loss of hydroxide ion and its subsequent nucleophilic reaction. Water is a much better leaving group than hydroxide, so that in acidic solution, where protonation of the oxygen can occur, water acts as the leaving group. Notice that because hydrochloric acid is completely dissociated in aqueous solution, hydronium ion, not hydrochloric acid, acts as the protonating agent. In acidic solution, the concentration of hydroxide ion is so low that, although hydroxide ion is a better nucleophile than water, it cannot compete with water for the nucleophilic site.

Problem 2.6
continued

It is always a good idea to keep track of lone pairs of electrons on heteroatoms, such as oxygen and sulfur, by drawing them in. As the mechanism proceeds, it also is a good idea to keep track of formal charges on the atoms of interest, in this case oxygen and sulfur.

Note that the first two steps of the mechanism represent a tautomerization, in which the overall result is movement of a proton from carbon to oxygen and movement of a double bond. Notice, also, that there are several intermediates in which sulfur has an expanded octet.

Take a look at some alternative mechanistic steps:

(i) One possibility is formation of **2-12** (by doubly protonating the sulfoxide oxygen of starting material), from which simultaneous elimina-

Problem 2.6
continued

tion of water and loss of a proton from the methylene group might be written. This would not be as good a step as those shown previously because the development of two adjacent positive centers is destabilizing.

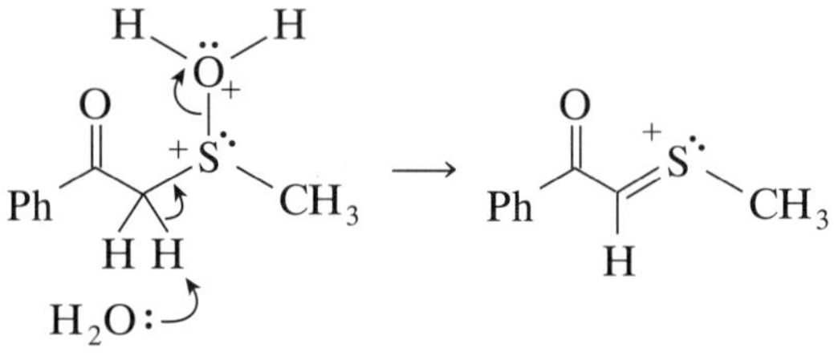

2-12

(ii) Removal of a proton always requires reaction with a base, which in this step could be water. Thus, the following representation is not strictly correct because no base is indicated to remove the proton; however, as stated previously, this would be considered acceptable by a number of instructors.

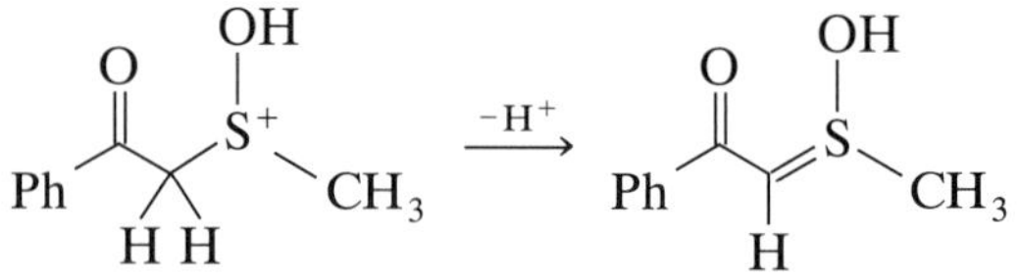

(iii) Showing **2-13** for removal of a proton is not accurate for two reasons. First, the arrow indicates electron movement in the wrong direction; this would produce H^- instead of H^+. (See Chapter 3 for further discussion of hydride loss.) Second, proton removal always requires reaction with a base, even if it is a weak base.

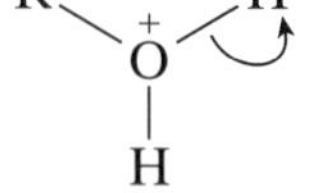

2-13

(iv) The following equation is an example of a [1,3] sigmatropic intramolecular shift of hydrogen. Chapter 6 discusses why this type of tautomeric reaction is unlikely to occur.

Problem 2.6
continued

3

Reactions of Nucleophiles
and Bases

The many reactions of nucleophiles and bases present an array that often is bewildering. These reactions usually are organized on the basis of the reacting group and the overall reaction (e.g., nucleophilic addition to carbonyl groups), which gives rise to a very large number of categories. Although there are a very large number of reactions, it soon becomes apparent that we can write mechanisms for most of them once we are familiar with a few general patterns and the general principles outlined in Chapter 2. As an example, consider that the hydrolysis of an ester in base is classified as a nucleophilic substitution at an aliphatic sp^2 carbon, whereas the reaction of hydrazine with a ketone is classified as a nucleophilic addition. However, as the following reactions show, both reactions involve nucleophilic addition to the carbonyl group, followed by loss of a leaving group.

Ester hydrolysis

Hydrazone formation

As you work your way through a number of reaction mechanisms, you will find that mechanistic patterns offer a way to organize organic reactions in a way that complements the organization based on functional groups.

This chapter includes examples of aliphatic nucleophilic substitution at both sp^3 and sp^2 centers, aromatic nucleophilic substitution, E2 elimination, nucleophilic addition to carbonyl compounds, 1,4-addition to α,β-unsaturated carbonyl compounds, and rearrangements promoted by base.

I. NUCLEOPHILIC SUBSTITUTION

The classification within this section is based on the structural (rather than the mechanistic) relationship between the starting materials and products. Mechanistically, all of the reactions considered in this section involve nucleophilic substitution as the first step, *except for* aromatic substitution via the aryne mechanism, which involves elimination followed by nucleophilic addition.

A. The S$_N$2 Reaction

The S$_N$2 reaction is a concerted bimolecular nucleophilic substitution at carbon. It involves an electrophilic carbon, a leaving group, and a nucleophile. The partial positive charge on the electrophilic carbon is due to the electron-withdrawing effect of the electronegative leaving group. This partial positive charge can be augmented by the presence of other electron-withdrawing groups attached to the electrophilic carbon, and the presence of such groups enhances reaction at the electrophilic center. For example, α-halocarbonyl groups react much faster than simple alkyl halides (see Examples 3.3 and 3.21).

The S_N2 reaction occurs only at sp^3-hybridized carbons. The relative reactivities of carbons in the S_N2 reaction are $CH_3 > 1° > 2° \gg 3°$, due to steric effects. Methyl, 1° carbons, and 1° and 2° carbons that also are allylic, benzylic, or α to a carbonyl group are especially reactive.

Example 3.1. *The S_N2 reaction: A concerted process.*

The electrons of the nucleophile interact with carbon at the same time that the leaving group takes both of the electrons in the bond between carbon and the leaving group. This particular example involves both a good leaving group and a good nucleophile.

$$^-CN \quad PhCH_2-OSO_2CF_3 \longrightarrow PhCH_2CN + {}^-OSO_2CF_3$$

Leaving Groups

Nucleophiles are discussed in Chapter 1, Section 9. Table 3.1 lists typical leaving groups and gives a qualitative assessment of their effectiveness. Usually, the less basic the substituent, the more easily it will act as a leaving group. This is because both basicity and leaving group ability are related to the stability of the anion involved. Frequently, these are both related to charge dispersal in the anion, with greater charge dispersal being associated with greater stability of the ion. In looking at Table 3.1, we see that the best leaving groups are those for which resonance, inductive effects, or size results in distribution of any negative charge.

Relative leaving group abilities also depend upon the solvent and the nature of the nucleophile. For example, negatively charged leaving groups will be stabilized by interactions with protic solvents, so that protic solvents will increase the rate of bond breaking for these groups. Although these effects are important in modifying reaction conditions and yields, they rarely are large enough to completely change the mechanism by which a reaction proceeds, and we will not consider them here in detail.

Hydride (H^-) rarely acts as a leaving group. Exceptions are the Cannizzaro reaction and hydride abstraction by carbocations. **Hint 3.1**

The S_N2 reaction rarely occurs with poor leaving groups. However, in other reactions, such as the nucleophilic substitution of carboxylic acid

TABLE 3.1   Leaving Group Abilities

Excellent

$$N_2, \ ^-OSO_2CF_3 \text{ (triflate)}, \ ^-O-\underset{\underset{O}{\|}}{\overset{\overset{O}{\|}}{S}}-\text{C}_6\text{H}_4-NO_2 \text{ (nosylate = Nos)},$$

$$^-O-\underset{\underset{O}{\|}}{\overset{\overset{O}{\|}}{S}}-\text{C}_6\text{H}_4-Br \text{ (brosylate = Bs)}, \ ^-O-\underset{\underset{O}{\|}}{\overset{\overset{O}{\|}}{S}}-\text{C}_6\text{H}_4-CH_3 \text{ (tosylate = Tos or Ts)},$$

$^-OSO_2CH_3$ (mesylate = Ms)

Good	Fair
I^-, Br^-, Cl^-, SR_2	$OH_2, NH_3, \ ^-OCOCH_3$ (acetate = OAc)

Poor	Very Poor
$F^-, \ ^-OH, \ ^-OR$	$^-NH_2, \ ^-NHR, \ ^-NR_2, R^-, H^-, Ar^-$

derivatives (see Sections 1.B and 3), reactions with a poor leaving group like OH^- or RNH^- are encountered more frequently. Hydroxide may act as a leaving group, but only when there is considerable driving force for the reaction, as in certain elimination reactions where the double bond formed is stabilized by resonance (see Ex. 3.10).

PROBLEM 3.1   **Explain why the following mechanistic step in the equilibrium between a protonated and an unprotonated alcohol is a poor one.**

Example 3.2.   *The S_N2 reaction of an alcohol requires prior protonation.*

The alcohol oxygen is protonated before substitution takes place. Thus, the leaving group is a water molecule, a fair leaving group, rather than the hydroxide ion, a poor leaving group.

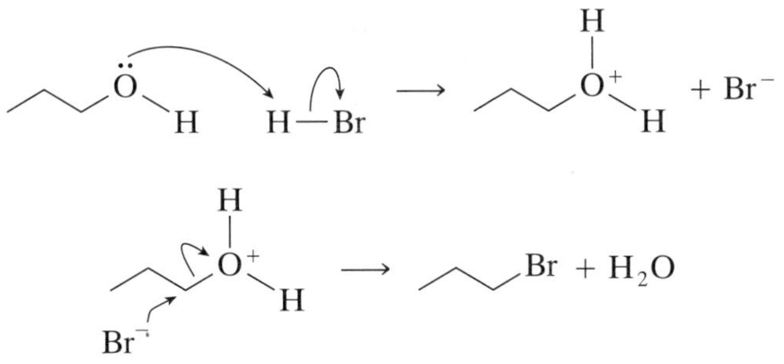

A poorer mechanistic option would show hydroxide as the leaving group:

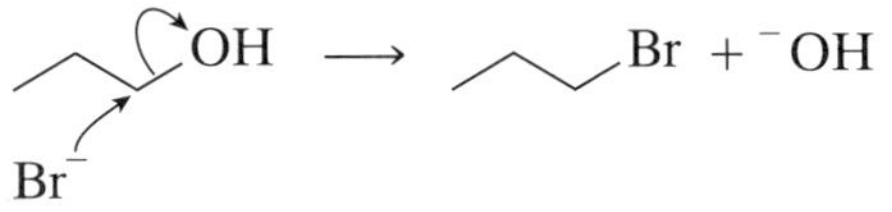

Stereochemistry

The S_N2 reaction always produces 100% inversion of configuration at the electrophilic carbon. Thus, as shown in Example 3.3, the nucleophile approaches the electrophilic carbon on the side opposite the leaving group (there is a 180° angle between the line of approach of the nucleophile and the bond to the leaving group).

Example 3.3. *Stereochemistry of the S_N2 reaction.*

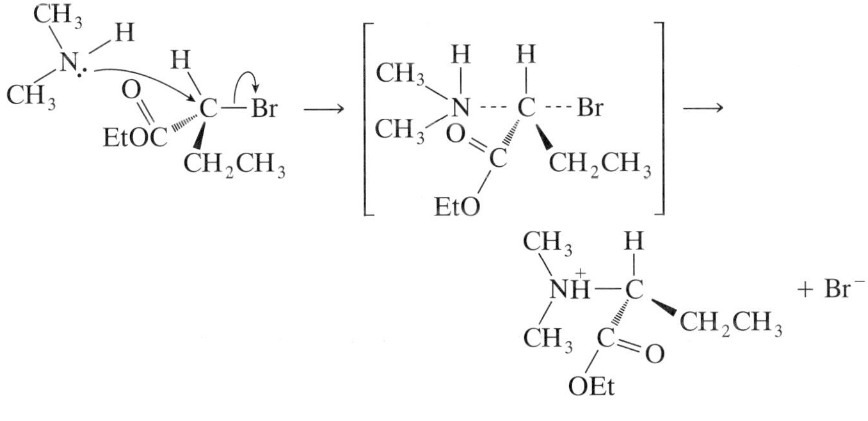

PROBLEM 3.2 **Consider the following synthesis, which involves alkylation of the phenolic oxygen (attachment of the benzyl group onto the oxygen).**

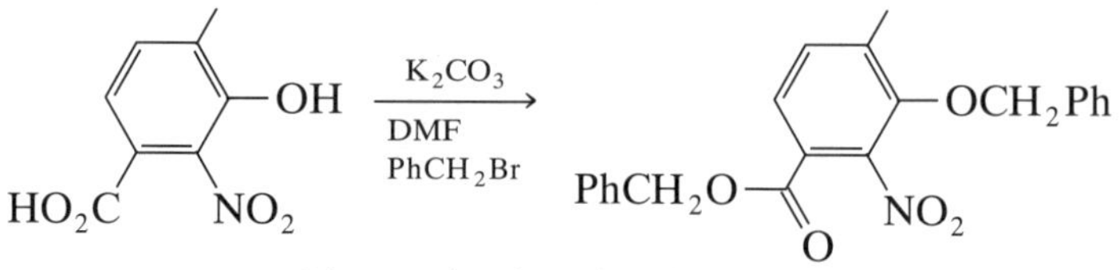

Propose a more reasonable mechanism for the alkylation than that shown in the following step. Formation of the product would involve deprotonation of the positively charged oxygen.

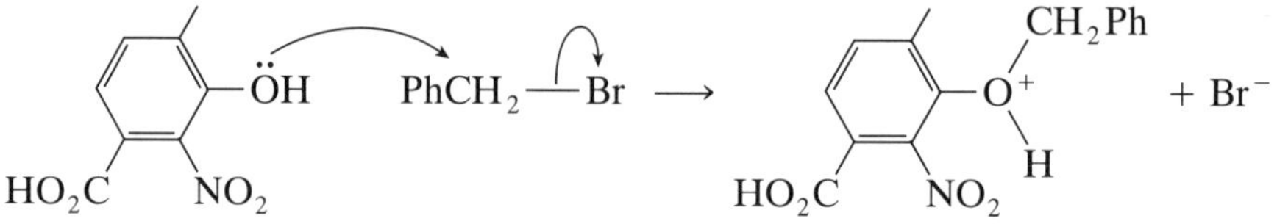

Pena, M. R.; Stille, J. K. *J. Am. Chem. Soc.* **1989**, *111*, 5417–5424.

PROBLEM 3.3 **Pick out the electrophile, nucleophile, and leaving group in each of the following reactions and write a mechanism for the formation of products.**

a.

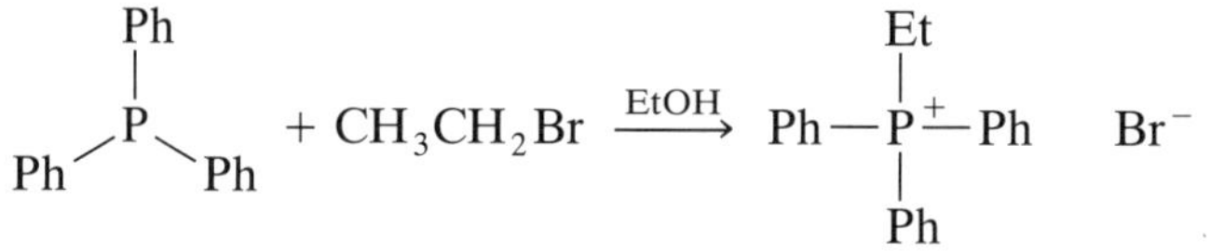

b. 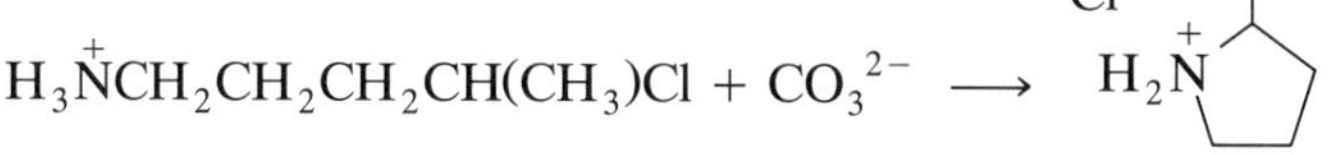 $H_3\overset{+}{N}CH_2CH_2CH_2CH(CH_3)Cl + CO_3^{2-} \longrightarrow$

c.

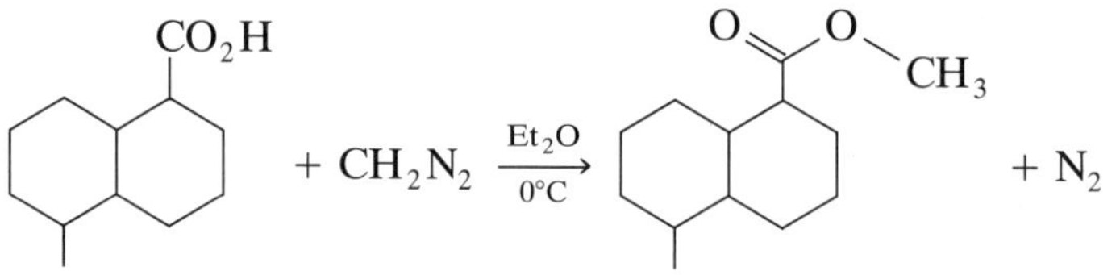

d.

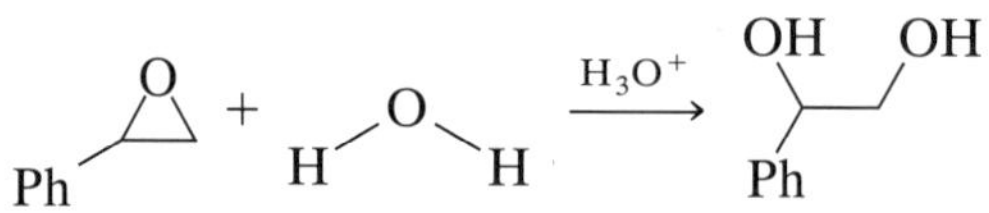

Neighboring Group Participation

On occasion, a molecule undergoing nucleophilic substitution may contain a nucleophilic group that participates in the reaction. This is known as the *neighboring group effect* and usually is revealed by *retention of stereochemistry* in the nucleophilic substitution reaction or by an *increase in the rate* of the reaction.

Example 3.4. *Neighboring group participation in the hydrolysis of ethyl 2-chloroethyl sulfide.*

The hydrolysis of the chlorosulfide proceeds to give the expected product. However, the reaction is 10,000 times faster than the reaction of the corresponding ether, $ClCH_2CH_2OCH_2CH_3$. This rate enhancement has been credited to the ready formation of a cyclic sulfonium ion due to intramolecular displacement of chloride by sulfur, followed by rapid nucleophilic reaction of water with the intermediate sulfonium ion.

Other groups that exhibit this behavior include thiol, sulfide, alkoxy (RO^-), ester, halogen, and phenyl.

Write step-by-step mechanisms for the following transformations: **PROBLEM 3.4**

PROBLEM 3.4
continued

b.

Wenkert, E.; Arrhenius, T. S.; Bookser, B.; Guo, M.; Mancini, P. *J. Org. Chem.* **1990**, *55*, 1185–1193.

B. Nucleophilic Substitution at Aliphatic sp^2 Carbon (Carbonyl Groups)

The familiar substitution reactions of derivatives of carboxylic acids with basic reagents illustrate nucleophilic substitution at aliphatic sp^2 carbons. (Substitution reactions of carboxylic acids, and their derivatives, with acidic reagents are covered in Chapter 4.) The mechanisms of these reactions involve two steps: (1) addition of the nucleophile to the carbonyl group and (2) elimination of some other group attached to that carbon. Common examples include the basic hydrolysis and aminolysis of acid chlorides, anhydrides, esters, and amides.

Example 3.5.   *Mechanism for hydrolysis of an ester in base.*

Unlike the one-step S_N2 reaction, the hydrolysis of esters in base is a two-step process. The net result is substitution, but the first step is nucleophilic addition to the carbonyl group, during which the carbonyl carbon becomes sp^3-hybridized. The second step is an elimination, in which the carbonyl group is regenerated as the carbon rehybridizes to sp^2.

This is followed by removal of a proton from the acid, by the methoxide ion, to yield methanol and the carboxylate ion:

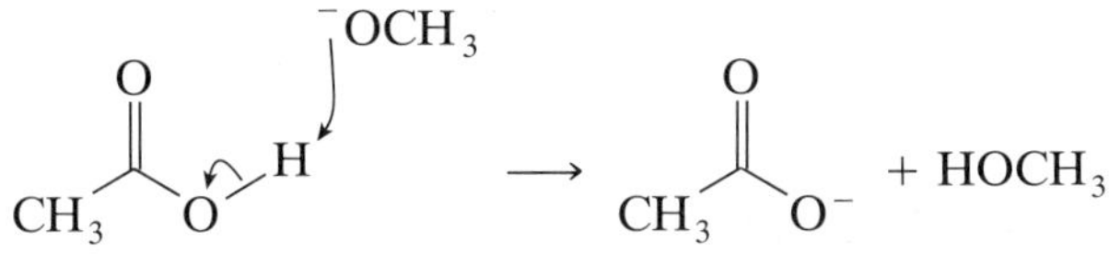

Another possible mechanism for this hydrolysis is an S_N2 reaction at the alkyl carbon of the ester:

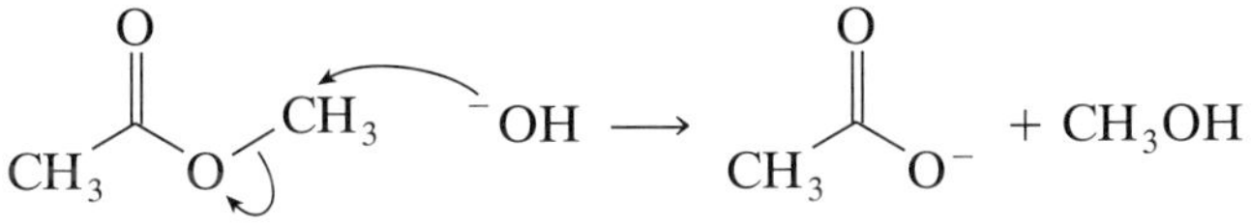

This single-step mechanism appears reasonable, because carboxylate is a fair leaving group and hydroxide is a very good nucleophile. However, labeling studies rule out this mechanism under common reaction conditions. The two-step mechanism must be favored because the higher mobility of the π electrons of the carbonyl group makes the carbonyl carbon especially electrophilic.

Hint 3.2

Direct nucleophilic substitution at an sp^2-hybridized center is not likely under common reaction conditions. Thus, nucleophilic substitution reactions at such centers usually are broken into two steps. (For exceptions to this hint, see Dietze, P.; Jencks, W. P. *J. Am. Chem. Soc.* **1989**, *111*, 5880–5886, and references cited therein.

There are several reasons why direct substitutions occur at sp^2-hybridized centers less readily than at sp^3 centers. First, because there is more *s* character in the bond to the leaving group, this bond is stronger than the corresponding bond to an sp^3-hybridized carbon. Second, the greater mobility of the π electrons at an sp^2 center increases the likelihood that the interaction will cause electron displacement. Third, because of the planar configuration of the substituents around an sp^2 center, there is strong steric interference to the approach of a nucleophile to the side opposite the leaving group. On the other hand, in the addition of a nucleophile to the carbonyl group, the nucleophile approaches perpendicular to the plane of the sp^2 orbitals so that there is maximum overlap with the π electron system. This means that the relatively unhindered addition step occurs in preference to direct substitution.

Thus, the mechanism for basic hydrolysis of an ester would *not be written* as follows:

PROBLEM 3.5   **Consider the mechanism shown for the following transformation. Propose a more reasonable alternative.**

$\longrightarrow$ Product

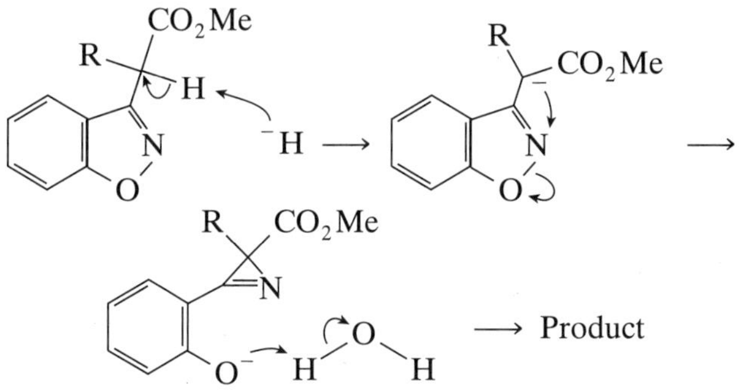

Write step-by-step mechanisms for the following transformations: **PROBLEM 3.6**

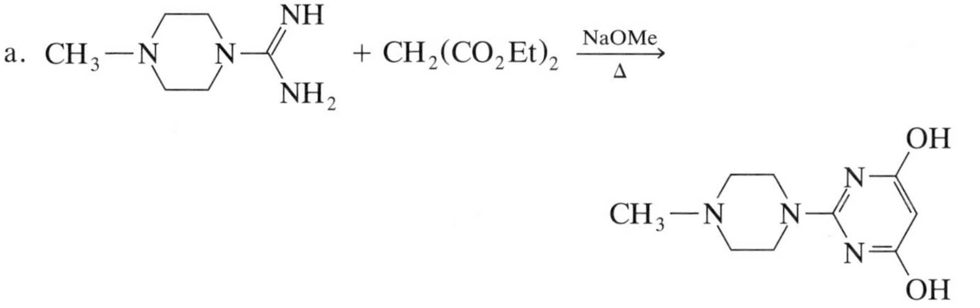

a.

Gueremy, C.; Audiau, F.; Renault, C.; Benavides, J.; Uzan, A.; Le Fur, *J. Med. Chem.* **1986**, *29*, 1394–1398.

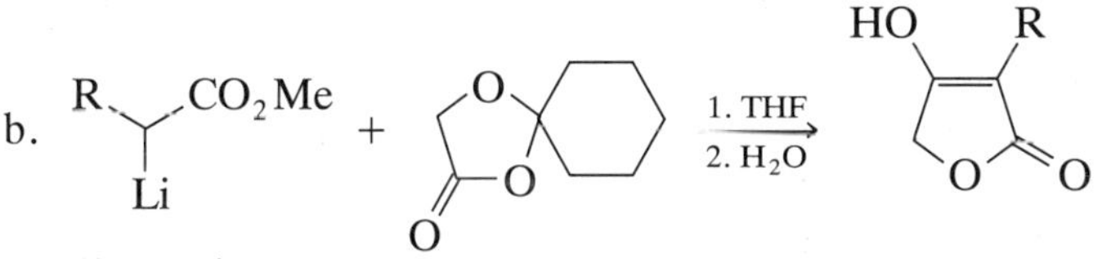

b.

62 mmol   25 mmol

Ramage, R.; Griffiths, G. J.; Shutt, F. E.; Sweeney, J. N. A. *J. Chem. Soc., Perkin Trans. I* **1984**, 1539–1545.

c. Critically evaluate the following partial mechanism for the reaction

PROBLEM 3.6
continued   given in part a:

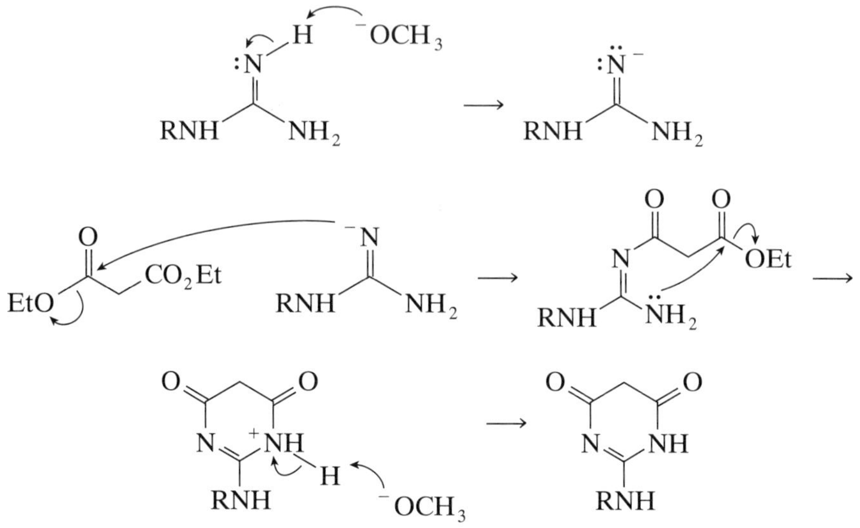

C. Nucleophilic Substitution at Aromatic Carbons

There are two mechanisms for nucleophilic aromatic substitution. Both occur in two important steps. In one mechanism, an addition is followed by an elimination. In the other mechanism, an elimination is followed by an addition.

Addition – Elimination Mechanism

The addition–elimination mechanism generally requires a ring activated by electron-withdrawing groups. These groups are especially effective at stabilizing the negative charge in the ring when they are located at positions *ortho* and/or *para* to the eventual leaving group.

Example 3.6. *Relative reactivity in the addition–elimination mechanism.*

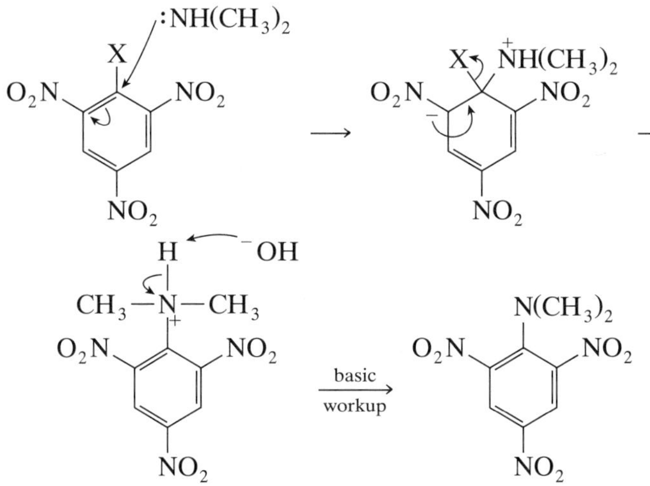

When X = halogen, the observed relative reactivities of the starting materials are F > Cl > Br > I. This indicates that the first step is rate-determining because the greater the electron-withdrawing power of the halogen (see Table 1.2), the more it increases the electrophilicity of the aromatic ring, making it more reactive to nucleophiles. If the second step were rate-determining, the relative reactivities would be reversed, because the relative abilities of the leaving groups are $I^- > Br^- > Cl^- > F^-$.

By drawing the appropriate resonance forms, show that the negative charge in the intermediate anion in Example 3.6 is stabilized by extensive electron delocalization.

PROBLEM 3.7

PROBLEM 3.8   **Write a step-by-step mechanism for the following transformation:**

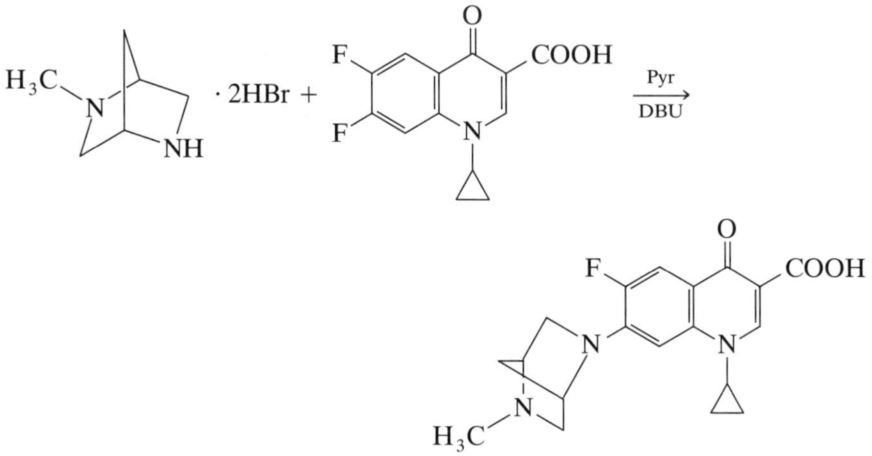

Braish, T. F.; Fox, D. E. *J. Org. Chem.* **1990**, *55*, 1684–1687. [This is the last step in the synthesis of danofloxicin, an antibacterial. Pyr (or Py) is a common acronym for pyridine; DBU is 1,8-diazabicyclo[5.4.0]undec-7-ene. A good reference for the translation of acronyms is Daub, G. H.; Leon, A. A.; Silverman, I. R.; Daub, G. W.; Walker, S. B. *Aldrichim. Acta* **1984**, *17*, 13–23.]

Elimination – Addition (Aryne) Mechanism

In reactions that proceed by the elimination–addition mechanism (often called the aryne mechanism), the bases used commonly are stronger than those used in reactions proceeding by the addition–elimination mechanism. Also, in this reaction, the aromatic ring does *not* need to be activated by electron-withdrawing substituents, although a reasonable leaving group (usually a halide) must be present.

Example 3.7.   *An elimination–addition mechanism—aryne intermediate.*

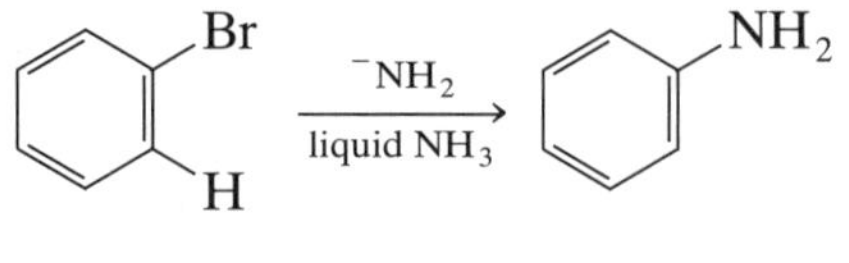

The following mechanism can be written for this reaction:

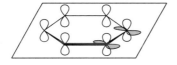

The intermediate with a triple bond is called benzyne. For substituted aromatic compounds, this type of intermediate is called an aryne. In benzyne, the ends of the triple bond are equivalent, and either can react with a nucleophile.

The triple bond in an aryne is not a normal triple bond. The six-membered ring does not allow the normal linear configuration of two sp-hybridized carbon atoms and their substituents. Thus, the carbons remain sp^2-hybridized, and the triple bond contains the σ bond, the π bond, and a third bond formed by overlap of the sp^2-hybridized orbitals that formerly bonded with the bromine and hydrogen atoms. This third bond is in the plane of the benzene ring and contains two electrons.

The rate-determining step can be either proton removal or departure of the leaving group, depending on the acidity of the proton and the ability of the leaving group. In many cases, the relative rates are so close that the reaction cannot be distinguished from a concerted process.

PROBLEM 3.9   Assume that in Example 3.7, the carbon bound to the bromine in bromobenzene is labeled by enrichment with ^{13}C. Where would this label be found in the product aniline?

PROBLEM 3.10   Write a step-by-step mechanism for the following transformation:

Bunnett, J. F.; Skorcz, J. A. *J. Org. Chem.* **1962**, *27*, 3836–3843.

2. ELIMINATIONS AT SATURATED CARBON

Important eliminations at saturated carbon are the E2 (bimolecular elimination) and Ei (intramolecular elimination) processes.

A. E2 Elimination

The E2 reaction is a concerted process, with a bimolecular rate-determining step. In this case, "concerted" means that bonding of the base with a proton, formation of a double bond, and departure of the leaving group all occur in one step.

Stereochemistry

The stereochemistry is usually *anti*, but in some cases is *syn*. The term *anti* means that the proton and leaving group depart from opposite sides of the bond, which then becomes a double bond. That is, the dihedral angle (measured at this bond) between their planes of departure is 180°. If they depart from the same side (the dihedral angle is 0°), the stereochemistry of the elimination is called *syn*.

syn elimination
(dihedral angle 0°)

anti elimination
(dihedral angle 180°)

Example 3.8. *An anti-E2 elimination.*

The dihedral angle between the proton and bromide is 180°, so this is an *anti* elimination.

Leaving Groups

The nature of the leaving group influences whether the reaction proceeds by an E2 mechanism. An excellent leaving group like $CH_3SO_3^-$ (mesylate) will favor competing reactions that proceed through a carbocation. Poor leaving groups, due to their failure to react, will allow competing reactions via anionic mechanisms.

Example 3.9. *In an acid-catalyzed elimination of water from an alcohol, water is the leaving group.*

The mechanism for an elimination step in the acid-catalyzed aldol condensation is written as follows:

The following step is less likely for the formation of an α,β-unsaturated aldehyde in acid (see Hint 2.5):

Example 3.10. *Under some conditions, hydroxide can act as a leaving group.*

A 3-hydroxyaldehyde (or ketone) will undergo elimination under basic conditions if the double bond being formed is especially stable, e.g., conjugated with an aromatic system. Such eliminations can occur under the reaction conditions of the base-promoted aldol condensation. An example is the formation of 3-phenyl-2-butenal by an E2 elimination from 3-hydroxy-3-phenylbutanal.

B. Ei Elimination

In another type of elimination reaction, called Ei or intramolecular, the base, which removes the proton, is another part of the same molecule. Such eliminations from amine oxides or sulfoxides have five-membered-ring transition states. These transition states are more stable with *syn* than with *anti* orientations of proton and leaving group, producing very high *syn* stereoselectivity.

Example 3.11. *An Ei reaction: Pyrolytic elimination from a sulfoxide.*

Curran, D. P.; Jacobs, P. B.; Elliott, R. L.; Kim, B. H. *J. Am. Chem. Soc.* **1987**, *109*, 5280–5282.

Write a mechanism for the following reaction. What is the other **PROBLEM 3.11**
product?

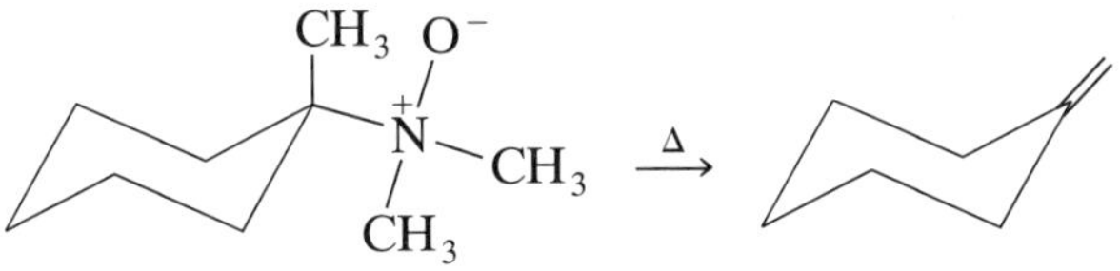

3. NUCLEOPHILIC ADDITION TO CARBONYL COMPOUNDS

Nucleophilic addition to the carbonyl groups of aldehydes and ketones occurs readily, and the carbonyl groups of carboxylic acids and their derivatives (acid chlorides, anhydrides, amides, and esters) also react with nucleophiles. In this section, the numerous nucleophilic addition reactions of carbonyl groups are organized first by the type of nucleophile (e.g., organometallics, nitrogen nucleophiles, carbon nucleophiles) and then according to the kind of carbonyl group.

Addition reactions with organometallic reagents usually are irreversible, but many other addition reactions are reversible. In these reversible additions, equilibrium may favor the starting materials. When the equilibrium does not favor product formation, the reaction can be made productive if the initial product is removed, either physically or by undergoing further reaction, as in Example 3.18.

A. Additions of Organometallic Reagents

Reactions of either Grignard or organolithium reagents with most aldehydes, ketones, or esters produce alcohols. Reactions of organolithium reagents with carboxylic acids, or of Grignard reagents with nitriles, produce ketones.

TABLE 3.2   Addition of Organometallic Reagents to Carbonyl Compounds and
Carboxylic Acid Derivatives

	$\overset{\displaystyle O}{\overset{\|}{R'CR''}}$	$\overset{\displaystyle O}{\overset{\|}{R'COEt}}$	$\overset{\displaystyle O}{\overset{\|}{R'CCl}}$	$\overset{\displaystyle O}{\overset{\|}{R'COH}}$	$R'C{\equiv}N$
RMgX	$R'\!-\!\overset{\overset{\textstyle OH}{\|}}{\underset{\underset{\textstyle R}{\|}}{C}}\!-\!R''$	$R'\!-\!\overset{\overset{\textstyle OH}{\|}}{\underset{\underset{\textstyle R}{\|}}{C}}\!-\!R$	$R'\!-\!\overset{\overset{\textstyle OH}{\|}}{\underset{\underset{\textstyle R}{\|}}{C}}\!-\!R$	$R'CO_2MgX$	$R'\!-\!\overset{\overset{\textstyle NH}{\|}}{C}\!-\!R \to \overset{\displaystyle O}{\overset{\|}{R'CR}}$
RLi	$R'\!-\!\overset{\overset{\textstyle OH}{\|}}{\underset{\underset{\textstyle R}{\|}}{C}}\!-\!R''$	$R'\!-\!\overset{\overset{\textstyle OH}{\|}}{\underset{\underset{\textstyle R}{\|}}{C}}\!-\!R, \ \overset{\displaystyle O}{\overset{\|}{R'CR}}$	—	$R'\!-\!\overset{\overset{\textstyle OH}{\|}}{C}HR$	—

The addition reactions of Grignard and organolithium reagents with carbonyl groups and carboxylic acid derivatives are summarized in Table 3.2.

Additions to Aldehydes and Ketones

The reactions of Grignard reagents (RMgX) and alkyllithium reagents (RLi) with aldehydes and ketones are similar. The mechanism is illustrated by the following example.

Example 3.12.   *The Grignard reaction of phenylmagnesium bromide with benzophenone.*

3-1

The electron pair in the carbon–magnesium bond of phenylmagnesium bromide is the nucleophile, and the carbonyl carbon of the ketone is the electrophile. Also, magnesium is an electrophile and the carbonyl oxygen is a nucleophile, so that the salt of an alcohol is the product of the reaction. The alcohol itself is generated by an acidic workup.

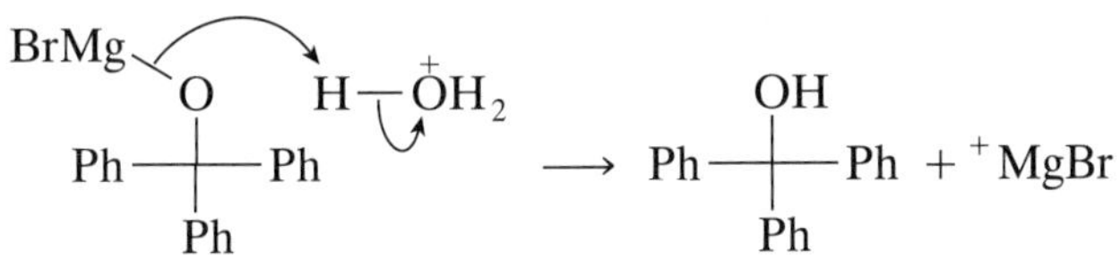

3-1

Additions to Carboyxlic Acid Derivatives

The stability of the intermediate formed by the addition of an organometallic reagent to a carboxylic acid derivative determines the product produced in the subsequent steps.

Hint 3.3

With organolithium reagents, stable intermediates are produced by addition to a carboxylic acid. On workup, these intermediates produce ketones. As with Grignard reagents, the reaction of organolithium reagents with esters produces tertiary alcohols because the intermediates decompose to ketones under the reaction conditions. The mechanisms for these processes are illustrated in the examples that follow.

Example 3.13. *Grignard addition to a nitrile.*

3-2

The protonation of the intermediate, **3-2**, to give **3-3** is similar to the protonation of **3-1** in the previous example.

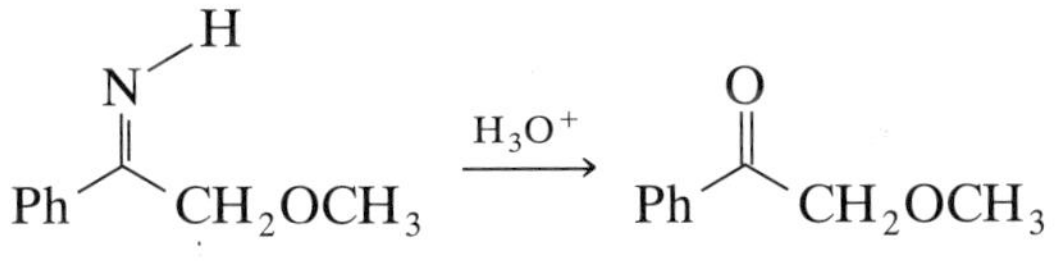

3-3   **3-4**

Then the initial product **3-3** can be hydrolyzed to the ketone **3-4**. (See the answer to Problem 3.14 for the mechanism of this reaction.)

Example 3.14.   *Addition of an organolithium reagent to a carboxylic acid.*

The reaction requires 2 mol of organolithium reagent per mole of acid. The first mole of organolithium reagent neutralizes the carboxylic acid, giving a salt.

The second mole adds to the carbonyl group to give a dilithium salt, **3-5**, which is stable under the reaction conditions. Sequential hydrolysis of each OLi group in acid, during workup, gives a dihydroxy compound, **3-6**, which is the hydrate of a ketone. A series of protonations and deprotonations transforms the hydrate into a species that can eliminate a molecule of water to form the ketone.

Example 3.15. *Grignard reaction of an ester.*

Esters react with 2 mol of Grignard reagent to give the salt of an alcohol. For example, reaction of ethyl benzoate with 2 mol of phenylmagnesium bromide gives a salt of triphenylmethanol. The first addition gives an intermediate, **3-7**, which is unstable under the reaction conditions.

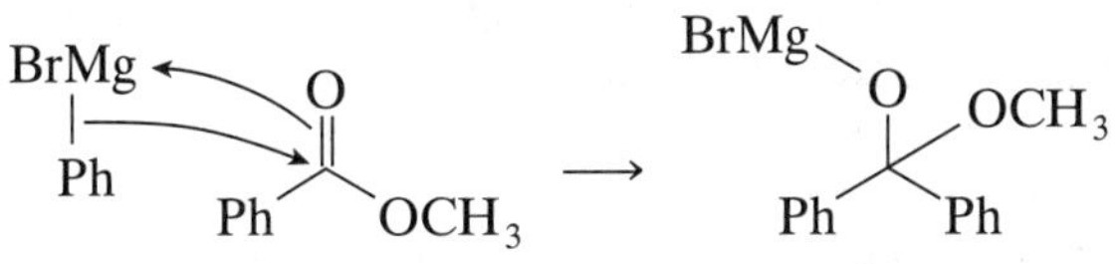

3-7

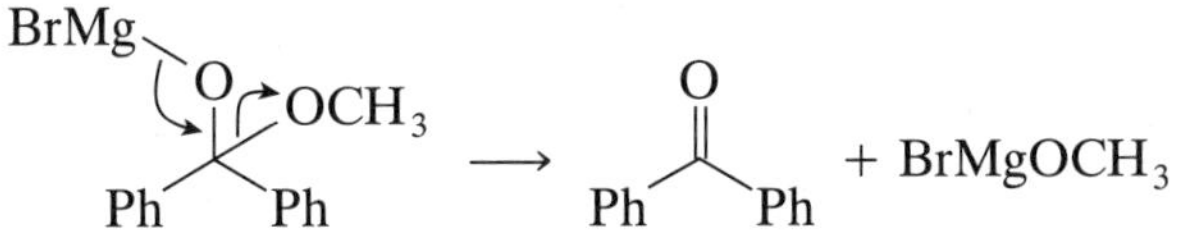

3-7

Another molecule of phenylmagnesium bromide now reacts with benzophenone, as shown in Example 3.12.

Write step-by-step mechanisms for the following reactions: **PROBLEM 3.12**

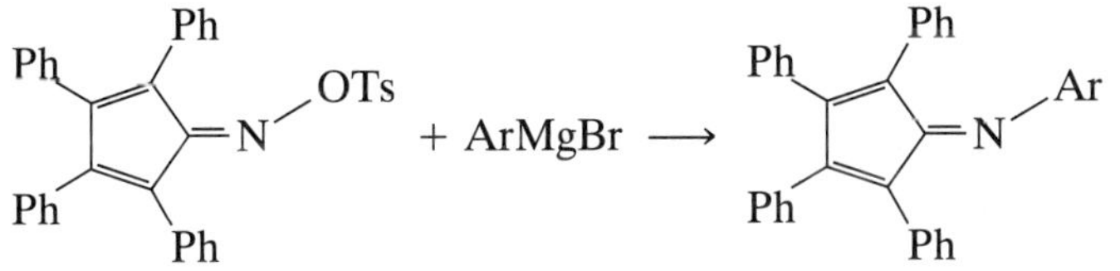

(Ar = Aryl)

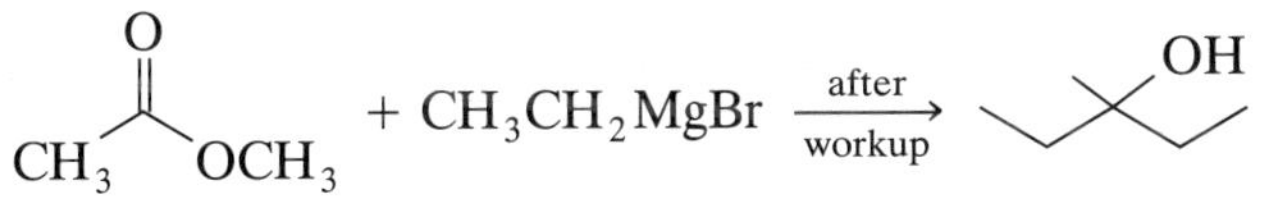

Hagopian, R. A.; Therien, M.J.; Murdoch, J. R. *J. Am. Chem. Soc.* **1984**, *106*, 5753–5754.

B. Reaction of Nitrogen-Containing Nucleophiles with Aldehydes and Ketones

A number of reactions of nitrogen-containing nucleophiles with aldehydes and ketones involve addition of the nitrogen to the carbon of the carbonyl group, followed by elimination of water to produce a double bond. Common examples are reactions of primary amines to produce substituted imines, reactions of secondary amines to produce enamines, reactions of hydrazine or substituted hydrazines to produce hydrazones, reactions of semicarbazides to give semicarbazones, and reactions of hydroxylamine to produce oximes. Usually these reactions are run with an acid catalyst.

In the synthesis of imines and enamines by this method, the water produced in the reaction must be removed azeotropically to drive the reaction to the right. In aqueous acid, equilibrium conditions favor the ketone rather than the imine. This relationship is the reason why Grignard reaction of a nitrile provides a good route to the synthesis of ketones. The intermediate imine formed is hydrolyzed easily to the corresponding ketone (e.g., the transformation of **3-3** to **3-4** in Example 3.13).

Example 3.16. *Mechanism for formation of a hydrazone.*

The first step in a mechanism for the following synthesis of a phenylhydrazone is an equilibrium protonation of the carbonyl oxygen.

$$PhNHNH_2 + \underset{}{\overset{O}{\underset{}{\parallel}}} \xrightarrow[NaOAc]{H_3O^+} \rangle=N-NH-Ph$$

The protonated carbonyl group then is more susceptible to reaction with a nucleophile than the neutral compound (note that the protonated carbonyl group is a resonance hybrid).

$$\rangle=O: \curvearrowright H-\underset{+}{O}H_2 \rightleftharpoons \rangle=\underset{+}{O}H \longleftrightarrow +\rangle-OH$$

The more nucleophilic nitrogen of the hydrazine reacts at the electrophilic carbon of the carbonyl group. Loss of a proton, facilitated by base ($^-$OAc) is followed by acid-catalyzed elimination of water.

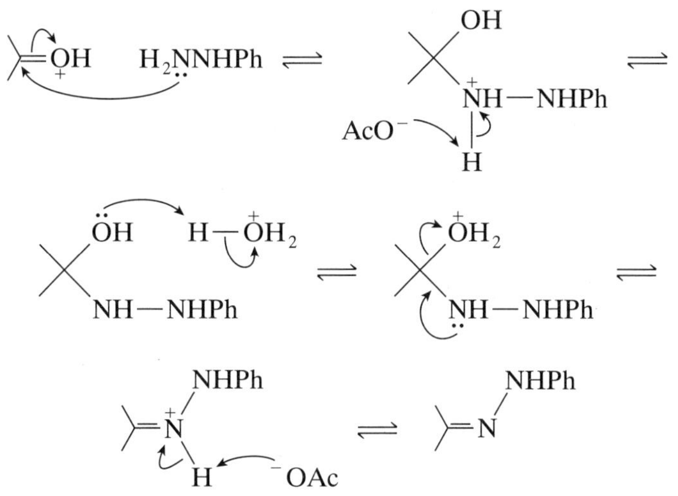

Explain why the nitrogen in phenylhydrazine that acts as the nucleophile is the nitrogen without the phenyl substituent.

PROBLEM 3.13

Write step-by-step mechanisms for the following transformations:

PROBLEM 3.14

a.

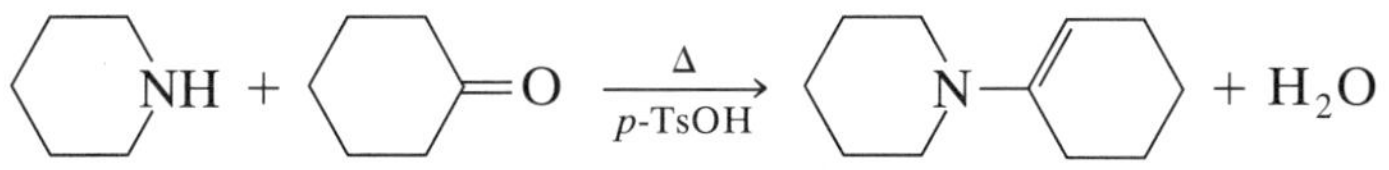

b.

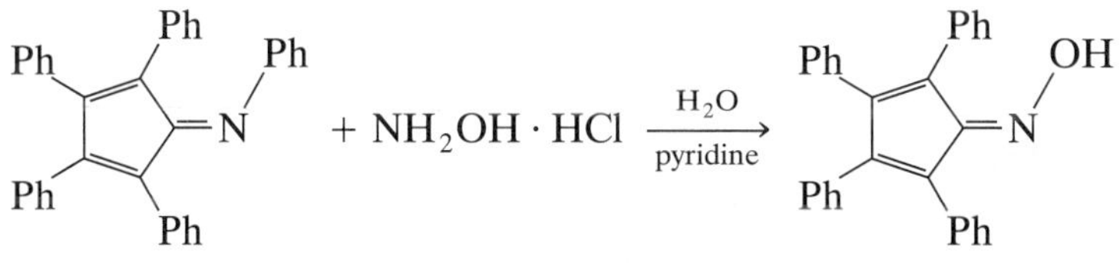

Hagopian, R. A.; Therian, J. J.; Murdoch, J. R. *J. Am. Chem. Soc.* **1984**, *106*, 5753–5754.

c.

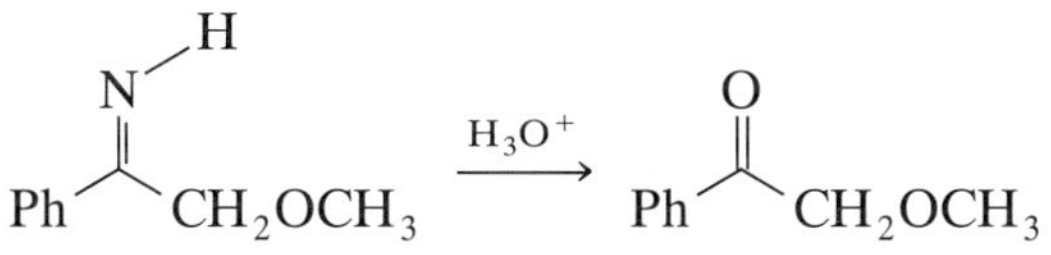

C. Reactions of Carbon Nucleophiles with Carbonyl Compounds

The Aldol Condensation

The aldol condensation involves the formation of an anion on a carbon α to an aldehyde or ketone carbonyl group, followed by nucleophilic reaction of that anion at the carbonyl group of another molecule. The reaction may involve the self-reaction of an aldehyde or ketone or the formation of the anion of one compound and reaction at the carbonyl of a different compound. The latter is called a mixed aldol condensation.

Example 3.17. *Condensation of acetophenone and benzaldehyde: Nucleophilic addition of an anion to a carbonyl group followed by an elimination.*

This reaction is a mixed aldol condensation of an aldehyde and a ketone.

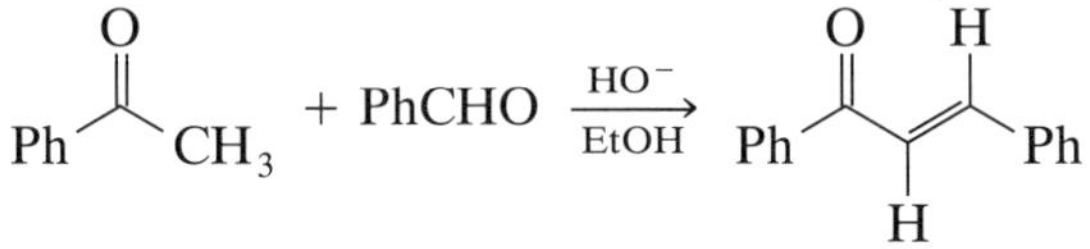

Consider a step-by-step mechanism for this process. The first step is removal of a proton from the carbon α to the carbonyl group of the ketone to give a resonance-stabilized anion. (Note that removal of the proton directly attached to the aldehyde carbonyl carbon does not give a resonance-stabilized anion, and there are no hydrogens on the carbon α to the aldehyde carbonyl.)

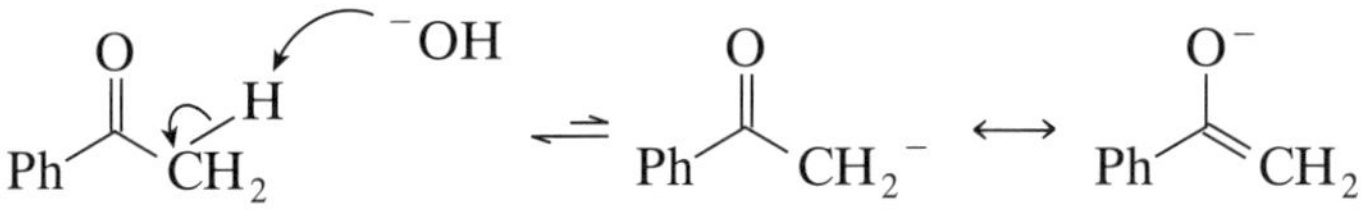

The equilibrium in this reaction favors starting material; in Problem 1.14.b, the equilibrium constant for this reaction was calculated to be approximately 10^{-9}. Nonetheless, the reaction continues to a stable product because the subsequent step has a much more favorable equilibrium constant. In this next step, the carbonyl group of the aldehyde undergoes nucleophilic addition by the enolate anion to give **3-8**:

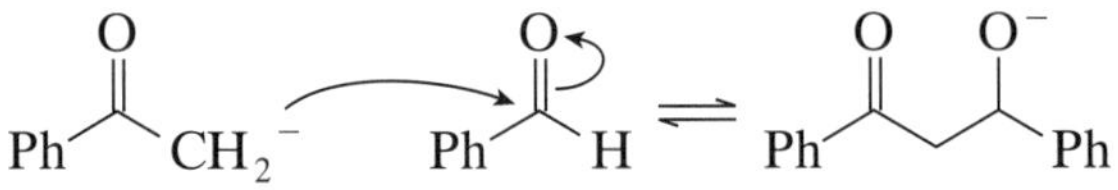

3-8

Why is there preferential reaction at the aldehyde carbon? In other words, why does the acetophenone anion react with the aldehyde instead of another acetophenone molecule? The ketone carbonyl is less reactive for two reasons. First, the tetrahedral intermediate formed by addition to the carbonyl group of a ketone is less stable than the intermediate formed by addition to an aldehyde because there is more steric interaction with the alkyl group of the ketone than with the corresponding hydrogen of the aldehyde. Second, inductive effects due to the two alkyl groups stabilize the carbonyl bond of a ketone relative to that of an aldehyde.

Anion **3-8** can remove a proton from the solvent, which often is ethanol.

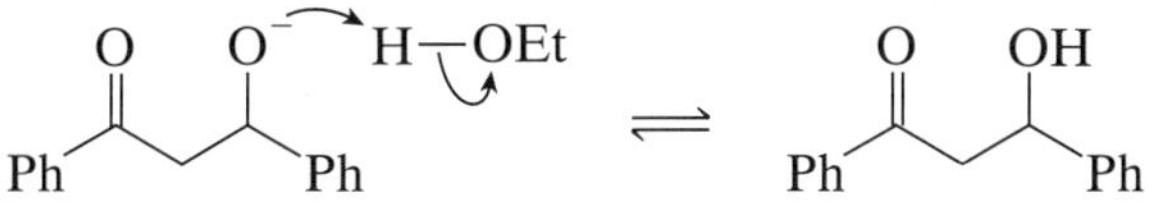

Finally, a base-promoted E2 elimination of water occurs to give the product. This elimination is driven energetically by the formation of a double bond, which is stabilized by conjugation with both a phenyl group and a carbonyl group.

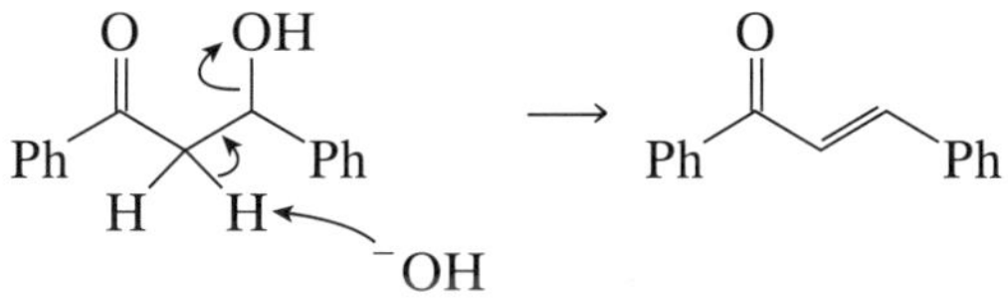

Write step-by-step mechanisms for the following reactions: **PROBLEM 3.15**

a.

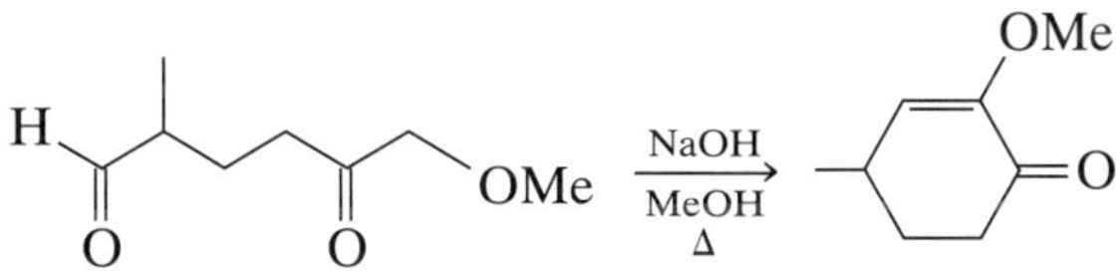

Gadwood, R. C.; Lett, R. M.; Wissinger, J. E. *J. Am. Chem. Soc.* **1984**, *106*, 3869–3870.

b. PhN=O + NCCH$_2$CO$_2$Et $\xrightarrow[\text{EtOH}]{\text{K}_2\text{CO}_3}$

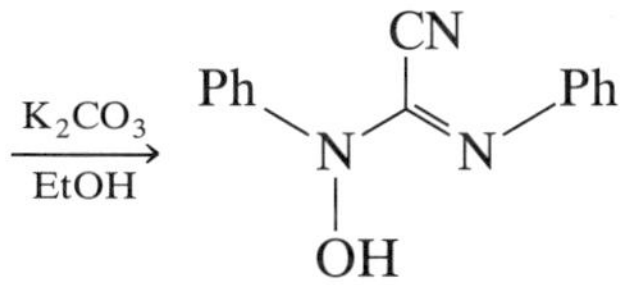

Bell, F. *J. Chem. Soc.* **1957**, 516–518.

The Michael Reaction and Other 1,4-Additions

The Michael reaction is addition of a carbon nucleophile to the β position of an α,β-unsaturated carbonyl compound or its equivalent. It also may be called a 1,4-addition reaction (the carbonyl oxygen is counted as 1 and the β-carbon as 4). The conjugation of the π bond with the carbonyl group imparts positive character to the β position, making it susceptible to reaction with a nucleophile. The product of this reaction, an enolate ion, also is stabilized by resonance.

When nucleophiles other than carbon add to α-β-unsaturated carbonyl compounds, the process is called a 1,4-addition.

Example 3.18.   *A typical Michael reaction.*

This example shows the addition of a fairly stable carbanion (stabilized by two adjacent carbonyl groups) to an α,β-unsaturated ketone.

Ramachadran, S.; Newman, M. S. *Org. Synth.* **1961**, *41*, 38.

The adduct formed initially is itself an enolate ion stabilized by resonance.

The enolate can remove a proton from the solvent to give the neutral product.

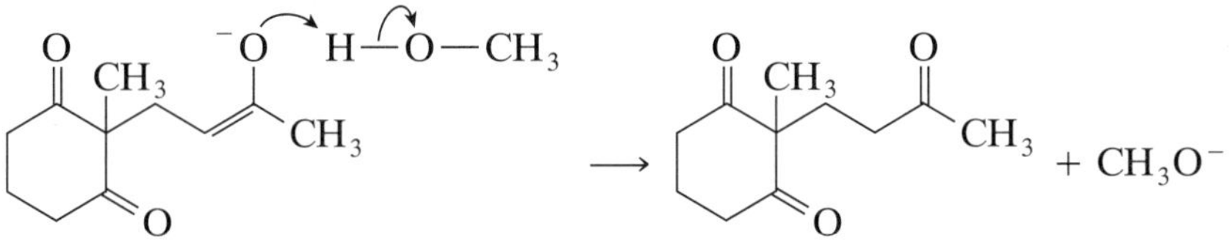

There are many instances in the literature where the Michael reaction is followed by subsequent steps. The following example is one of them.

Example 3.19. *1,4-Addition followed by subsequent reaction.*

The overall reaction is

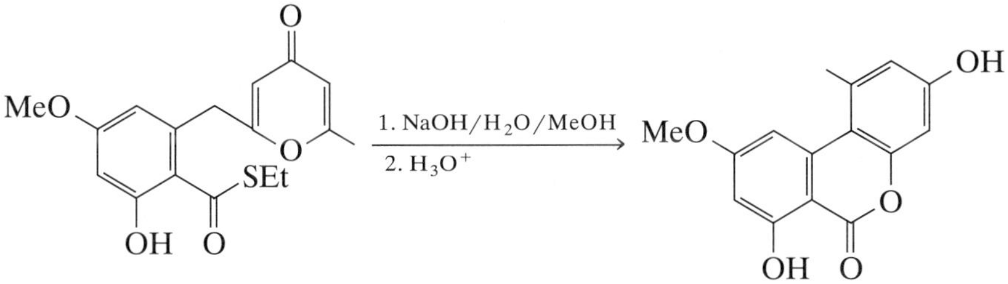

Abell, C.; Bush, B. D.; Staunton, J. *J. Chem. Soc., Chem. Commun.* **1986**, 15–17.

Analysis of the starting material indicates an acidic phenolic hydroxyl, a thioester susceptible to base-promoted hydrolysis, and an α,β-doubly unsaturated ketone that could undergo 1,4-addition followed by subsequent reaction. From the structure of the product, it is clear that both the thioester and the α,β-unsaturated ketone undergo reaction. Because hydroxide is the base, a proton will be removed readily from the phenolic hydroxyl group, forming **3-9-1**.

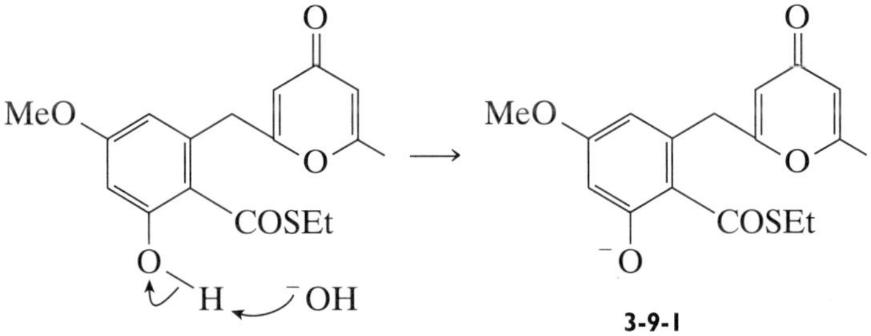

3-9-1

Whether the ester or one of the positions β to the carbonyl group reacts first cannot be ascertained from the data given. Thus, both possibilities are discussed.

Mechanism 1   A drawing of another resonance form of **3-9-1**, **3-9-2**, shows that the oxygen of the thioester has negative character and could act as an intramolecular nucleophile:

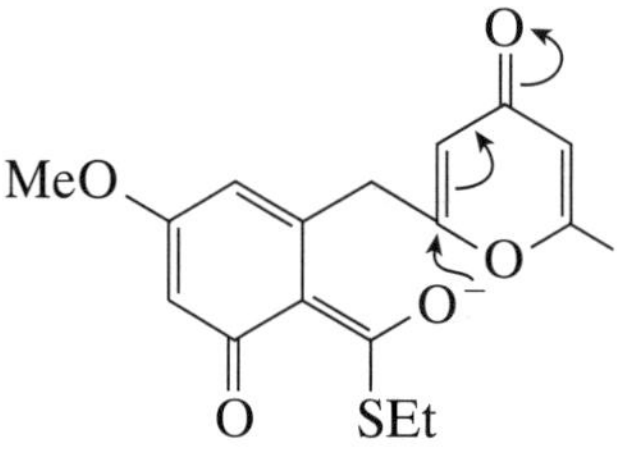

3-9-2	**3-10**

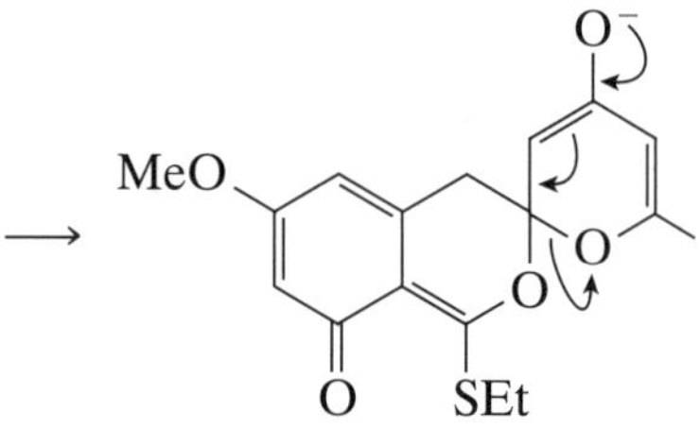

Comparison of **3-10** with the product reveals a central ring with the same atomic skeleton as the product, but a right-hand ring that does not. Thus, the latter opens to an enolate ion, **3-11**.

The enolate **3-11** can remove a proton from solvent to give **3-12**, which can undergo addition of hydroxide to give a resonance-stabilized anion, **3-13**.

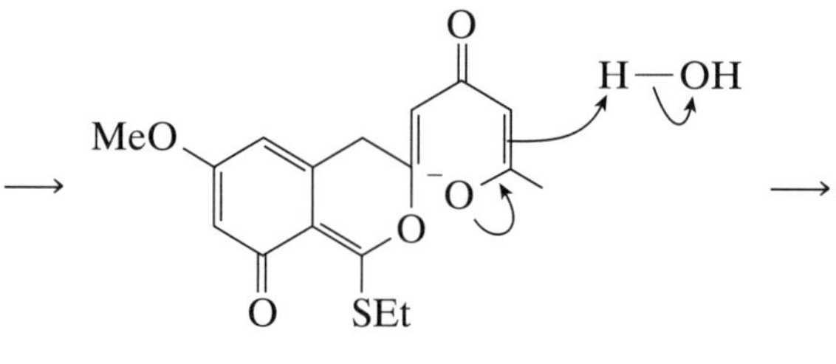

3-11

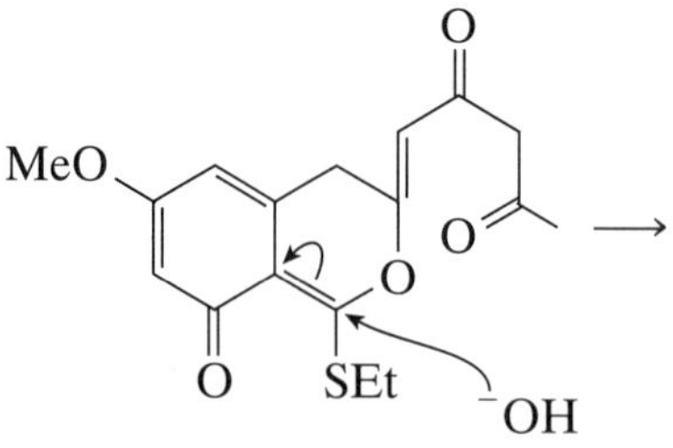

3-12 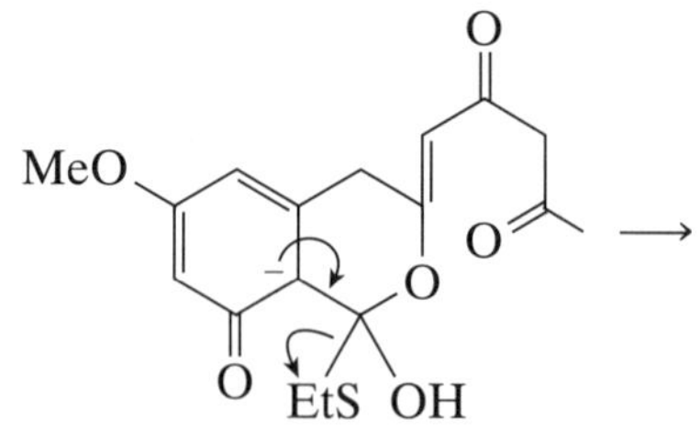   **3-13**

3-14

Ion **3-13** can lose ethyl thiolate to give **3-14**.

Intermediate **3-14** contains a number of acidic protons. Removal of some of these would give anions that probably react to give side products, and removal of others may result in unproductive equilibria. Removal of the proton shown gives a resonance-stabilized anion, **3-15**, which can react with the terminal carbonyl group of the side chain to form the tricyclic structure **3-16**.

3-15   **3-16**

The intermediate **3-16** removes a proton from solvent to give **3-17**, and **3-17** undergoes elimination of water to give **3-18**.

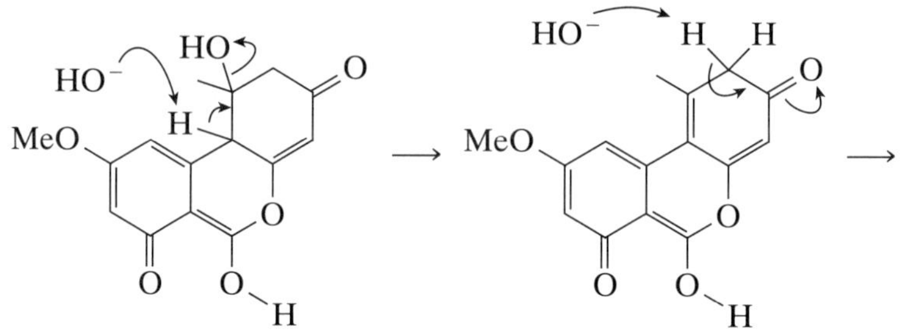

3-17   **3-18**

Removal of a proton from the right-hand ring of **3-18** gives a phenolate ion, **3-19**.

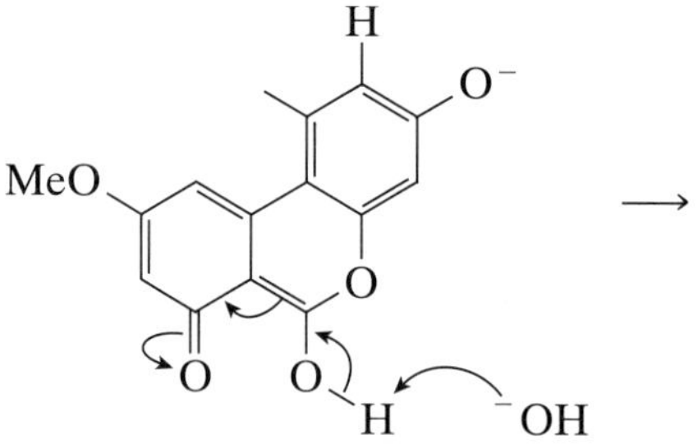

3-19

Removal of another proton from **3-19** gives a diphenolate, **3-20**, which can be protonated to give the product upon workup in aqueous acid.

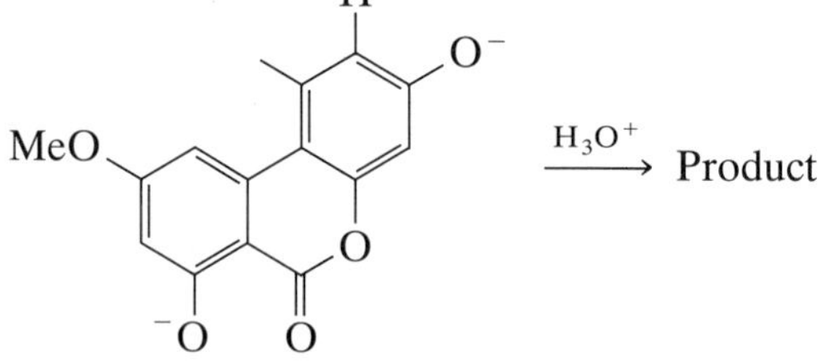

3-20

In this reaction, as in many others, the exact timing of steps, especially proton transfers, is difficult to anticipate. For example, the proton removed from **3-19** actually may be removed in an earlier step.

Mechanism 2 In this mechanism, formation of the right-hand ring occurs before formation of the middle ring. After the formation of **3-9**, there is 1,4-addition of hydroxide ion to the α,β-unsaturated ketone.

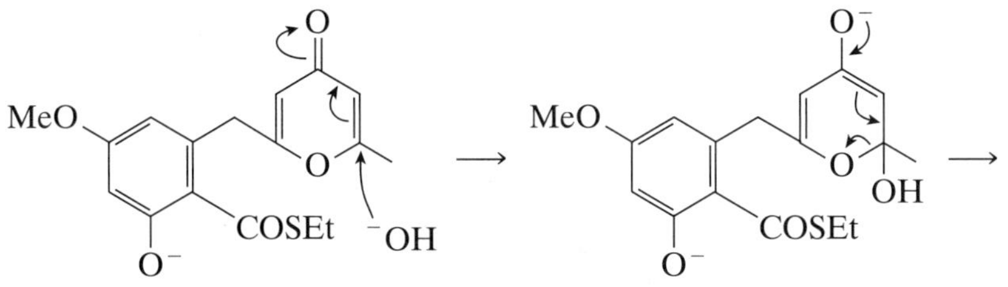

3-9   **3-21**

Ring opening of **3-21** follows to give a new enolate, **3-22**, which can be protonated at carbon by water to give **3-23**.

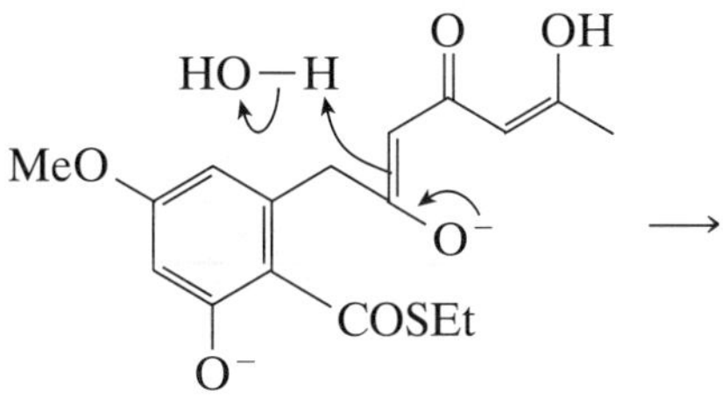

3-22

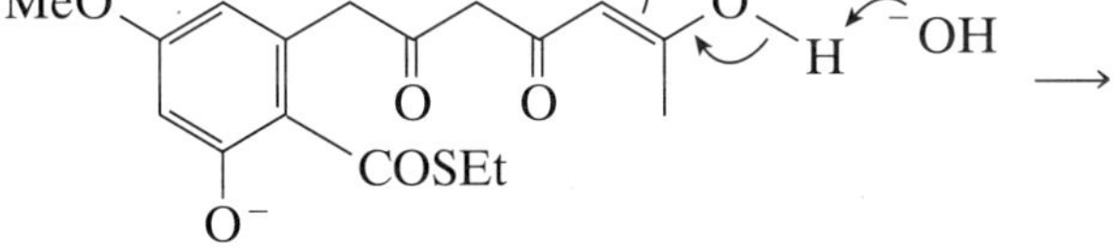

3-23

Base-promoted tautomerization of the enol in **3-23** gives **3-24**. An alternative route to the methyl ketone **3-24** starts with reaction of hydroxide on the other carbon β to the ketone in the right-hand ring of **3-9**. These steps follow a course analogous to that depicted (**3-21** to **3-22** to **3-23**).

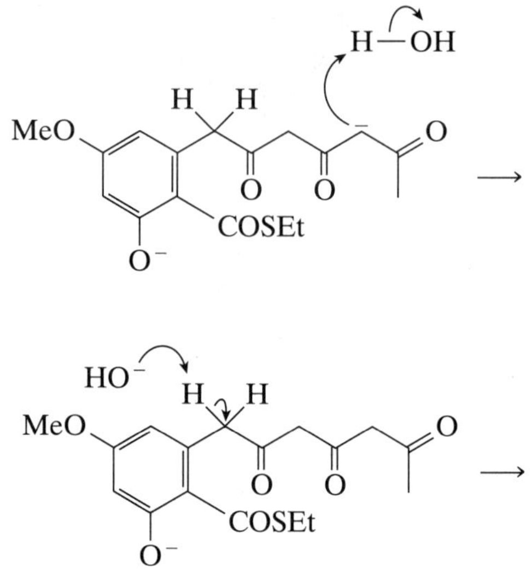

3-24

The mechanism then continues with an intramolecular aldol condensation.

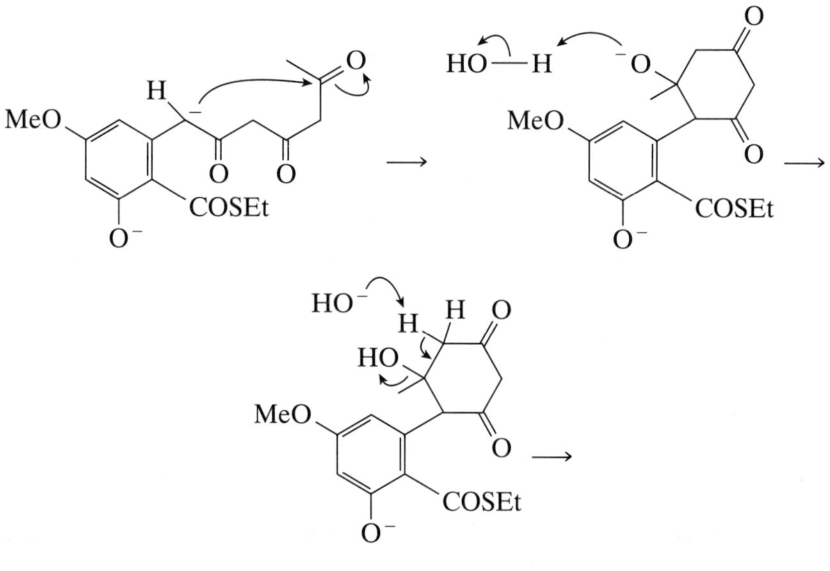

3-25

The resulting β-hydroxy ketone, **3-25**, can eliminate water to give **3-26**. Base-promoted tautomerization of one of the protons in the box in **3-26** gives the enol, and removal of the proton in the circle then gives the phenolate ion, **3-27**.

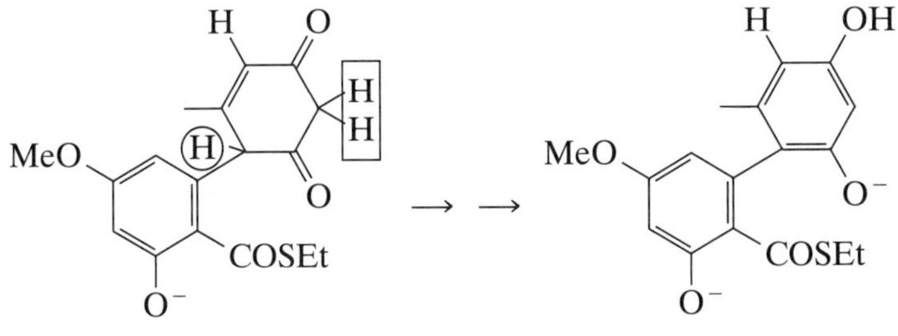

3-26   **3-27**

The nucleophilic phenolate oxygen atom of **3-27** adds to the carbon of the thioester group; then ethylthiolate is eliminated.

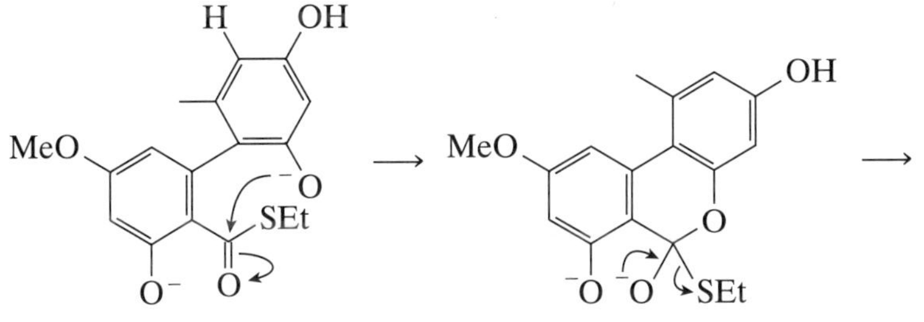

3-28

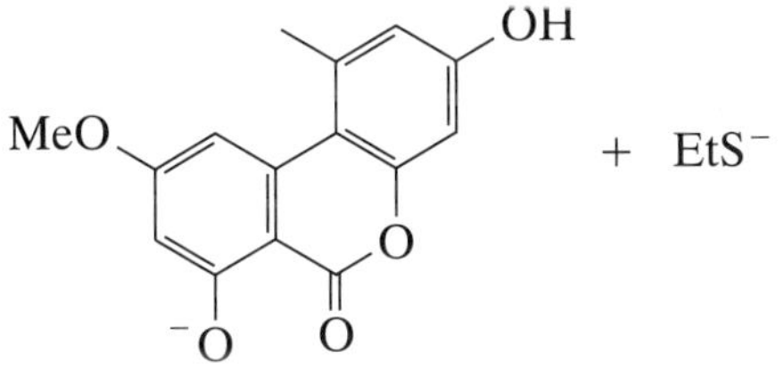

+ EtS⁻

3-29

Once the phenolate ion **3-27** has reacted, the phenol in the right-hand ring of either **3-28** or **3-29** reacts with hydroxide to give a new phenolate in this ring. One of these possibilities is represented next. (The phenolate ion in the right-hand ring of **3-27** reduces the acidity of the other phenolic group in that ring. Thus, we anticipate that the proton of the second phenolic group is removed in a later step.)

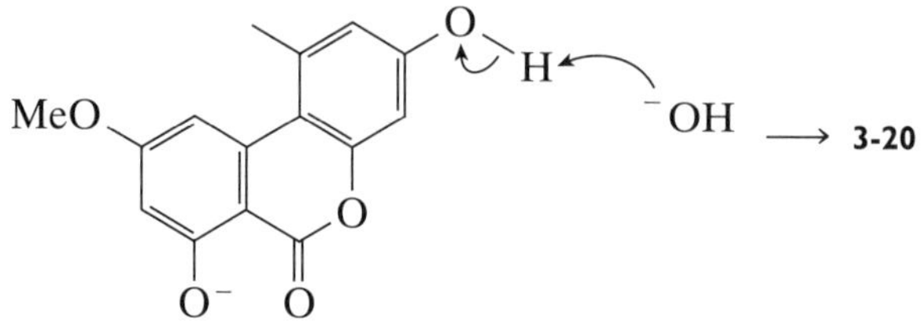

3-29

Both mechanisms for the reaction seem reasonable. The authors of the paper cited (Abell *et al.* 1986) showed that **3-30** also cyclizes to the product in excellent yield. Note that **3-30** is the phenol corresponding to the intermediate phenolate **3-24** of mechanism 2. This evidence does not prove that **3-24** is an intermediate in the reaction, but does support it as a viable possibility.

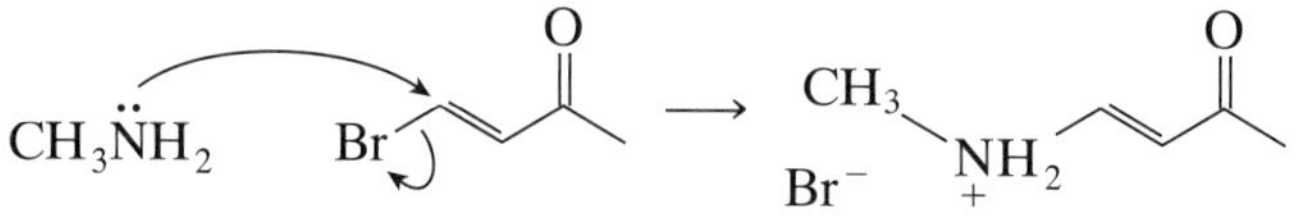

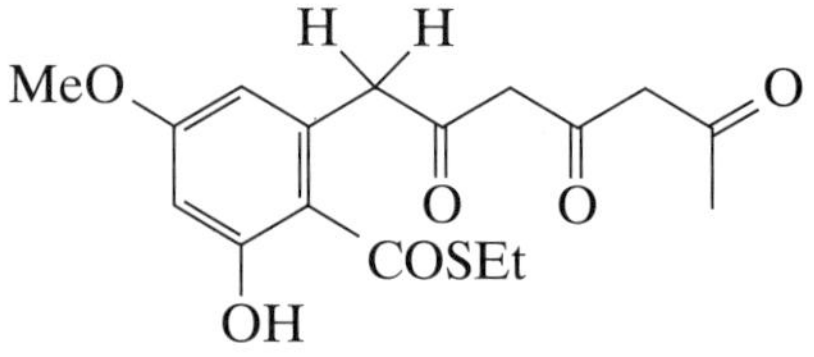

3-30

PROBLEM 3.16   **Why is the following mechanistic step unlikely? How would you change the mechanism to make it more reasonable?**

Write step-by-step mechanisms for the following transformations. **PROBLEM 3.17**

a.

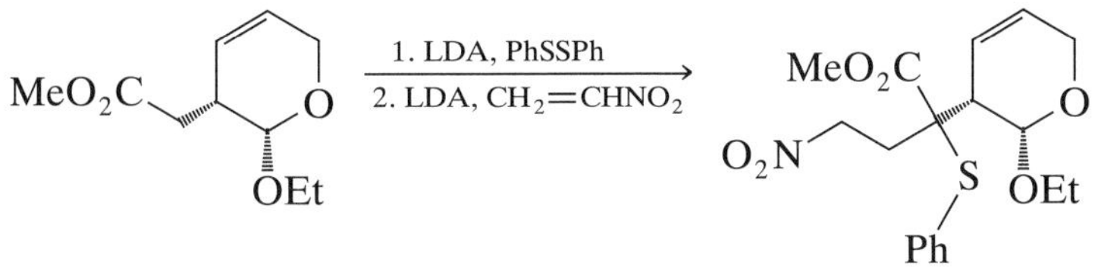

Curran, D. P.; Jacobs, P. B.; Elliott, R. L.; Kim, B. H. *J. Am. Chem. Soc.* **1987**, *109*, 5280–5282.

Lithium diisopropylamide (LDA) is a strong base, but not a good nucleophile because of steric inhibition by the two isopropyl groups directly attached to the nitrogen anion.

b.

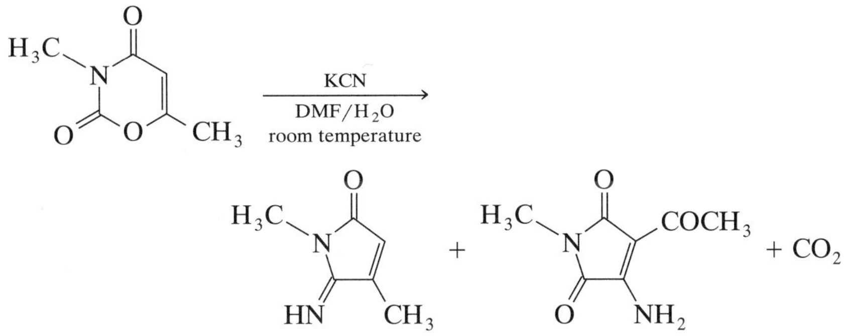

Yogo, M.; Hirota, K.; Maki, Y. *J. Chem. Soc., Perkin Trans. I* **1984**, 2097–2102.

4. BASE-PROMOTED REARRANGEMENTS

A. The Favorskii Rearrangement

A typical Favorskii rearrangement involves reaction of an α-halo ketone with a base to give an ester or carboxylic acid, as in the following example:

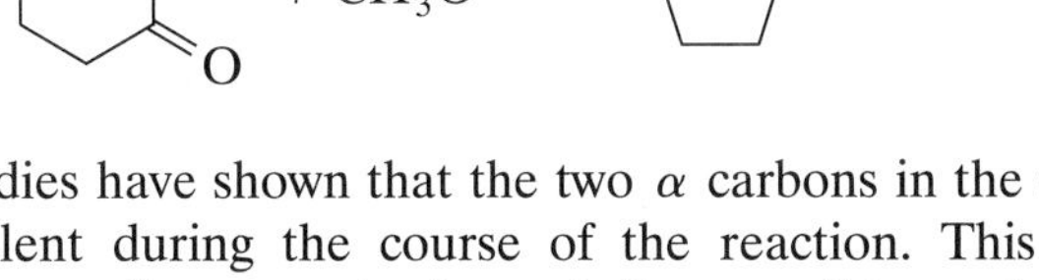

Labeling studies have shown that the two α carbons in the starting ketone become equivalent during the course of the reaction. This means that a symmetrical intermediate must be formed. One possible mechanism, which is consistent with this result, is as follows:

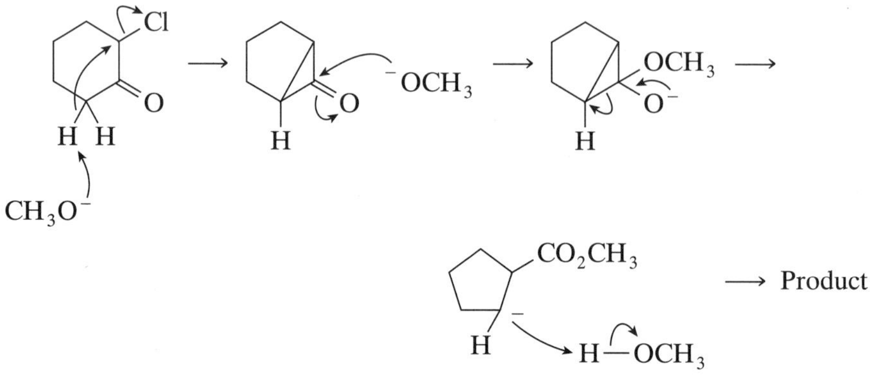

B. The Benzilic Acid Rearrangement

This is a rearrangement of an α-diketone, in base, to give an α-hydroxy-carboxylic acid. The reaction gets its name from the reaction of benzil to give benzilic acid:

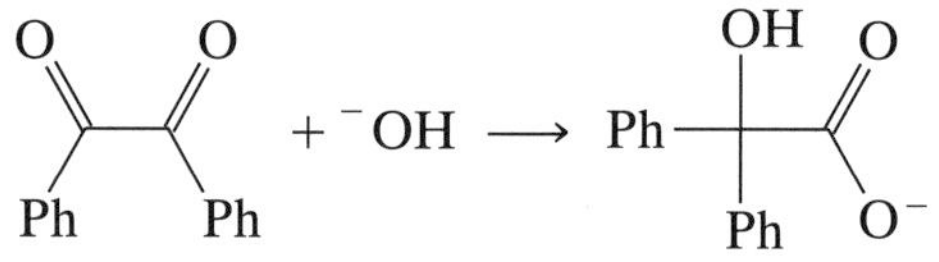

The mechanism involves nucleophilic addition of the base to one carbonyl group, followed by transfer of the substituent on that carbon to the adjacent carbon:

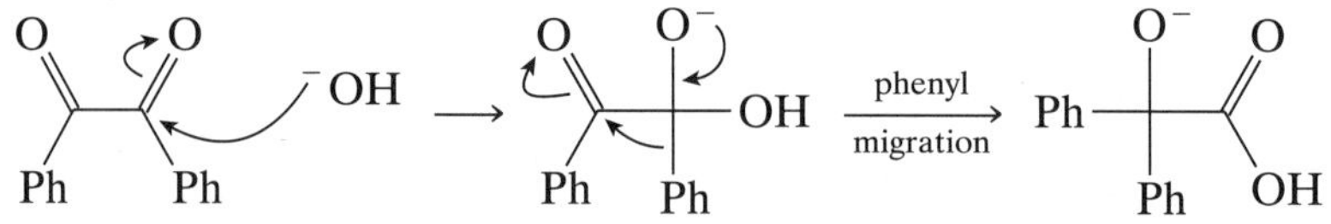

The final steps, under the reaction conditions, are protonation of the alkoxide and deprotonation of the carboxylic acid to give the corresponding carboxylate salt.

Write step-by-step mechanisms for the following transformations: **PROBLEM 3.18**

a.

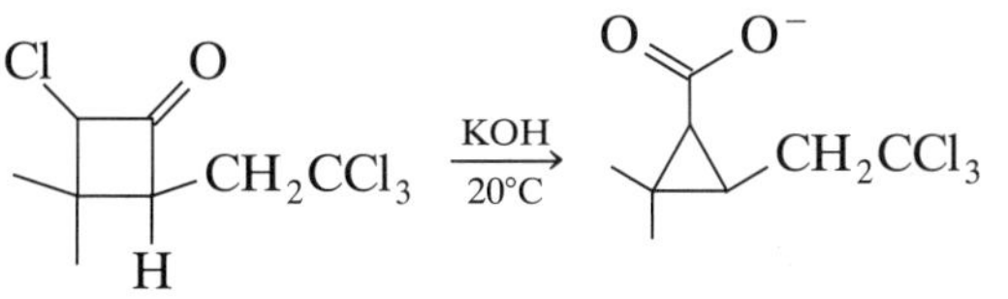

Martin, P.; Greuter, H.; Bellus, D. *J. Am. Chem. Soc.* **1979**, *101*, 5853–5854.

b.

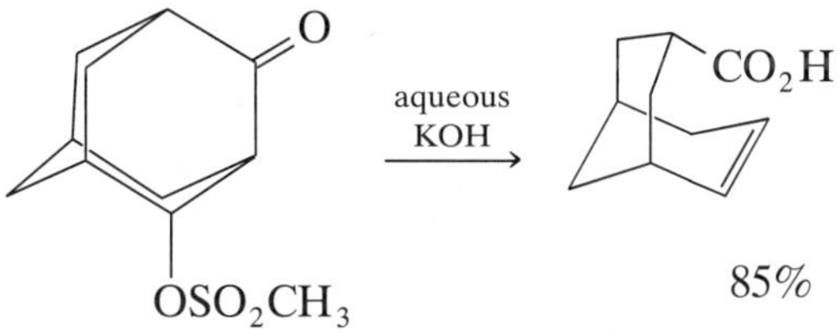

85%

Sasaki, T.; Eguchi, S.; Toru, T. *J. Am. Chem. Soc.* **1969**, *91*, 3390.

c.

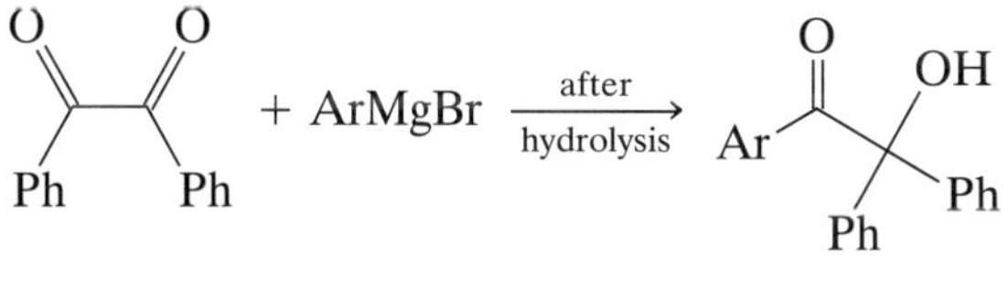

March, J. *Advanced Organic Chemistry*, 3rd ed.; Wiley: New York, 1985; p. 970.

Propose a mechanism for the following transformation and offer an **PROBLEM 3.19**
explanation for the difference in stereochemistry obtained when the
reaction is run in methanol and in the ether dimethoxyethane (DME).

PROBLEM 3.19
continued

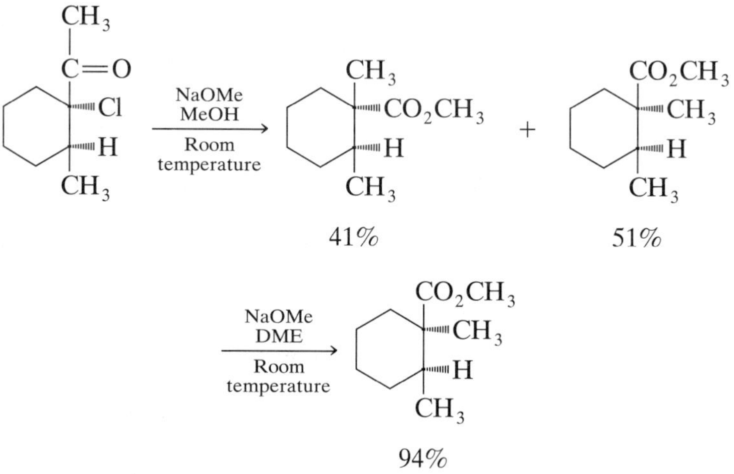

41%   51%

94%

House, H. O.; Gilmore, W. F. *J. Am. Chem. Soc.* **1961**, *83*, 3980.

5. ADDITIONAL MECHANISMS IN BASIC MEDIA

Example 3.20. *Nucleophilic addition followed by rearrangement.*

Write a mechanism for the following transformation:

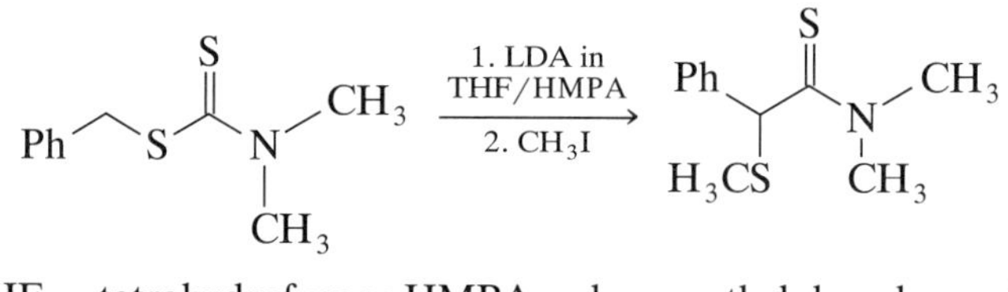

(THF = tetrahydrofuran; HMPA = hexamethylphosphoramide)

First, number the atoms in starting material and product to ascertain how the atoms have been reorganized.

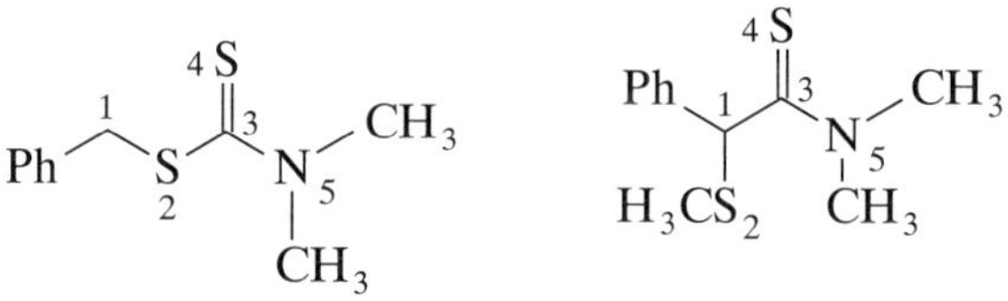

Numbering indicates that C-1 has become attached to C-3 and that the methyl group on S-2 must come from methyl iodide. Focusing attention on positions 1 and 3 of the starting material reveals the following: (1) the protons at position 1 are acidic because they are benzylic, that is, if a proton is removed from this position, the resulting anion is stabilized by resonance; (2) position 3, a thiocarbonyl carbon, is an electrophile and should react with nucleophiles. Thus, the first step of the reaction might be as follows:

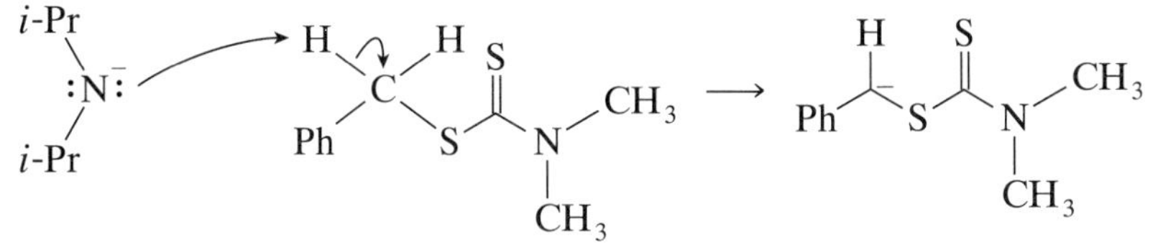

The next step then would be nucleophilic reaction of the carbanion with the electrophilic carbon of the thiocarbonyl group. This reaction joins carbons 1 and 3, as was predicted from the numbering scheme.

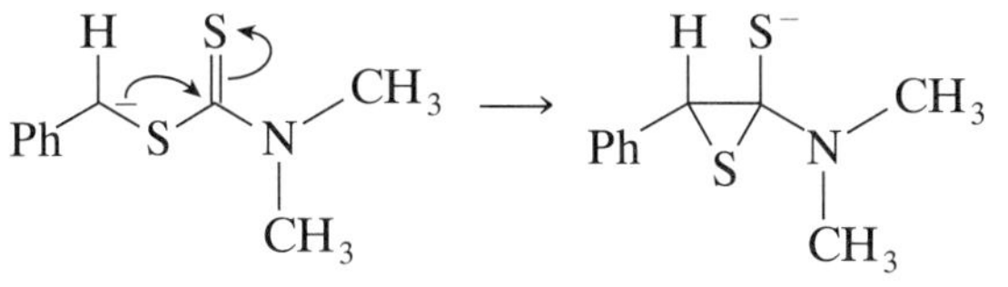

The resulting three-membered-ring intermediate (thiirane) is not stable under the reaction conditions. We know this because there is no three-membered ring in the product! The ring strain in a three-membered ring and the negatively charged sulfur facilitate the ring opening. There are three possible bonds that could be broken when the electron pair on the thiolate makes a π bond with the carbon to which it is attached. Each possibility gives an anion whose stability can be approximated by comparing the relative strengths of the corresponding acids (formed when each anion is protonated). These values can be approximated by choosing compounds from Appendix C with structures as close as possible to the structural features of interest. Breaking the C—N bond would give the dimethylamide ion (the pK_a's of aniline and diisopropylamine are 31 and 36, respectively); breaking the C—C bond would give back the starting material (the pK_a of toluene is 43); and breaking the C—S bond would give a new thiolate ion (the pK_a of ethanethiol is 11). Thus, on the basis of the thermodynamic stability of the product, the C—S bond of the ring would break. In fact, breaking the C—S bond leads to an anion that is related structurally to the final product of the reaction. However, keep in mind that although thermodynamics often is helpful, it does not always predict the outcome of a reaction.

3-31

The original transformation is completed by an S_N2 reaction of the thiolate ion, **3-31**, with methyl iodide.

Example 3.21.   *A combination of proton exchange, nucleophilic addition, and nucleophilic substitution.*

Write a mechanism for the following transformation:

py = pyridine

When a reaction involves only part of a large molecule, such as the steroids in these reactions, it is common to abbreviate the structure. In the structures that follow, the wavy lines indicate the location of the A and B rings that are left out.

First, consider the reaction medium. In pyridine, cyanide is very basic and is also an excellent nucleophile. Because nucleophilic substitution at a position α to a carbonyl is facile, one possible step is nucleophilic substitution of Br by CN. This would be an S_N2 reaction with 100% inversion.

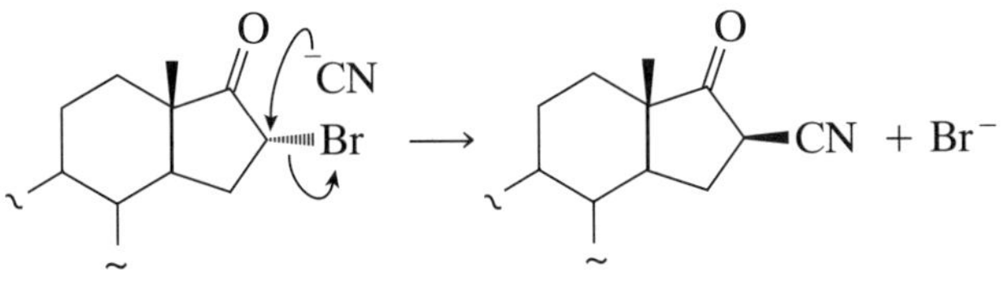

However, there is no reasonable pathway from the product of this reaction to the final product.

Another possible step is a nucleophilic reaction of cyanide at the electrophilic carbonyl carbon:

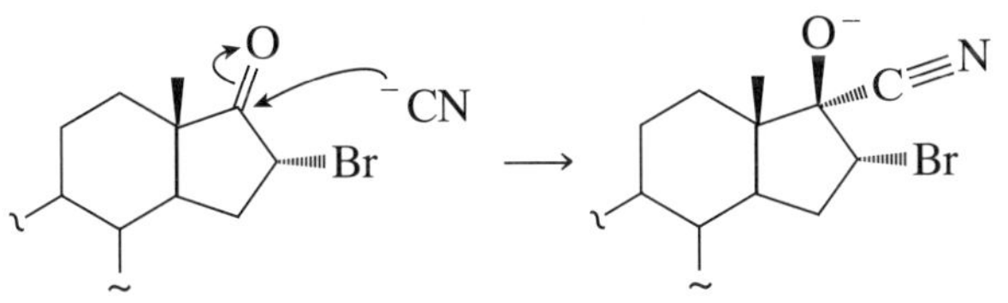

The cyanide approach has been directed so that the alkoxide produced is *anti* to the halide. In this position the alkoxide is situated in the most favorable orientation for backside nucleophilic reaction at the carbon bearing the bromo group to give an epoxide:

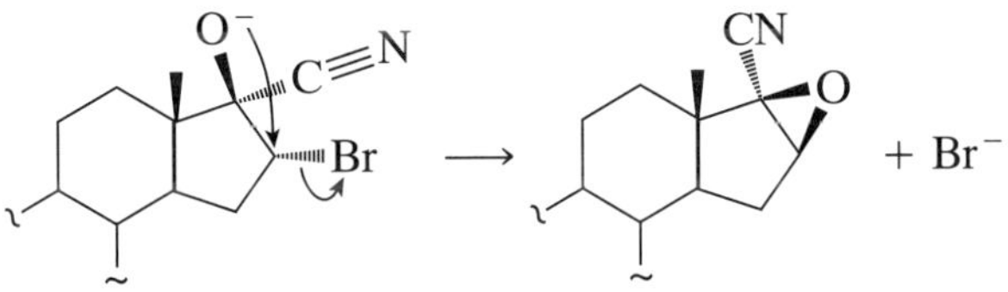

Wrong isomer!

However, this reaction leads to the wrong stereochemistry for the product.

A third possible mechanism can explain the stereochemical result. In this mechanism, a proton is removed from and then returned to the α carbon, such that the starting material is "epimerized" before cyanide reacts. (Epimerization is a change in stereochemistry at one carbon atom.)

The epimerization mechanism is supported by the following findings: the starting material and the epimeric bromo compound are interconverted under the reaction conditions, and both isomers give a 75% yield of the epoxide product.

Numazawa, M.; Satoh, M.; Satoh, S.; Nagaoka, M.; Osawa, Y. *J. Org. Chem.* **1986**, *51*, 1360–1362.

PROBLEM 3.20   **Write reasonable step-by-step mechanisms for the following transformations:**

a.

Note: The sulfone is first treated with excess butyllithium to form a dianion and then the α-chlorocarbonyl compound is added.

Eisch, J. J.; Dua, S. K.; Behrooz, M. *J. Org. Chem.* **1985**, *50*, 3674–3676.

PROBLEM 3.20
continued

b.

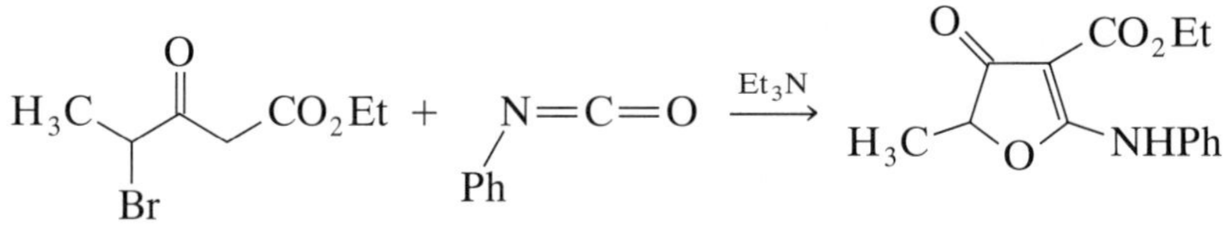

Mack, R. A.; Zazulak, W. I.; Radov, L. A.; Baer, J. E.; Stewart, J. D.; Elzer, P. H.; Kinsolving, C. R.; Georgiev, V. S. *J. Med. Chem.* **1988**, *31*, 1910–1918.

c.

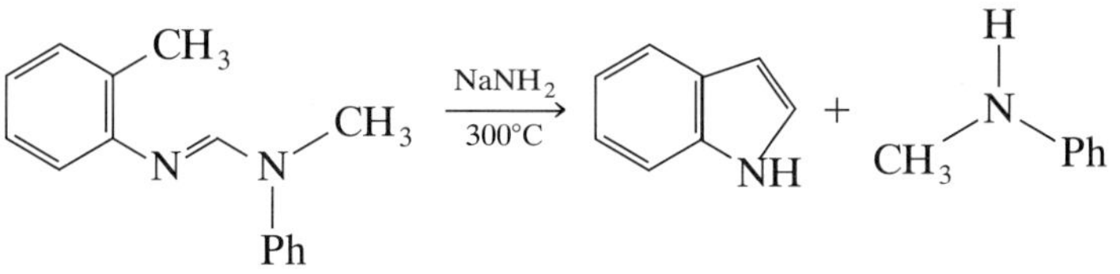

Lorenz, R. R.; Tullar, B. F.; Koelsch, C. F.; Archer, S. *J. Org. Chem.* **1965**, *30*, 2531–2533.

d.

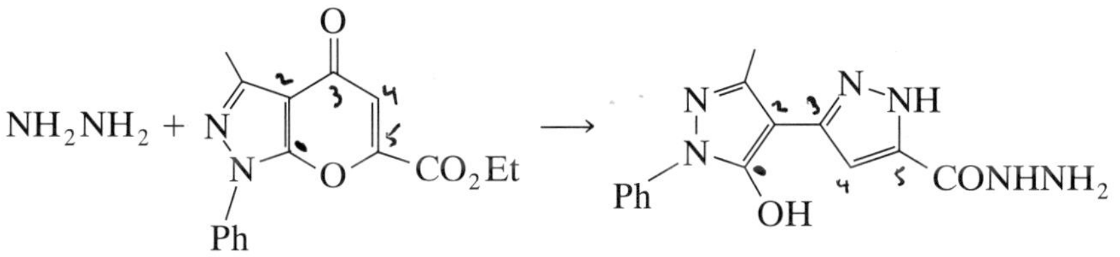

Khan, M. A.; Cosenza, A. G. *Afinidad* **1988**, *45*, 173–174; *Chem. Abstr.* **1988**, *109*, 128893.

e.

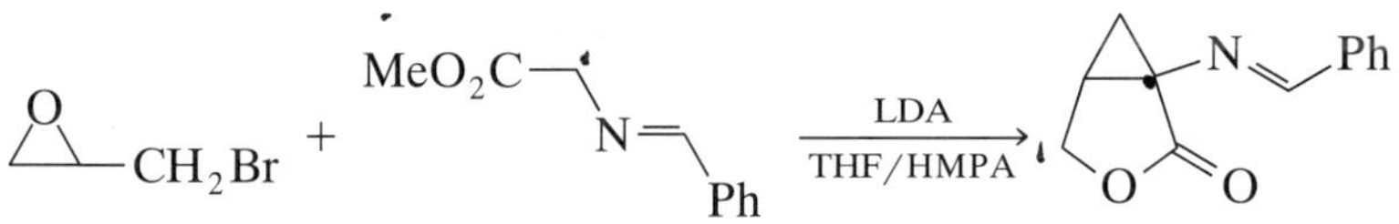

Bland, J.; Shah, A.; Bortolussi, A.; Stammer, C. H. *J. Org. Chem.* **1988**, *53*, 992–995.

Consider the following reaction: **PROBLEM 3.21**

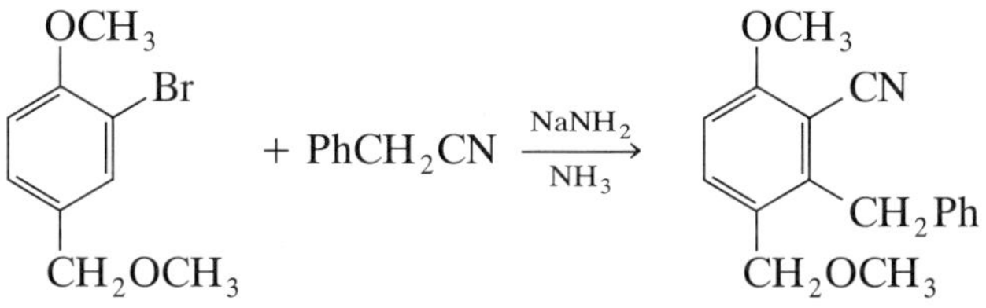

43% yield

PROBLEM 3.21
continued

Two mechanisms for the reaction are written next. Both proceed through formation of the anion of phenylacetonitrile:

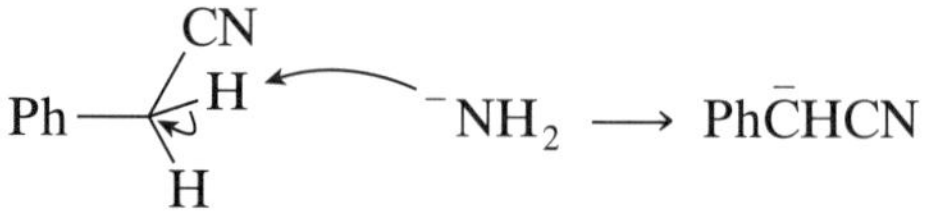

Decide which is the better mechanism and discuss the reasons for your choice.

Mechanism 1

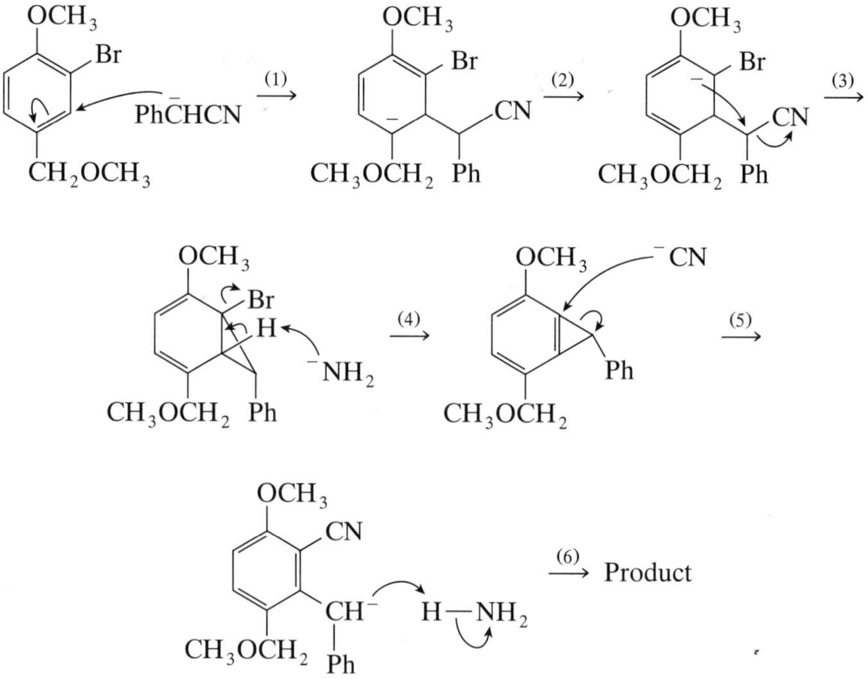

Mechanism 2

PROBLEM 3.21
continued

OCH$_3$
Br
H $^-$NH$_2$
CH$_2$OCH$_3$

$\xrightarrow{(7)}$

OCH$_3$
PhC̄HCN
CH$_2$OCH$_3$

$\xrightarrow{(8)}$

OCH$_3$
C≡N
CH$_3$OCH$_2$ Ph

$\xrightarrow{(9)}$

OCH$_3$
N$^-$
Ph
CH$_3$OCH$_2$

$\xrightarrow{(10)}$

OCH$_3$
CN
H—OH
CHPh
CH$_2$OCH$_3$

$\xrightarrow{(11)}$ Product

Khanapure, S. P.; Crenshaw, L.; Reddy, R. T.; Biehl, E. R. *J. Org. Chem.* **1988**, *53*, 4915–4919.

The following esterification is an example of the Mitsunobu reaction.
Notice that there is inversion of configuration at the asymmetric
carbon bearing the alcohol group in the starting material.

PROBLEM 3.22

EtO$_2$C
N═N
CO$_2$Et + Ph$_3$P + PhCO$_2$H + H—OH
CH$_3$
C$_6$H$_{13}$ $\xrightarrow{Et_2O}$

EtO$_2$C H
N—N
H CO$_2$Et + Ph$_3$PO + PhCO$_2$—H
CH$_3$
C$_6$H$_{13}$

Mitsunobu, O.; Eguchi, M. *Bull. Chem. Soc. Jpn.* **1971**, *44*, 3427–3430. For a review article on
the versatility of the reaction, see Mitsunobu, O. *Synthesis*, **1981**, 1–28.

The reaction mechanism is believed to proceed according to the following
outline: Triphenylphosphine reacts with the diethyl azodicarboxylate to give
an intermediate, which is then protonated by the carboxylic acid to form a
neutral salt. This salt then reacts with the alcohol to form dicarboethoxyhy-
drazine and a new salt. This salt then reacts further to give a triphenylphos-

PROBLEM 3.22   phine oxide and the ester. Using this outline as a guide, write a mechanism
continued   for the reaction.

ANSWERS TO PROBLEMS

Problem 3.1   First, although the atoms are balanced on both sides of the equation, the
                   charges are not. Also, there is no way to write a reasonable electronic state
                   for the oxygen species on the right. If the electrons were to flow as written,
                   the oxygen on the right would have a double positive charge (highly
                   unlikely for an electronegative element) and only six electrons lacking two
                   for an octet.

                   A proton, H^+, should be shown on the right side of the equation, rather
                   than a hydride, H^- (see Hint 2.2). Also, as stated in Hint 3.1, hydride is an
                   extremely poor leaving group. The curved arrow, on the left side of the
                   equation, is pointed in the wrong direction (see Hint 2.3). This should be
                   apparent from the fact that the positively charged oxygen atom is far more
                   electronegative than hydrogen. Thus, the arrow should point toward oxy-
                   gen, not away from it.

                   Loss of a proton is often written in mechanisms simply as $-H^+$. We will
                   not do this because protons are always solvated in solution. Therefore, in
                   this book, with the exception of Chapter 7, the loss of a proton will always
                   be shown as assisted by a base. However, the base need not be a strong
                   base. Thus, for the transformation indicated in the problem, the following
                   could be written:

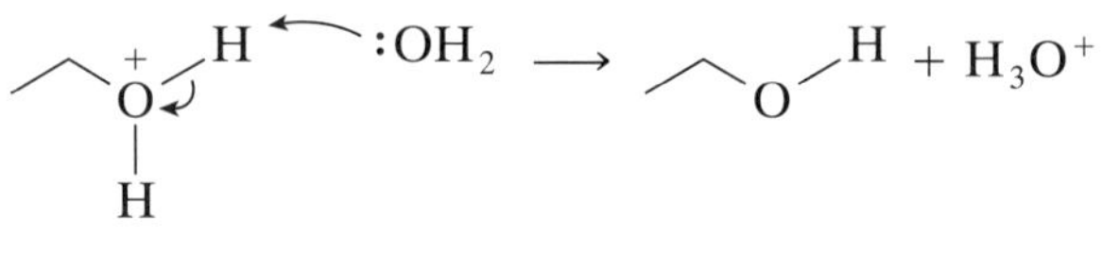

Problem 3.2   According to Appendix C, the pK_a of bicarbonate is 10.2, whereas the pK_a
                   of phenol is 10.0 and the pK_a's of *m*-nitrophenol and *p*-nitrophenol are
                   8.3 and 7.2 (i.e., an increase of 1.8 and 2.8 pH units), respectively. The
                   combined effect of the substituents would make this phenol sufficiently
                   acidic to be converted almost entirely to the corresponding phenoxide ion,
                   which will act as a nucleophile. (The negatively charged phenoxide ion is a
                   much better nucleophile than neutral phenol.)

Problem 3.2
continued

   The carboxylic acid group in the starting material will also be converted to a salt in carbonate solution. Although phenoxide is a better nucleophile than carboxylate ion, this phenoxide may not react as rapidly as the carboxylate with benzyl bromide because it is much more sterically hindered (by the *ortho* methyl and nitro groups). S_N2 reactions are slowed considerably by steric hindrance.

Problem 3.3

a. The methylene carbon of the ethyl bromide is the electrophile, the lone pair of electrons on the phosphorus of triphenylphosphine is the nucleophile, and the bromide ion is the leaving group.

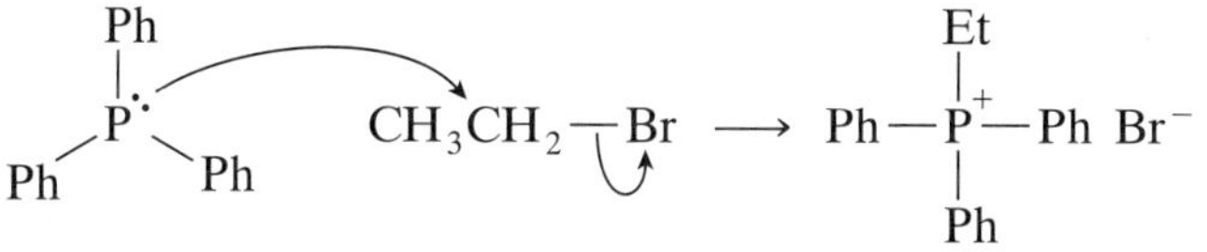

The product salt is stable, and no further reaction takes place. If you wrote that ethoxide was formed and acted as a nucleophile to give further reaction, you neglected to consider the relative acidities of HBr and EtOH. (See Table 1.3 and Appendix C.) The relative pK_a's indicate that the following reaction does not occur:

$$Br^- + EtOH \rightleftharpoons HBr + EtO^-$$

HBr has a pK_a of -9.0 and EtOH has a pK_a of 16. Thus, the K for this reaction is 10^{-25}!

   Prior ionization of ethyl bromide to the carbocation is unlikely, because the primary carbocation is very unstable.

b. Base is present to neutralize the amine salt, giving free amine. The free amine is the nucleophile, the carbon bearing the chlorine is the electrophile, and chloride ion is the leaving group. The driving force for this intramolecular substitution reaction is greater than that of an intermolecular reaction, because of entropic considerations.

Problem 3.3
continued

Under the basic reaction conditions, the salt shown and the neutral product would be in equilibrium.

c. The overall reaction can be separated into two sequential substitution reactions. In the first step, the proton of the carboxylic acid is the electrophile, the carbon of diazomethane is the nucleophile, and the carboxylate anion is the leaving group. In the second step, the carboxylate ion is the nucleophile, the methyl group of the methyldiazonium ion is the electrophile, and nitrogen is the leaving group. Loss of the small stable nitrogen molecule provides a lot of driving force to the reaction. (See Hint 2.14.)

d. In acid, the first step is protonation of the oxygen of the epoxide in order to convert it into a better leaving group. This is followed by ring opening to the more stable carbocation (the one stabilized by conjugation with the phenyl group), followed by nucleophilic reaction of water at the positive carbon to give the product. Thus, in this reaction, the oxygen of water is the nucleophile, the more highly substituted carbon is the electrophile, and the protonated epoxide oxygen is the leaving group.

Problem 3.3
continued

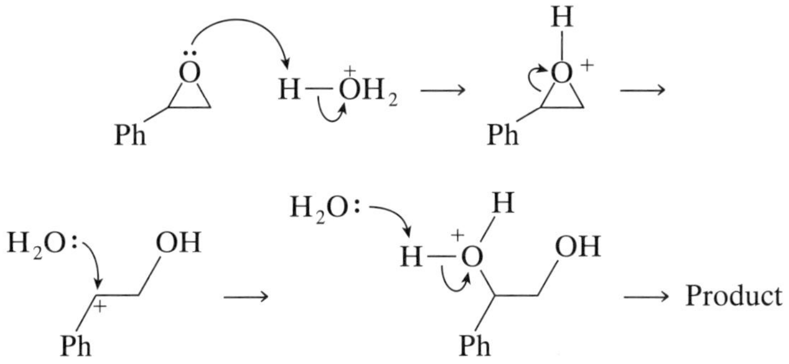

If this reaction were run in base, the following mechanism would apply:

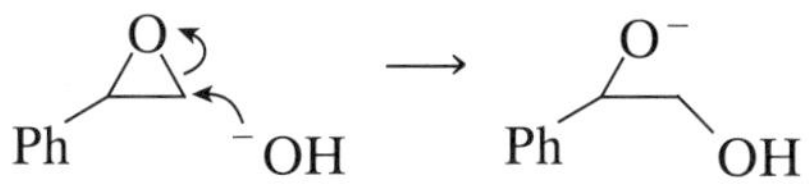

This is one of the few examples where RO⁻ acts as a leaving group. The reason this reaction takes place is that it opens a highly strained three-membered ring. Note that in base the nucleophile reacts at the less substituted carbon. This is because the S_N2 reaction is sensitive to steric effects.

Water is not a strong enough nucleophile to open an epoxide in the absence of acid, so that in neutral water the following mechanistic step is invalid:

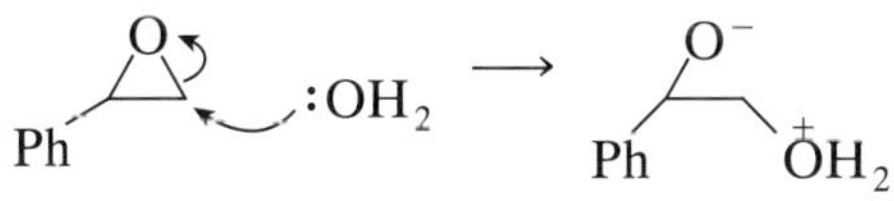

a. The carbon skeleton has rearranged in this transformation. Moreover, numbering of corresponding atoms in product and starting material indicates that it is not the ethyl group, but the nitrogen, that moves. In other

Problem 3.4

words, carbon-3 of the ethyl group is attached to carbon-2 in both the starting material and the product. This focuses attention on the nitrogen, which is a nucleophile. Because chlorine is not present in the product, an intramolecular nucleophilic substitution is a likely possibility. This intramolecular S_N2 reaction is an example of neighboring group participation. This reaction gives a three-membered-ring intermediate, which can open in a new direction to give the product. In this second S_N2 reaction, hydroxide is the nucleophile and the CH_2 group of the three-membered ring is the electrophile. Notice that, as in the previous problem, S_N2 reaction of the nucleophile on the three-membered ring occurs at the less hindered carbon.

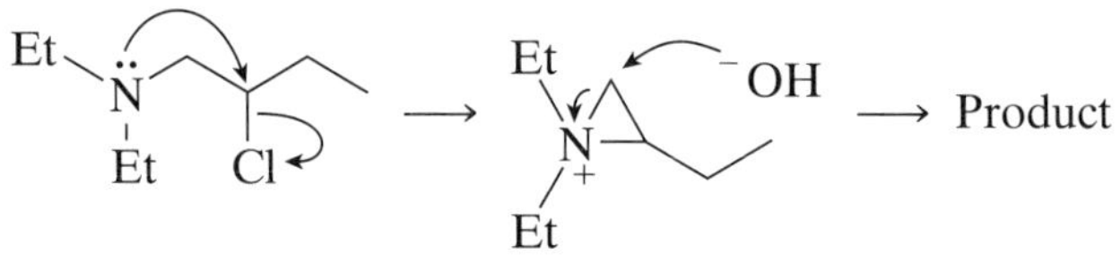

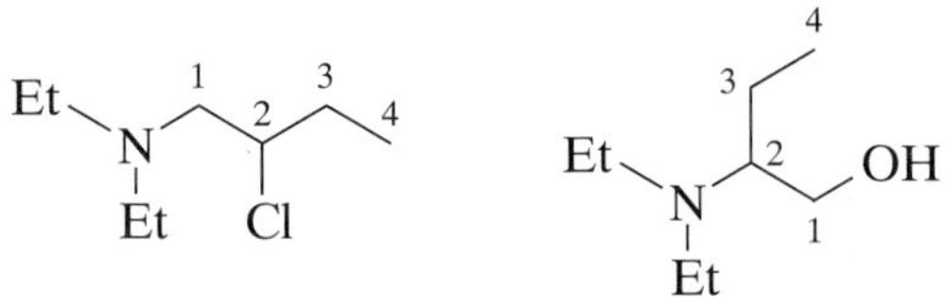

Some steps in an alternative mechanism, written by a student, are shown here:

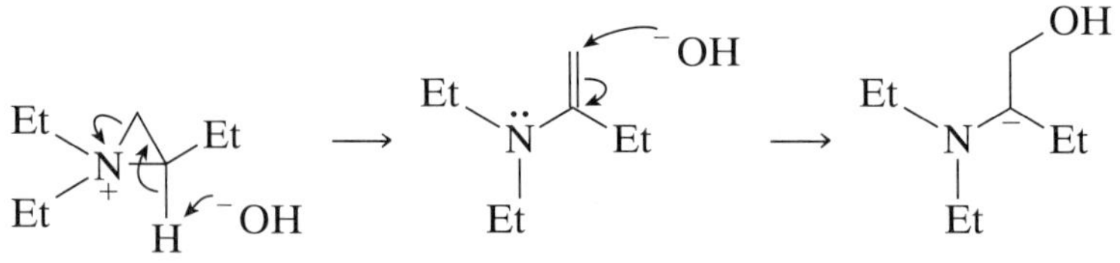

An unlikely step is the addition of hydroxide to the double bond. Double bonds of enamines, like this one, tend to be nucleophilic rather than electrophilic. That is, the resonance interaction of the lone pair of electrons on nitrogen with the double bond is more important than the inductive withdrawal of electrons by the nitrogen. Another way of looking at this is to realize that the final carbanion is not stabilized by resonance and, thus, is not likely to be formed in this manner.

Problem 3.4
continued

b. The first step is removal of the most acidic proton, the one on the central carbon of the isopropyl group. The anion produced is stabilized by conjugation with both the carbonyl group and the new double bond to the isopropyl group. Removal of no other proton would produce a resonance-stabilized anion. The second step is nucleophilic reaction of the anion with the electrophilic carbon, the one activated by two bromines. Bromide ion acts as the leaving group.

The mechanism shown violates Hint 3.2, because the second step shows nucleophilic substitution at an sp^2-hybridized nitrogen occurring as a single concerted process. Change it to two steps, namely, addition followed by elimination:

Problem 3.5

a. Because a strong base is present, the mechanism is not written with one of the neutral guanidino nitrogens acting as the nucleophile. Instead, the first step is removal of a proton from the guanidino NH_2 group. This gives a

Problem 3.6

resonance-stabilized anion, **3-32**. Removal of the proton from the imino nitrogen would produce a less stable anion because it is not resonance stabilized (see Problem 1.13.d). In the following, R = the *N*-methylpyrazine ring.

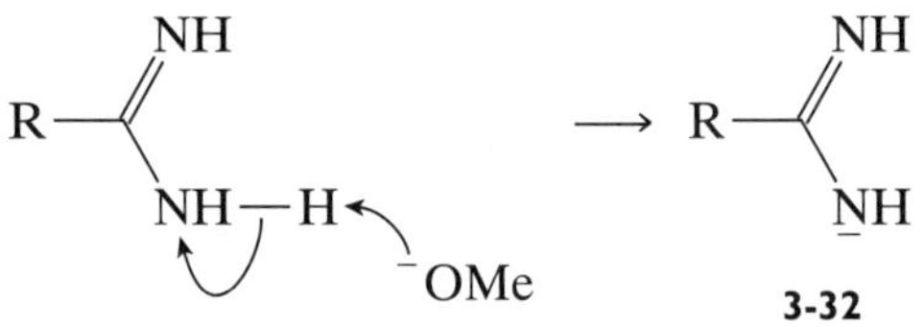

3-32

Another reasonable step that can take place under the reaction conditions is removal of one of the protons on the carbon α to the carbonyl groups in diethyl malonate. However, this reaction does not lead to the product and is an example of an unproductive step. Another unproductive reaction is ester interchange in diethyl malonate. In this case, sodium methoxide would react with the ethyl ester to produce a methyl ester. However, this reaction is inconsequential because methyl and ethyl esters have similar reactivity and the alkyl oxygens with their substituents are lost in the course of the reaction.

The nucleophilic nitrogen of **3-32** adds to the carbonyl group, and then ethoxide ion is lost.

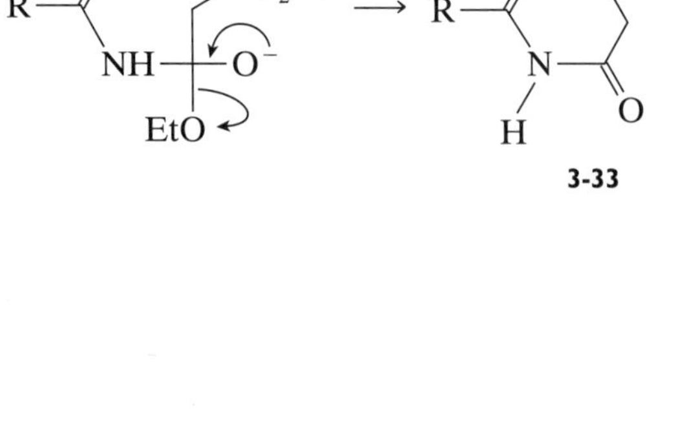

3-33

Intermediate **3-33** reacts by a route completely analogous to the previous steps to give **3-35**, a tautomer of the product. That is, the proton removed from the amino nitrogen of **3-33** leads to a resonance-stabilized anion, **3-34**. (The anion formed by removal of a proton from the imide nitrogen would not be resonance stabilized.) The nucleophilic anion, **3-34**, adds to the remaining ester carbonyl. Elimination of ethoxide then gives **3-35**.

Problem 3.6
continued

Intermediate **3-35** undergoes two tautomerizations to give the product. The first tautomerization involves removal of the proton on nitrogen, because this gives an anion that is considerably more delocalized than the anion that would be produced by removal of a proton from carbon. Either enolate oxygen of **3-36** can pick up a proton from either ethanol or methanol. Finally, removal of a second acidic proton forms the product phenolate, **3-37**.

In fact, the reactions shown for this mechanism are reversible, and an important driving force for the reaction is production of the stable phenolate salt. The neutral product would be obtained by acidification of the reaction mixture upon workup.

Note: In a tautomerism like that shown for **3-36**, transfer of a proton generally is written as an intermolecular process, not as an intramolecular process as pictured here:

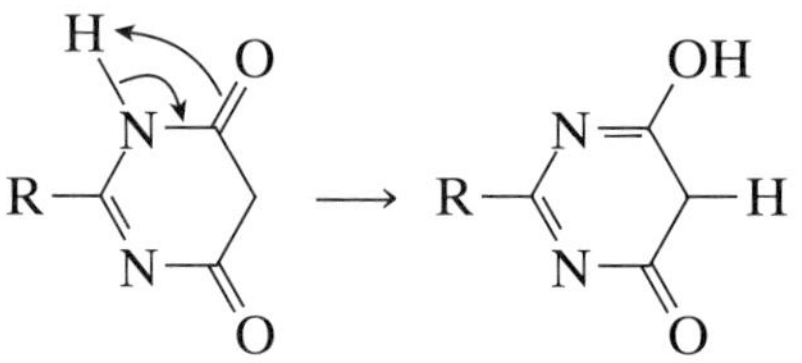

3-35

b. Because the product contains no saturated carbon chain longer than one carbon, it is unlikely that the carbons of the six-membered ring are part of the product. Furthermore, because the functionality attached to the six-membered ring is an acetal and this functional group is easily cleaved (e.g., during aqueous workup), it is probable that cyclohexanone is the other product. Also, the methoxy group of the ester is not present in the product, which suggests a nucleophilic substitution at the carbonyl of the ester.

Numbering of starting material and product also aids in analysis. The relationship between the starting materials and the product is seen more clearly if **3-39-1**, a tautomer of the product, is depicted.

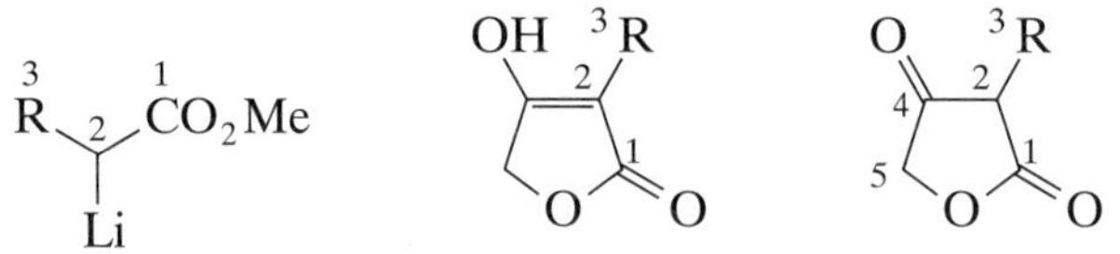

3-38   **3-39-1**

It is reasonable to assume that three carbons of the product **3-39-1** arise from the lithio reagent and that the other two carbons are introduced from **3-40**. The simplest assumption is that carbons 1, 2, and 3 in **3-39-1** are derived from the lithium reagent and carbons 4 and 5 and oxygen-6 are derived from **3-40**. If, instead, the carbons of the lithio reagent become carbons 1, 2, and 3 of structure **3-39-2**, carbons 4 and 5 from the reagent **3-40** would no longer be bonded to one another. It is difficult to write a

Problem 3.6
continued

mechanism that would account for this bonding change. If the lithio reagent is incorporated as in **3-39-2**, the ring oxygen (O-6) could be derived from either of the ring oxygens in **3-40**, as illustrated in the following structures.

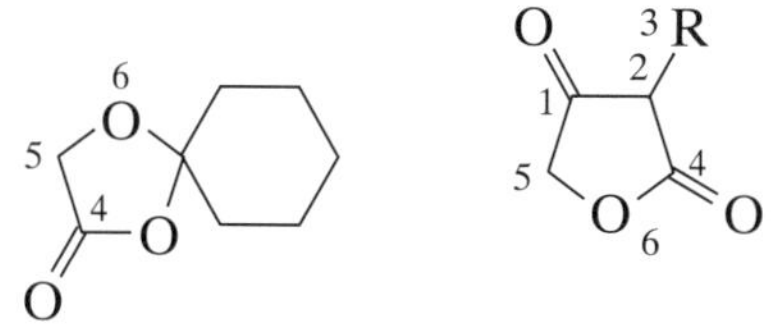

3-40   **3-39-2**

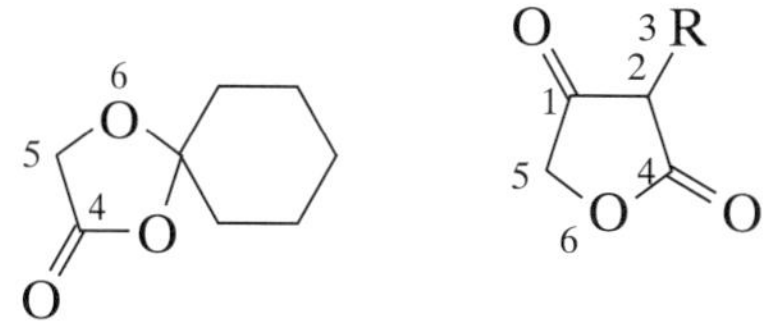

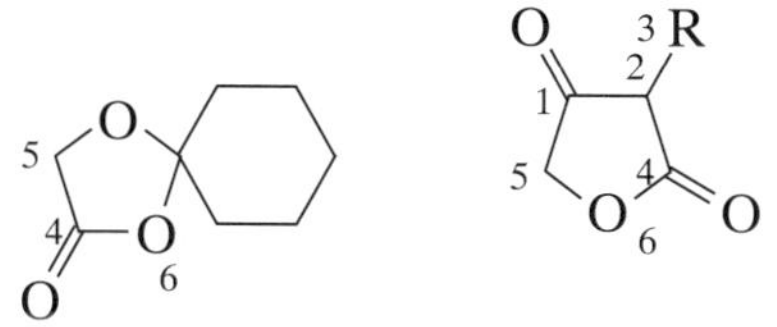

Occam's razor then leads us to assume that the source of the carbons is as shown in structure **3-39-1**.

Analysis by numbering the atoms, the presence of a nucleophile (the lithium reagent) and an electrophile (the ester carbonyl group of **3-40**), all suggest nucleophilic substitution of the ester as a likely first step.

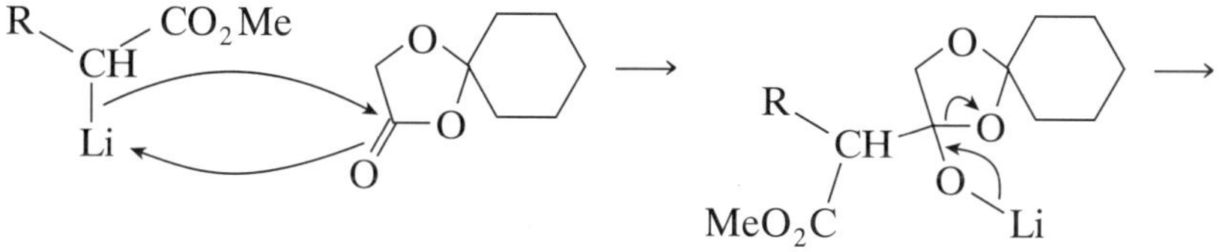

The elimination part of this initial nucleophilic substitution might be concerted with the loss of cyclohexanone, or it might occur in a separate step, as follows:

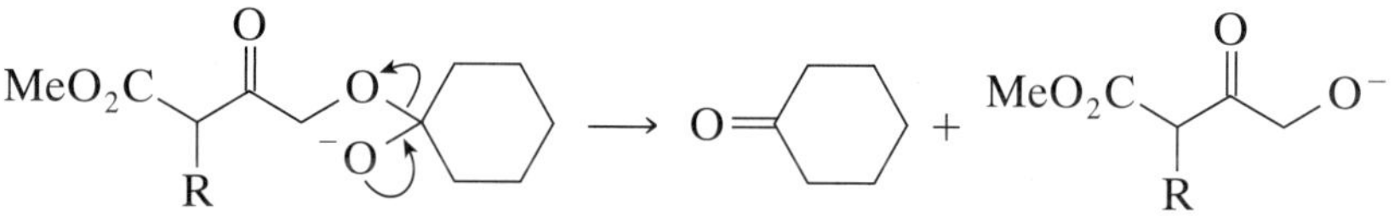

3-41

An alternative elimination with breaking of a carbon–carbon bond instead of a carbon–oxygen bond, is less likely because the carbanion formed would be much less stable than the alkoxide ion.

The alkoxide ion, **3-41**, reacts intramolecularly with the ester to produce a lactone.

3-41

The remaining steps show tautomerization of the ketone.

Because there is a considerable excess of the lithium reagent, it is possible that the acidic proton α to both the ester and the ketone is removed to form the dianion **3-42** and that this is the species that undergoes cyclization.

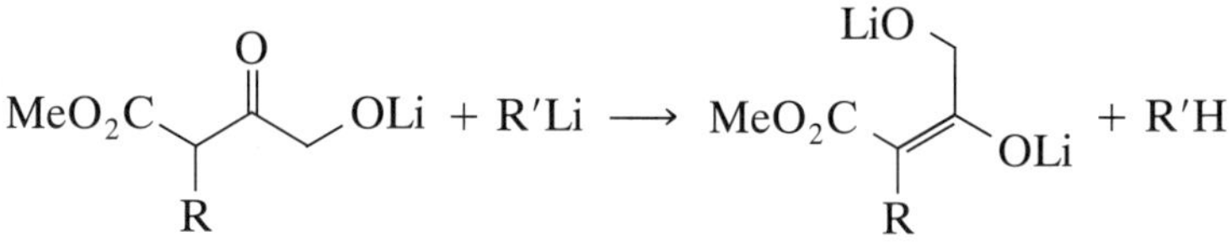

Problem 3.6
continued

3-42

However, this reaction would be slower than the reaction of the alkoxide **3-41** because of the delocalization of charge shown in **3-43**.

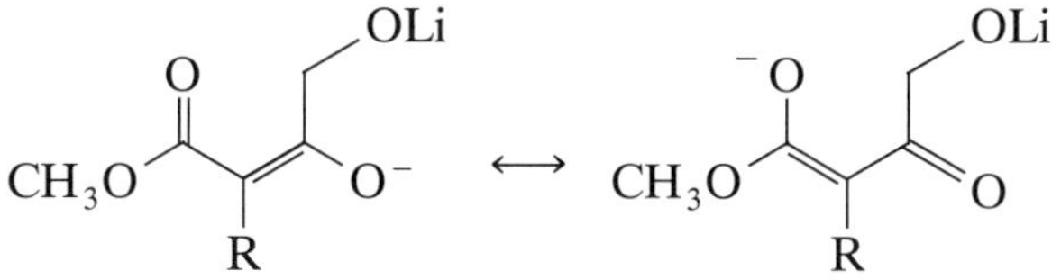

3-43

Because of repulsion between the two centers of negative charge, it may be that the cyclization reaction proceeds through the alkoxide **3-41** even if the dianion **3-42** is the predominant species in the reaction mixture.

c. The first step should be removal of a proton from an amino group because this gives a resonance-stabilized anion. (Use the amino group that becomes substituted during the course of the reaction.)

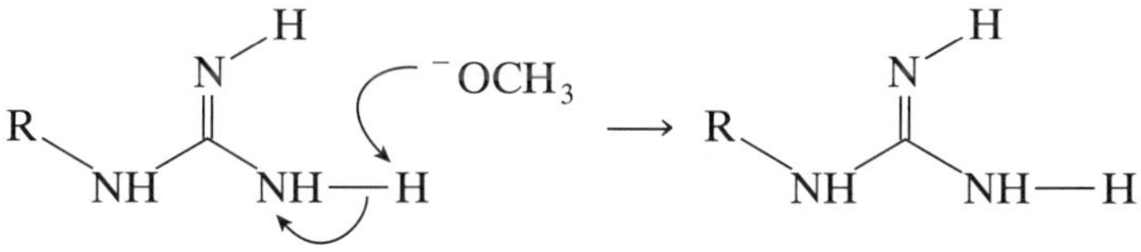

Substitution of the ethoxy group by the guanidino group should be a two-step process: addition followed by elimination:

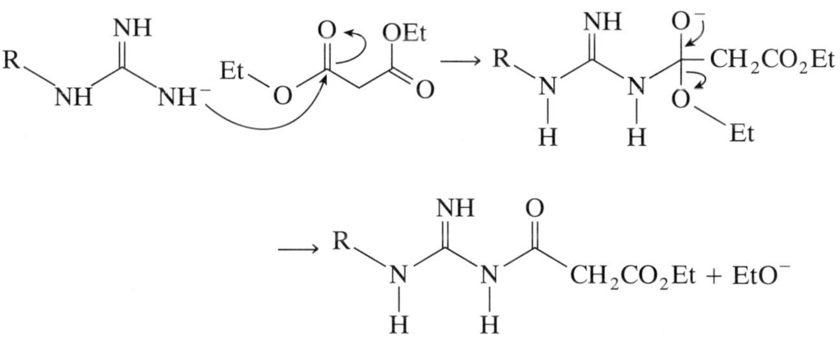

Problem 3.6
continued

In the next step, because of the basicity of the medium and the acidic protons that are present, the amino proton is removed prior to nucleophilic substitution of the second ester carbonyl. Because of the resonance stabilization possible for the resulting anion, the amino proton, not the imino proton, reacts. Removal of a proton prior to cyclization also eliminates the need for the last step shown in the problem.

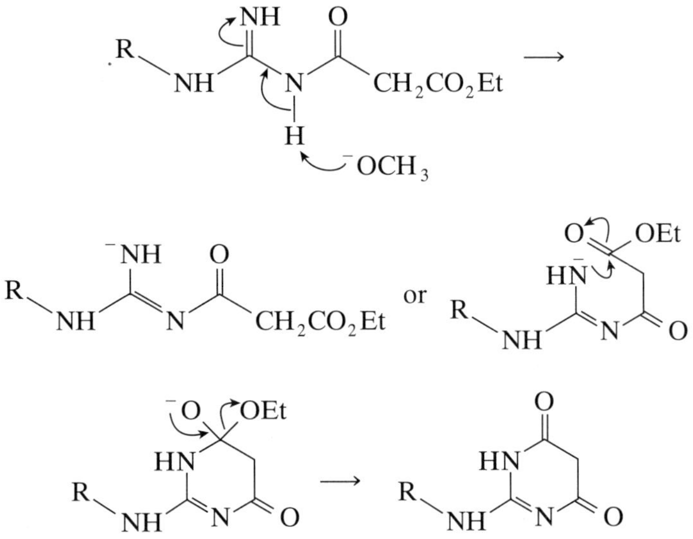

Problem 3.7

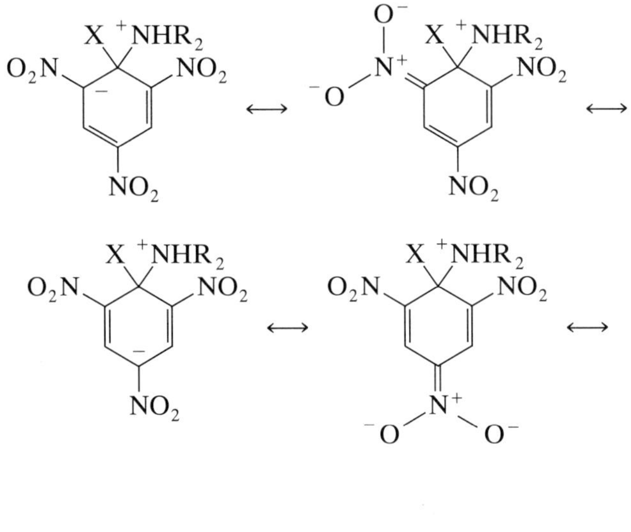

Problem 3.7
continued

Resonance forms, in which the nitro group is shown in its alternative forms,

do not add to the stability of the anion relative to the stability of the starting material because both the starting material and the intermediate have such resonance forms. Thus, these forms often are omitted from answers to questions like this one.

The first steps, which are not shown, are the removal of the protons from the protonated amine starting material. Those steps look similar to the final step of the mechanism shown.

Problem 3.8

Problem 3.8
continued

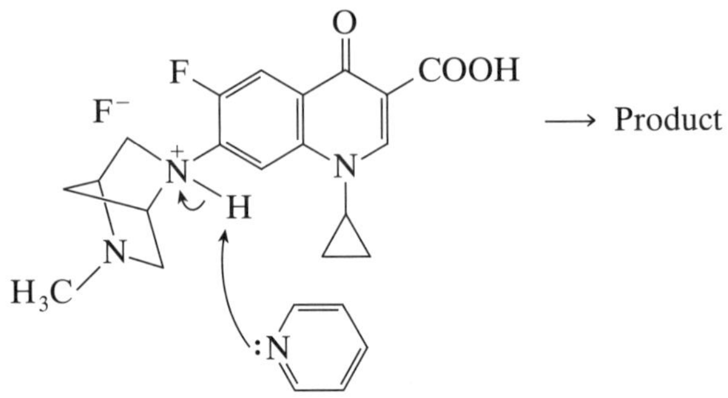

→ Product

Substitution of the particular fluorine shown is favored because the intermediate anion formed is stabilized by delocalization of the charge on the keto oxygen (resonance). Reaction at the carbon bearing the other fluorine would result in an intermediate in which the negative charge could not be delocalized onto this oxygen.

Initial formation of an aryne, followed by nucleophilic attack, is not a likely mechanism. The most important factors that detract from such a mechanism are the following: (1) the base is not a strong base, and (2) the carbon–fluorine bond is very strong.

Problem 3.9

Because the aryne intermediate is symmetrical, half of the ^{13}C label will be on the carbon bearing the amino group and the other half will be on the carbon *ortho* to the amino group.

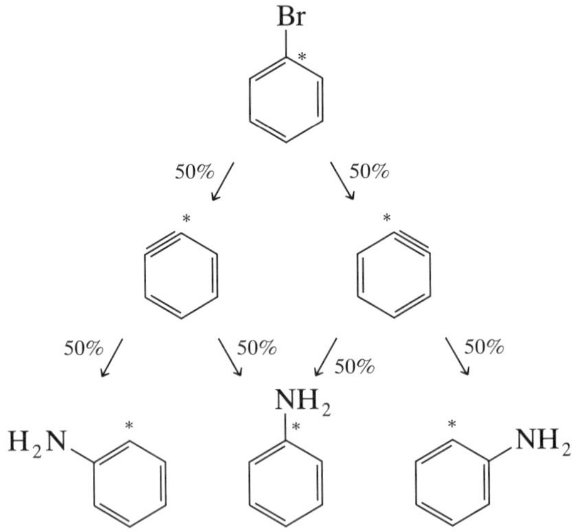

The two outer products at the base of the pyramid are the same: Both are aniline with the label in the *ortho* position.

The most acidic proton in the molecule is on the carbon α to the nitrile. This proton is removed first. The second step is elimination of HCl to give an aryne. (As indicated in Chapter 3, Section 1.C, this might also be a two-step process in which removal of the proton is followed by loss of the chloride ion.)

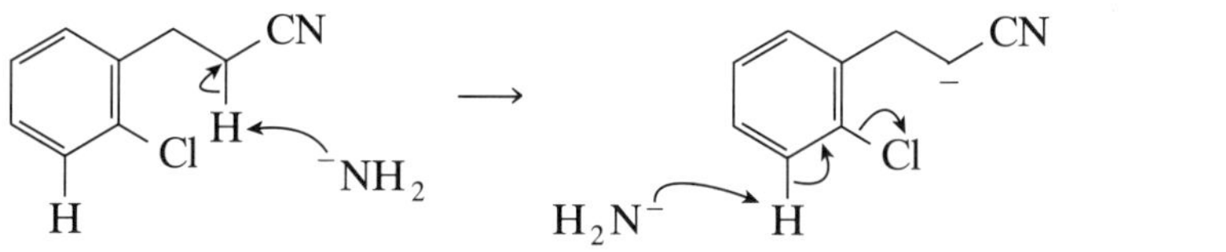

Problem 3.10

The aryne intermediate is usually written with a triple bond and a delocalized aromatic system, as shown in **3-44**. The anion in the side chain reacts as a nucleophile with the electrophilic aryne. The resulting anion, **3-45**, can remove a proton from ammonia to give **3-46**. Because the product has been reached, we usually stop writing the reaction mechanism at this point. However, in the reaction mixture, amide will remove a proton from the carbon α to the cyano group of **3-46**. Only during workup will the anion be protonated to give back **3-46**.

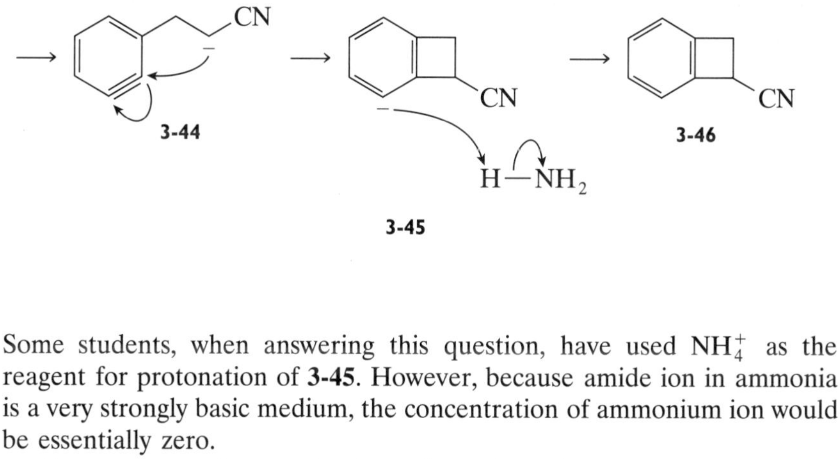

Some students, when answering this question, have used NH_4^+ as the reagent for protonation of **3-45**. However, because amide ion in ammonia is a very strongly basic medium, the concentration of ammonium ion would be essentially zero.

The representation, **3-47**, shown here is equivalent to the structure usually written, **3-44**; however, keep in mind that the extra π bond drawn, the one highlighted, is not an ordinary π bond. It is formed by the overlap

Problem 3.10
continued

of two sp^2 orbitals, one on each carbon. This bond is perpendicular to the other π bonds shown in **3-47**.

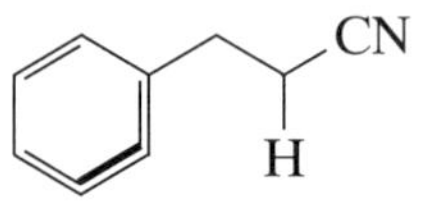

3-47

Problem 3.11

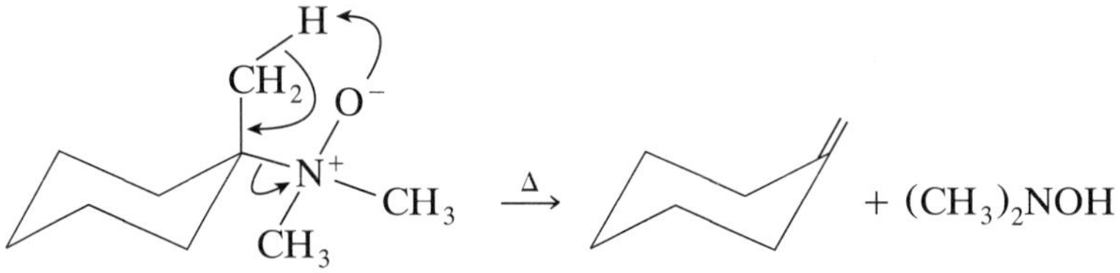

An interesting aspect of this elimination reaction is that it gives only the isomer with the exocyclic double bond. This results from the strict stereochemical requirements of the five-membered-ring transition state: all of the atoms must lie in the same plane. This rules out the alternate reaction, removal of a proton from a ring carbon, because too much distortion of the cyclohexane ring would be required.

Problem 3.12

a. This reaction is analogous to that in Example 3.15. However, we can also apply our knowledge that esters undergo nucleophilic substitution. Therefore, the initial reaction of the Grignard reagent will give a ketone:

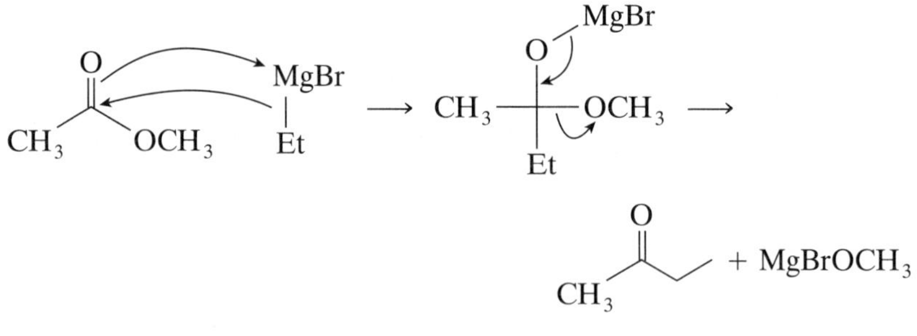

The ketone then reacts with another mole of Grignard reagent to give the salt of an alcohol, which will be converted to the alcohol during acidic workup.

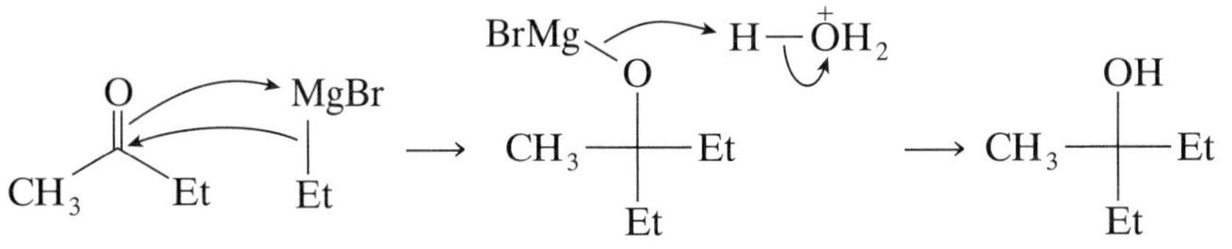

Problem 3.12
continued

b. This reaction is an unusual addition to a carbonyl derivative. Normally, the nucleophile would react at the carbon atom of the C=N group, but because the resonance form with a positive charge on this carbon would impart antiaromatic character to the ring, the electrophilic character of this carbon is decreased and the aryl group reacts at the unsaturated nitrogen instead. The direction of addition is also influenced by stabilization of the intermediate **3-48**. To the extent that the carbon of the C—Mg bond in **3-48** is negative, it can increase the aromatic character, and thus the stability, of the five-membered ring system.

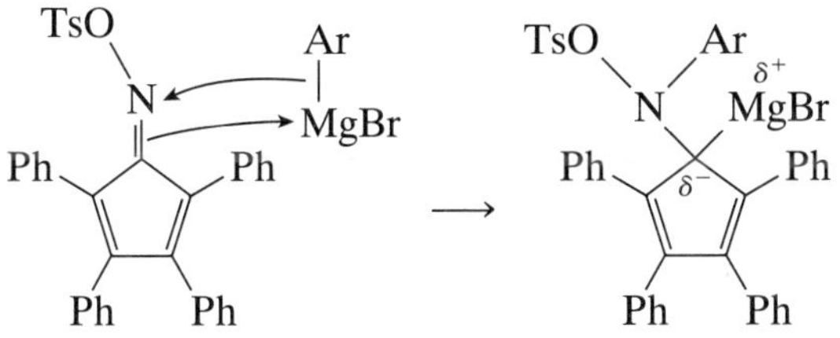

3-48

Subsequent elimination of the tosylate group gives the product:

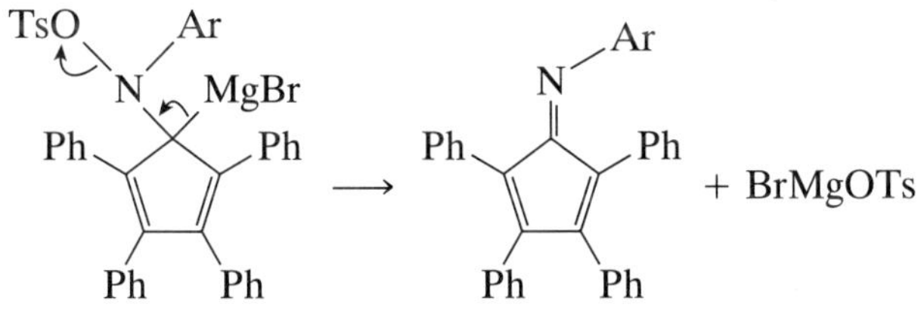

Problem 3.12
continued

A less likely mechanism would be direct displacement of the tosylate anion by the Grignard reagent because this would be an S_N2 reaction at an sp^2 atom (see Hint 3.2).

Problem 3.13

The nitrogen attached to the phenyl ring is less nucleophilic because the lone pair of electrons is delocalized onto the aromatic ring.

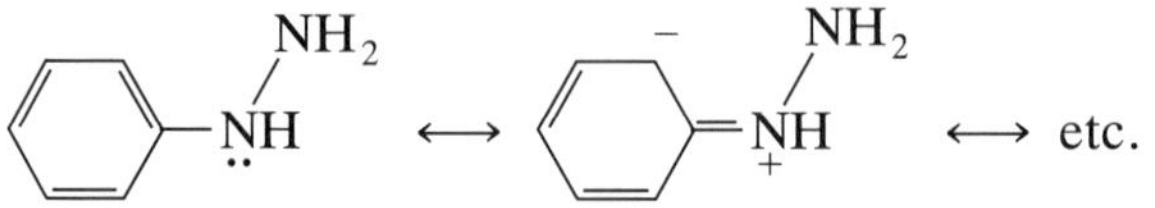

Problem 3.14

a. This is another example of the reaction of an amine with a carbonyl compound in the presence of an acid catalyst. The first steps are protonation of the carbonyl group, nucleophilic addition of the amine, and deprotonation of the nitrogen to give intermediate **3-49**.

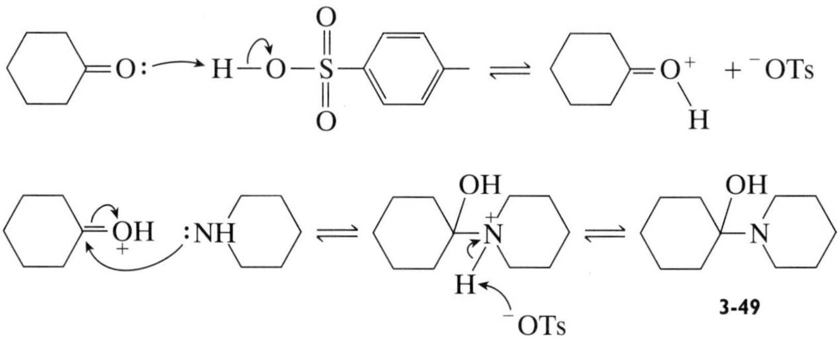

Unlike the intermediate in Example 3.16, intermediate **3-49** has no proton on the nitrogen. Thus, it is not possible to form an imine via loss of a water molecule. Instead, dehydration occurs via loss of a proton from carbon to give the product, which is called an enamine.

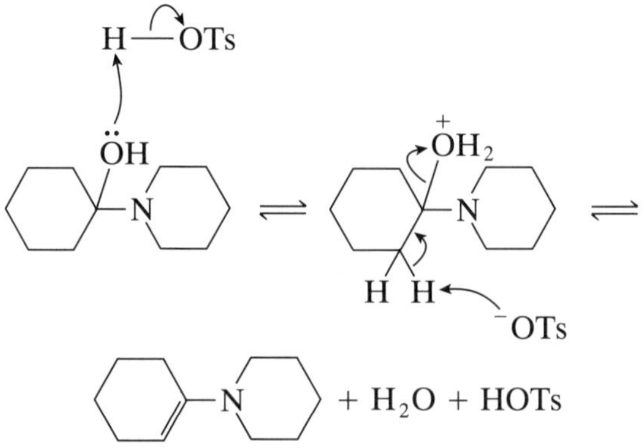

Problem 3.14
continued

Notice that in the last step, the catalyst is regenerated and water is produced. The equilibria in this problem favor toward starting material. However, the enamine can be obtained in significant amounts if the product water is removed from the reaction mixture as it is produced.

b. Analysis of the starting materials and products reveals that the $=$NPh group has been replaced by $=$NOH. An outline of steps can be formulated on the basis of this observation: (1) reaction of hydroxylamine at the carbon of the C$=$N and (2) elimination of the NPh group. Details to be worked out include identifying the actual nucleophile and electrophile in (1) and the actual species eliminated in (2). These can be ascertained by considering the relative acidities and basicities of the species involved.

   Because the nitrogen in hydroxylamine hydrochloride has no lone pairs of electrons, the salt cannot be the nucleophile. The pyridine in the reaction mixture can remove a proton from the nitrogen of hydroxylamine hydrochloride. This equilibrium favors starting material (note the relative pK_a's in the following), but excess pyridine will release some hydroxylamine.

$$\text{Py-N: } + \overset{+}{N}H_3OH \rightleftharpoons \text{Py-}\overset{+}{N}H + :NH_2OH$$

$$pK_a = 8.03 \qquad\qquad pK_a = 5.2$$

Moreover, because the salts of pyridine and hydroxylamine are sources of protons, it is unlikely that the anion derived from hydroxylamine, $^-$NHOH, or an anion like **3-50** will be formed as an intermediate. The anion in **3-50** can be compared to the anion formed when aniline ($pK_a = 30$) acts as an acid. Such a strong base will not be produced in significant amounts in a medium in which there is protonated amine (see pK_a's mentioned previously).

3-50

Problem 3.14
continued

Thus, the nitrogen of the imine will be protonated prior to nucleophilic reaction at the carbon.

The electrophilic protonated imine reacts with nucleophilic hydroxylamine.

Before the phenyl-substituted nitrogen acts as a leaving group, it too is protonated to avoid the poor leaving group PhNH$^-$.

Problem 3.14
continued

c. In the first step, the basic nitrogen of the imine is protonated. This converts the molecule into a better electrophile, and water, acting as a nucleophile, adds at the positive carbon.

The resulting intermediate can be deprotonated at oxygen and protonated at nitrogen.

The result is to convert the nitrogen into a better leaving group.

The oxygen in the molecule provides the driving force for the reaction by stabilizing the positive charge. Deprotonation gives the product.

Problem 3.15   a. Although there are three carbons from which a proton could be removed to produce an enolate ion, only one of the possibilities leads to the product shown.

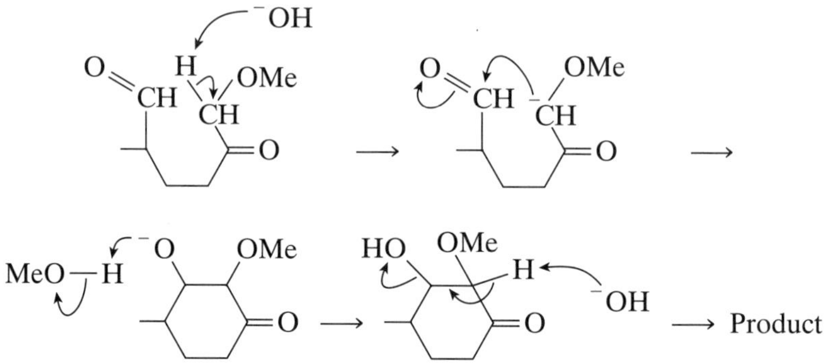

A common shortcut that students take is to write the following mechanistic step for loss of water:

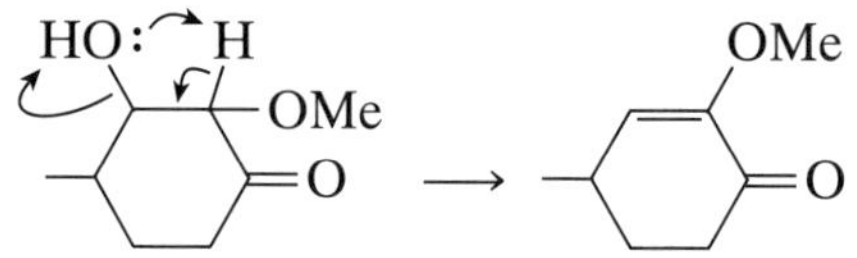

Because hydroxide ion is a much stronger base than the alcohol used as a base in this step, the elimination using hydroxide ion is a better step.

b. The first step of this reaction is removal of the very acidic proton α to both a cyano and a carbethoxy group. There are hydroxide ions present in 95% ethanolic solutions of carbonate, so that either hydroxide ion or carbonate ion can be used as the base.

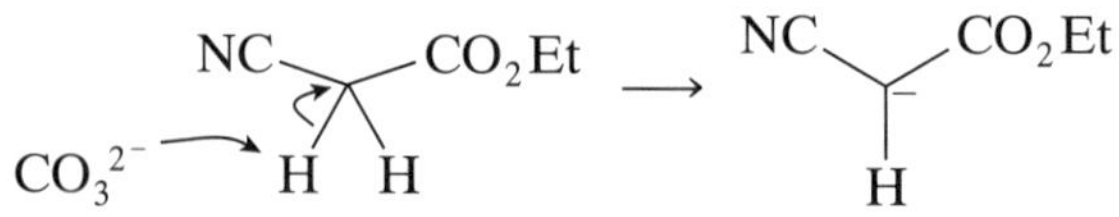

The resulting anion can then carry out a nucleophilic reaction with the electrophilic nitrogen of the nitroso group of nitrosobenzene.

Problem 3.15
continued

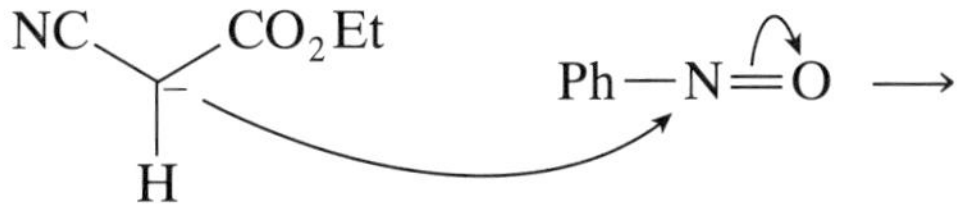

The resulting oxyanion can remove a proton from solvent.

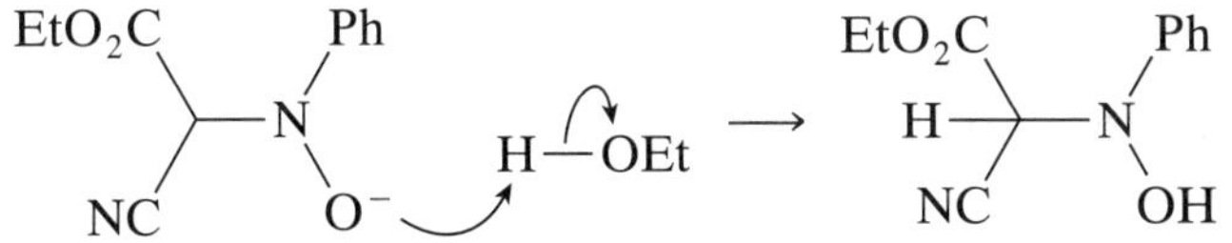

3-51

After the first addition, the reaction repeats itself with removal of the second α proton in **3-51** and addition to a second molecule of nitrosobenzene.

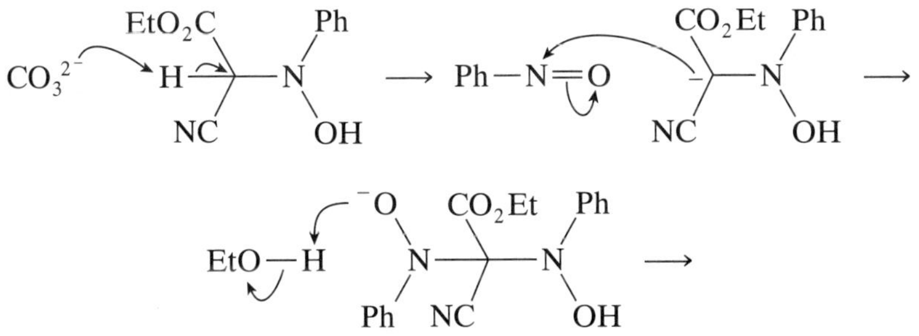

Finally, hydroxide can add to the carbethoxy group, and the intermediate undergoes elimination to give the products.

Product, **3-52**, a half ester of carbonic acid, is unstable and would decompose to CO_2 and HOEt under the reaction conditions.

Another possible mechanism for the final stages of the reaction involves an intramolecular nucleophilic reaction:

The fact that **3-53** is not necessary does not eliminate it as a possible intermediate in the reaction.

A student wrote the following as a mechanism for the final elimination step:

The student's mechanism involves the elimination of hydroxide ion and the formation of the following cation:

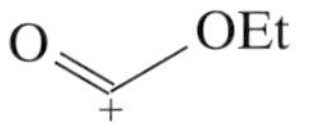

Problem 3.15
continued

Formation of this cation, a very strong acid, would not be expected in a basic medium. (See Hint 2.5.) If the structures of the eliminated fragments had been drawn, this unlikely step might not have been suggested.

This mechanism shows a direct nucleophilic substitution at an sp^2-hybridized carbon, which is unlikely. An alternative is addition of the amine to the α,β-unsaturated system, followed by elimination of bromide.

Problem 3.16

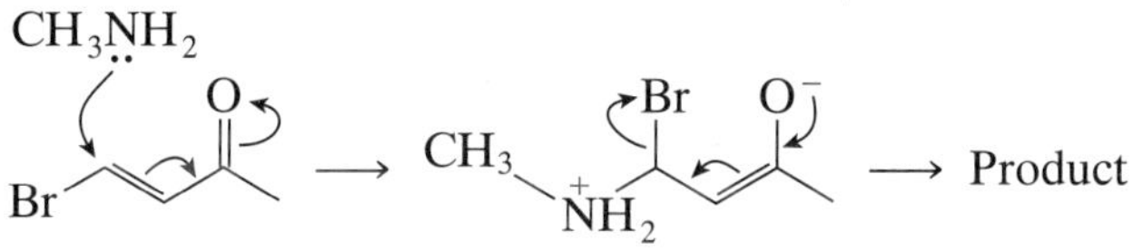

This addition follows the usual course for a 1,4-addition reaction, but the subsequent addition of a proton at the 1 position (oxygen of the carbonyl group) is replaced by elimination of bromide ion.

a. Notice that both new groups in the molecule are attached to the carbon next to the carbonyl group. Thus, the first steps are removal of a proton from the carbon α to the carbonyl group to give an anion, followed by nucleophilic substitution effected by that anion.

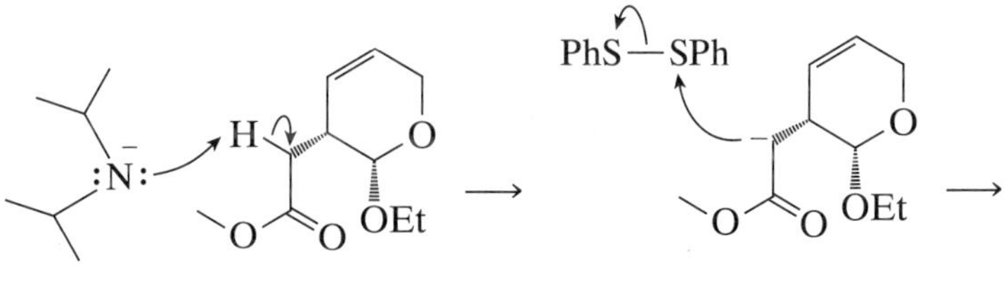

Problem 3.17

Then a second anion is formed, which adds to the β carbon of the α,β-unsaturated nitro compound. The nitro group can stabilize the intermediate anion by resonance, analogous to a carbonyl group.

3-54

The diisopropylamine formed when LDA acts as a base is much less acidic than the proton α to the nitro group (see Appendix C). Thus, protonation of the anion **3-54** must take place during workup.

b. Cyanide is a good nucleophile. A 1,4-addition to the α,β-unsaturated carbonyl puts the cyano group in the right place for subsequent reaction.

Formation of the other product involves addition to a carbonyl group as the first step.

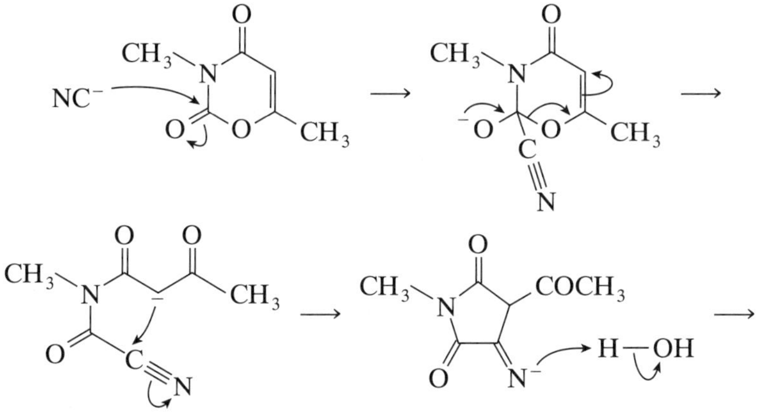

Problem 3.17
continued

The final steps are tautomerization.

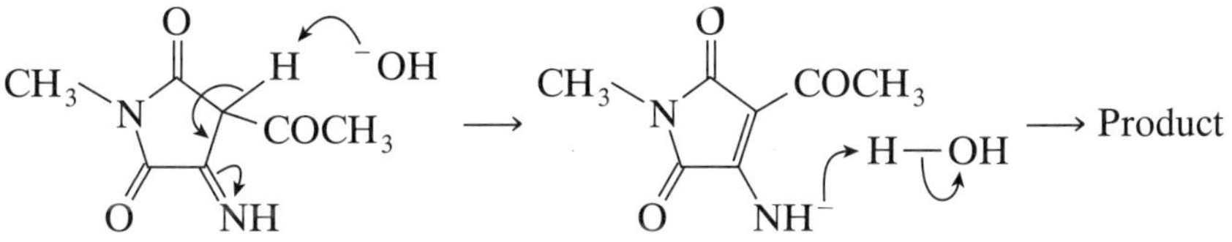

Problem 3.18

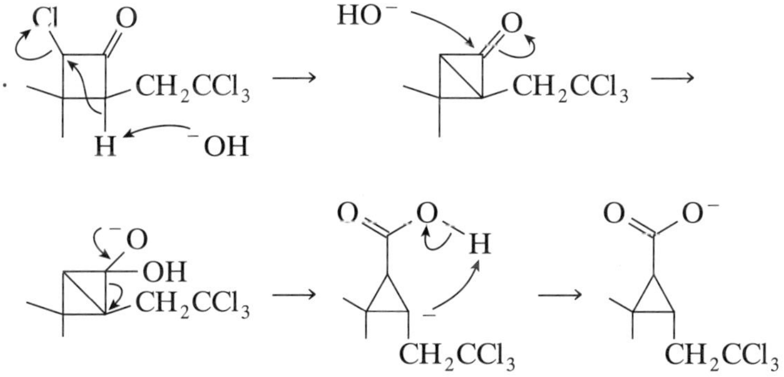

The last step shows intramolecular transfer of a proton because five-membered cyclic transition states are readily achieved. The process could also be represented by two intermolecular steps

Problem 3.18
continued

b.

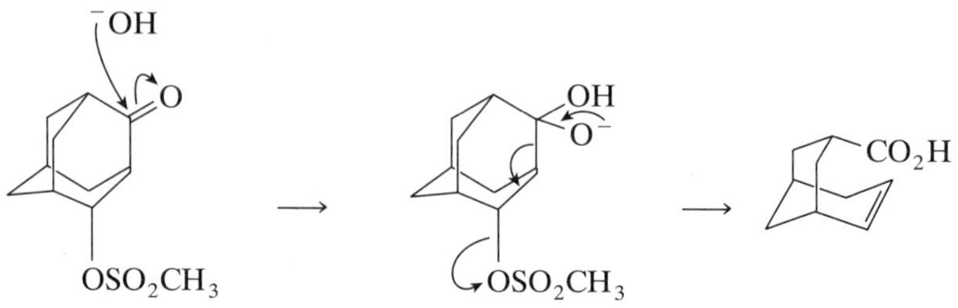

c.

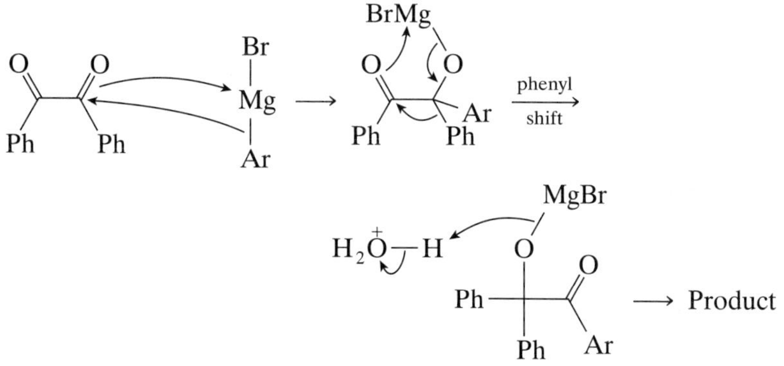

Problem 3.19

The stereospecific reaction occurring in dimethoxyethane could arise by a concerted mechanism such as the following:

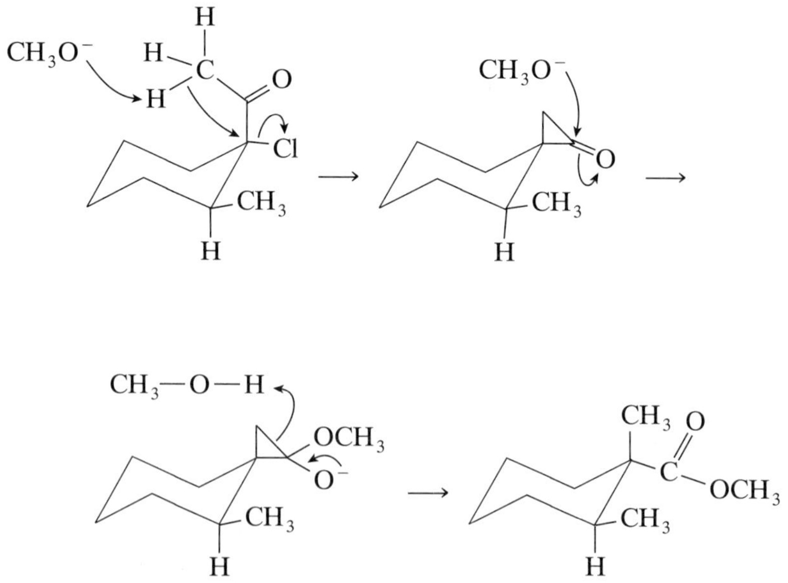

The loss of stereochemistry when the reaction is carried out in methanol may be the result of a stepwise mechanism:

Problem 3.19
continued

CH_3O^-

H H O H Cl CH_3 H $\longrightarrow$

$\bar{C}H_2$ O Cl CH_3 H $\longleftrightarrow$

H_2C O^- Cl CH_3 H $\longrightarrow$

H_2C O^- $+$ CH_3 H $\longleftrightarrow$

H_2C O $+$ CH_3 H $\rightleftharpoons$

O $+$ CH_2 CH_3 H

O CH_3 H

O CH_3 H

Product Ester $\longleftarrow$ ^-OMe $\longleftarrow$

Development of a carbocation center due to loss of chloride ion can account for the loss of stereochemistry at this carbon. It seems reasonable to write this step after removal of a proton to give the anion because, if chloride loss occurs before anion formation, the carbocation would be expected to react with solvent before the anion could be formed. In that case one might expect to isolate some of the compound in which chloride had been replaced by methoxyl. The effect of the solvent on the stereochemistry of the reaction is due to its ability to solvate the intermediate charged species involved in a stepwise mechanism.

a. The reaction mechanism requires two separate nucleophilic steps, i.e., the mechanism requires two nucleophiles rather than one. The reaction can-

Problem 3.20

Problem 3.20
continued

not be run with the α-chloro compound in the presence of *n*-butyllithium, because reactions between these reagents would give several important side products. Thus, **3-55**, the dilithium derivative of the starting sulfone, is formed first, and then the chloro compound is added.

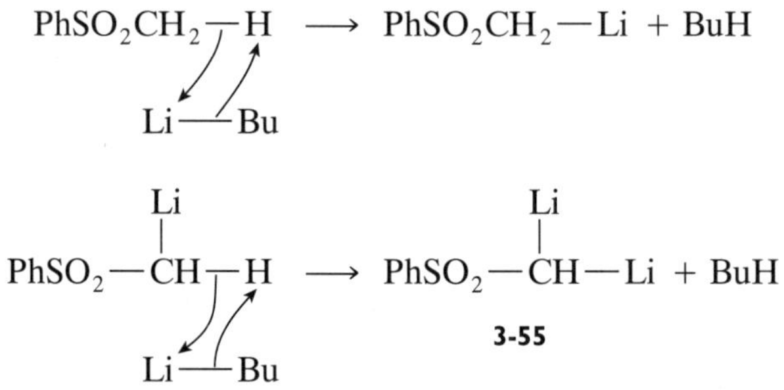

$$PhSO_2CH_2 \!-\! H \longrightarrow PhSO_2CH_2 \!-\! Li + BuH$$

$$Li \!-\! Bu$$

$$PhSO_2 \!-\! \overset{\overset{\displaystyle Li}{|}}{C}H \!-\! H \longrightarrow PhSO_2 \!-\! \overset{\overset{\displaystyle Li}{|}}{C}H \!-\! Li + BuH$$

3-55

$$Li \!-\! Bu$$

There are two electrophilic positions in the chloro compound; the carbonyl carbon and the carbon α to it, which bears chloride as a leaving group. Nucleophilic reaction at either position by **3-55** is a reasonable reaction, and we will illustrate both possibilities. Reaction at the carbonyl gives **3-56**, which can close to a cyclopropane.

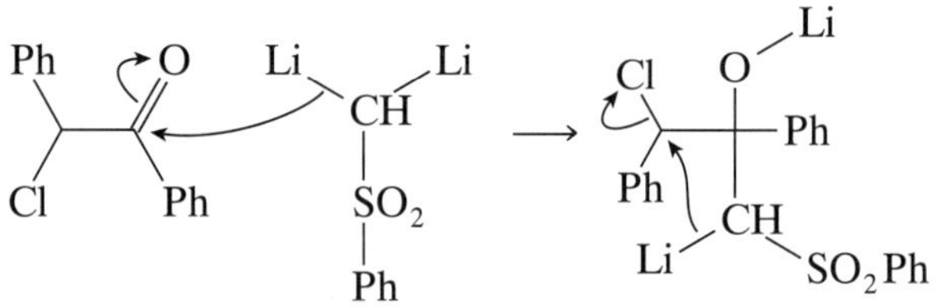

3-56

The cyclopropane ring then opens to give the product.

Problem 3.20
continued

3-57

The other mechanism involves **3-55** as the nucleophile in the S_N2 displacement at the highly reactive chloro-substituted carbon α to the carbonyl. The remaining anion, **3-58**, reacts with the carbonyl group to give **3-57**.

3-55   **3-58**

The authors of the paper cited favor the first mechanism by analogy to the reaction of **3-55** with 1-chloro-2,3-epoxypropane.

Notice how this reaction resembles Example 3.20. In Example 3.20, the benzylic carbon is inserted between the S and C=S. In this reaction, the carbon introduced by the phenylsulfonyl anion is inserted between the benzoyl group and the α carbon of the starting carbonyl compound. In both cases, the "insertion" is effected by forming one bond to close a three-membered ring and then breaking a different bond to open the three-membered ring.

b. This reaction is run in the presence of base. The most acidic hydrogen in the starting materials is on the carbon between the ester and ketone functional groups. If that proton is removed, the resulting anion can act as a nucleophile and add to the carbonyl group of the isocyanate. The oxyanion formed (stabilized by resonance with the nitrogen) can undergo an intramolecular nucleophilic substitution to produce the five-membered ring. Base-catalyzed tautomerization gives the final product.

Problem 3.20
continued

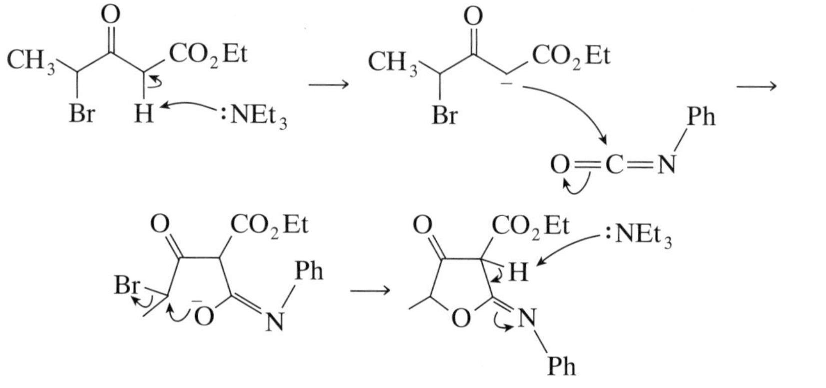

3-59

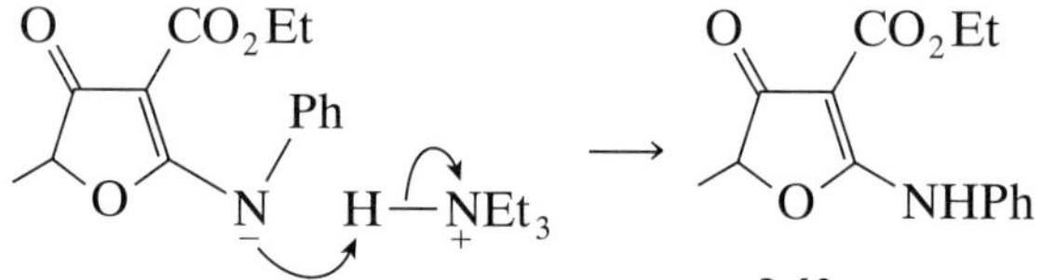

3-60

The anionic intermediate through which the tautomers **3-59** and **3-60** interconvert is a resonance hybrid:

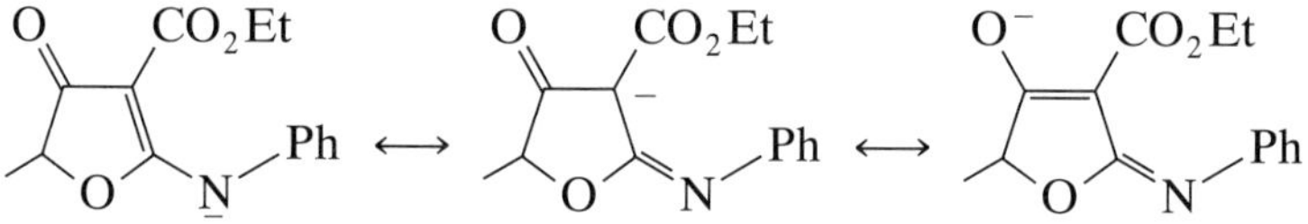

A much less likely first step is nucleophilic reaction of the carbonyl oxygen of the isocyanate with the carbon attached to bromine:

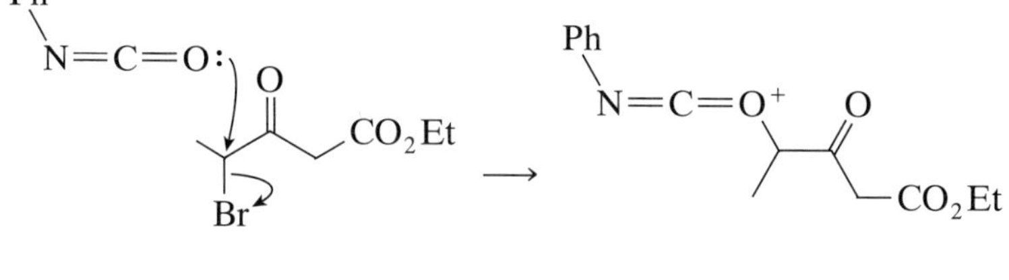

Problem 3.20
continued

Although carbonyl groups will act as bases with strong acids, their nucleophilicity generally is quite low. We can get a rough idea of the basicity of the carbonyl group, relative to the acidity of the proton actually removed, from Appendix C. The basicity of acetone is estimated from the pK_a of its conjugate acid (-2.85). The pK_a of the proton should be similar to that of ethyl acetoacetate (11). Thus, the ketoester is much more acidic than the carbonyl group is basic, and it is much more likely that the proton would be removed. Nonetheless, a positive feature of this mechanism is that there is a viable mechanism leading to product. (The next step would be removal of a proton on the carbon between the ketone and ester.)

c. In this reaction, the amide anion, a very strong base, removes a benzylic proton. Cyclization, followed by loss of methylphenylamide, and tautomerization lead to the product. Notice that nucleophilic reaction of the benzylic anion and loss of methylphenylamide are separate steps, in accord with Hint 3.2.

The anion corresponding to the product is considerably more stable than either the amide ion or the phenylamide ion, so protonation of the final anion will take place during workup.

As in Problem 3.19.b, it is necessary to be able to distinguish between tautomers and resonance forms. The following two structures are tautomers, so a mechanism needs to be written for their interconversion.

d. Note that 2 mol of hydrazine have reacted to give the product. It also appears that the introduction of each mole is independent of the other and, thus, each mechanism can be shown separately. The hydrazinolysis of the ester is shown first.

A mechanism for reaction with the other mole of hydrazine follows. The reaction sequence shown is preferred because the anion produced by the first step, addition at the β position, is resonance-stabilized. The anion produced by reaction of hydrazine at the keto carbonyl carbon would not be resonance stabilized.

Problem 3.20
continued

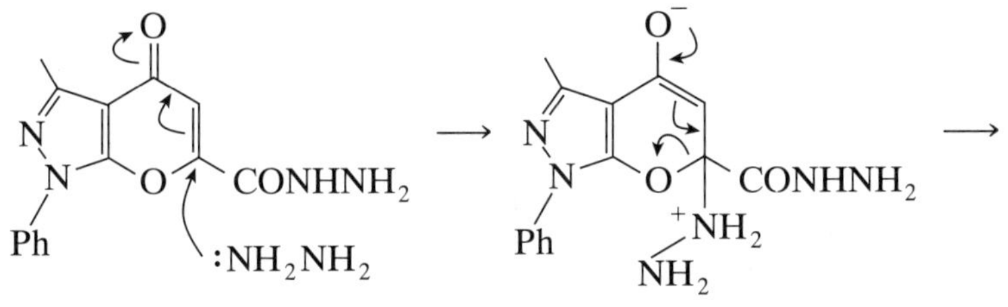

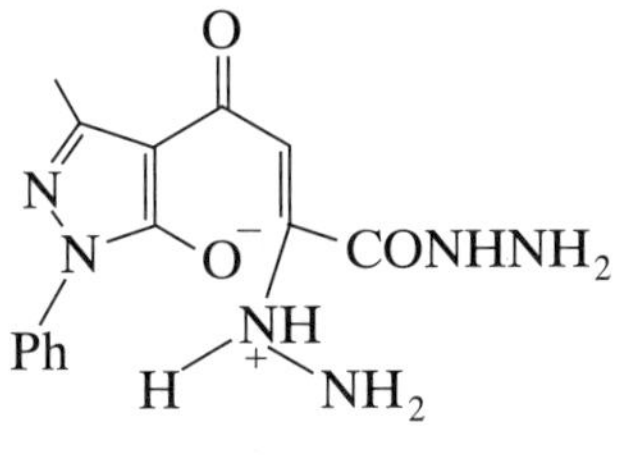

3-61

The enolate ion, **3-61**, will pick up a proton from solvent, and the hydrazinium ion will lose a proton to a base, such as a molecule of hydrazine. The distance between the groups in the molecule makes intramolecular transfer of a proton quite unlikely. The nucleophilic hydrazine group in the neutral intermediate, **3-62**, can react intramolecularly with the electrophilic carbon of the carbonyl group.

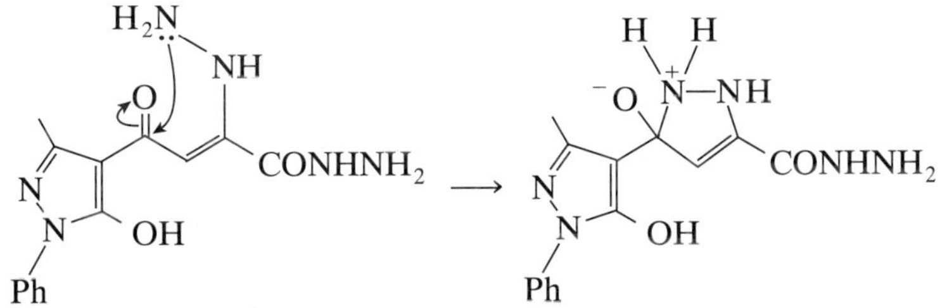

3-62

Again, the positive nitrogen loses a proton and the negative oxygen picks up a proton. In the final step, base-promoted elimination of water occurs. The driving force for loss of water is formation of an aromatic ring.

Problem 3.20
continued

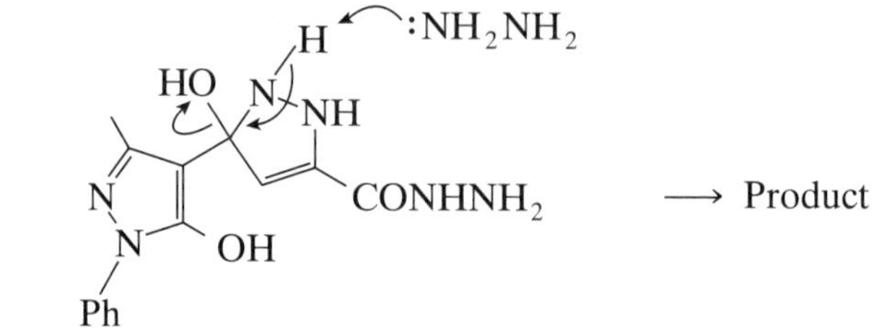

Do not write intramolecular loss of water in the last step. The mechanism shown is better for two reasons. First, the external base, hydrazine, is a much better base than the hydroxyl group. Second, most eliminations go best when the proton being removed and the leaving group are *anti* to one another.

The following would not be a good step:

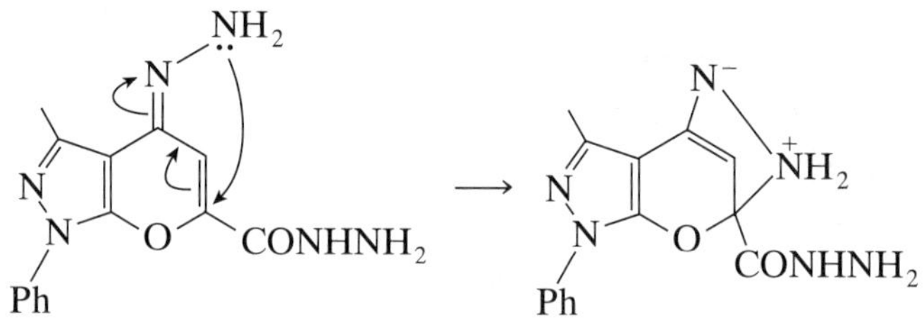

All of the atoms involved in this step lie in a plane (the C=N nitrogen is sp^2-hybridized), but because the conjugated double bond system is locked into the transoid arrangement due to the position of the C=C double bond in the six-membered ring, the amino group cannot reach close enough to cyclize in the manner shown. A further point to consider in this type of reaction is that the approaching nucleophile needs to interact with the *p* orbitals of the double bond system, so the most favorable approach of the nucleophile is perpendicular to the plane of the conjugated system.

e. Bromine is not present in the product, so that one reaction probably is nucleophilic substitution at the carbon bearing the bromine. A simple analysis (by numbering or by inspection) also reveals that the methylene group in the imine ester is substituted twice in the reaction. Thus, removal of a proton from the methylene group is a good first step, followed by a nucleophilic substitution reaction.

Problem 3.20
continued

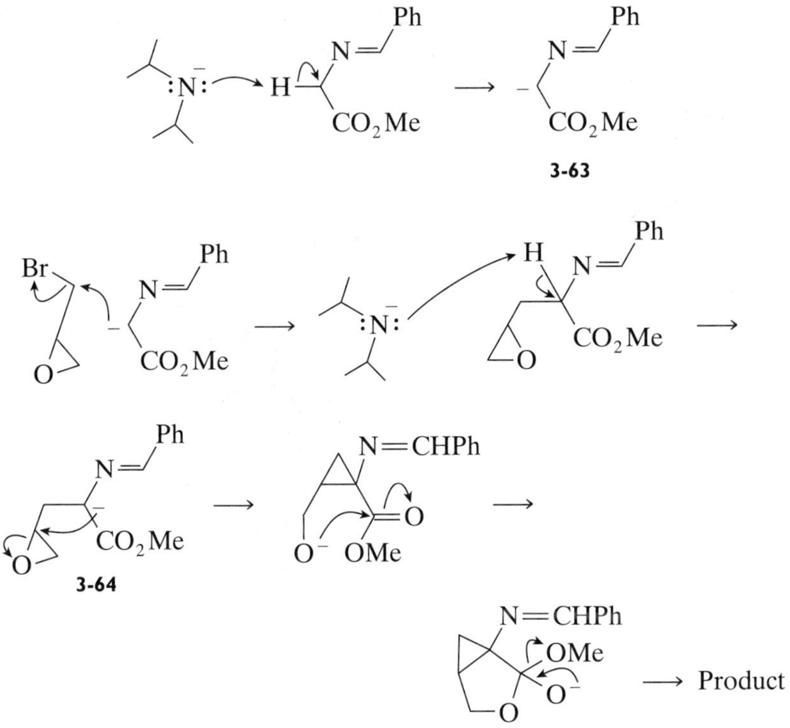

3-63

3-64

$\longrightarrow$ Product

Another possible nucleophilic reaction of anion **3-64** would be at the other electrophilic carbon of the epoxide:

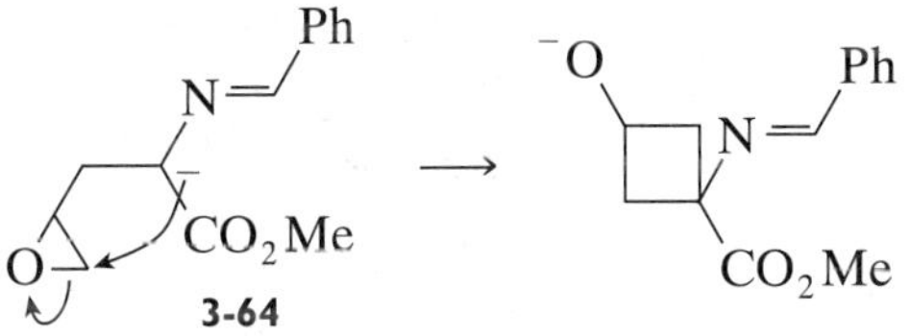

3-64

This reaction might have been favored for two reasons. First, reaction with the epoxide, an S_N2 reaction, should occur best at the least hindered carbon. Second, formation of a four-membered ring would be favored by enthalpy because it would have less strain energy than the three-membered ring. The fact that the three-membered ring is actually formed must mean that the reaction is directed by entropy.

Another possible mechanism starts with nucleophilic reaction by **3-63** at the epoxide. The alkoxide ion then displaces bromide to produce a new epoxide.

Proton removal followed by nucleophilic reaction gives a cyclopropane.

The nucleophilic alkoxide, **3-65**, reacts intramolecularly with the carbon of the ester functional group. The resulting intermediate loses methoxide to give the product.

Initial reaction of **3-64** at the CH of the epoxide (rather than the CH$_2$) is at the more hindered carbon, which is not preferred for an S$_N$2 reaction.

Problem 3.20
continued

The reaction conditions, sodamide in ammonia, as well as the lack of electron-withdrawing groups directly attached to the aromatic ring, suggest the aryne mechanism (mechanism 2) rather than the nucleophilic aromatic substitution (mechanism 1). There are several additional problems with mechanism 1.

Problem 3.21

Step 1, the nucleophilic addition of the anion to the aromatic ring, is less likely than the aryne formation (step 7) because the intermediate anion is not very stable. The methoxy and bromine substituents can remove electron character from the ring only by inductive effects. Nucleophilic aromatic substitution (except at elevated temperatures) ordinarily requires substituents that withdraw electrons by resonance. The usual position for nucleophilic reaction in this mechanism is at the carbon bearing the leaving group, in this case, bromide. However, in some other mechanism, involving several steps, addition of the nucleophile at other positions might be acceptable.

For step 2, the arrow between the two structures should be replaced with a double-headed arrow, indicating that these are resonance structures.

In step 3, cyanide can act as a leaving group, even though it is not a very good one. Furthermore, the product contains a cyano group. Thus, another problem with this step is that if cyanide leaves the molecule, it will be diluted by the solvent to such a low concentration that subsequent addition will occur very slowly. (However, that does not mean that it cannot happen.)

In step 4, the elimination of HBr is reasonable because it gives an aromatic system and because *syn*-E2 elimination is a common reaction.

In step 5, direct substitution by cyanide ion at an *sp*^{2} center, as shown, is an unlikely process.

In step 6, removal of a proton from ammonia, by an anion that is much more stable than the amide ion, is very unlikely; this step would have a very unfavorable equilibrium. In other words, the product-forming step, like step 11, would occur on workup.

Other comments:

1. Writing the aryne formation (step 7) as a two-step process would be acceptable.

2. Direct substitution of cyanide for bromine on the ring is not a good mechanistic step. Initially, nucleophilic substitution at an *sp*^{2}-hybridized carbon is unlikely.

Problem 3.21
continued

3. An aryne intermediate is unlikely to react as a nucleophile. Because of the high *s* character in the orbitals forming the third bond, arynes tend to be electrophilic, not nucleophilic.

4. Other mechanisms, that involve formation of a carbanion in the ring also are unlikely because this carbanion is not stabilized by strongly electron-withdrawing substituents on the ring.

Problem 3.22

There are two apparent ways that triphenylphosphine can react with the diethyl azocarboxylate. One is nucleophilic reaction at the electrophilic carbonyl carbon, and the other is 1,4-addition. Because the ester groups are intact in the hydrazine product, the 1,4-addition is more likely. The intermediate anion can be protonated by the carboxylic acid to give a salt.

The electrophilic phosphorus atom then undergoes nucleophilic reaction with the alcohol. An intriguing aspect of this reaction is the fate of the proton on the alcohol. There is no strong base present in the reaction mixture. A good possibility is that the carboxylate anion removes the proton from the alcohol as it is reacting with the phosphorus.

How do we rationalize what appears to be a trimolecular reaction? Because the solvent is a nonpolar aprotic solvent, the phosphonium carboxylate must be present as an ion pair and can be considered as a single entity. The carboxylate ion also may be properly situated to remove the proton.

Problem 3.22
continued

Finally, the carboxylate anion acts as a nucleophile to displace triphenylphosphine oxide and give the inverted product.

An alternate mechanism, in which the alcohol reacts with the phosphonium salt without assistance from the carboxylate anion, is less likely because the intermediate produced has two positive centers adjacent to each other.

Reactions Involving Acids and Other Electrophiles

Acids and electrophiles are electron-deficient species. According to the Lewis concept, all electrophiles (e.g., cations, carbenes, metal ions) are acids by definition. However, from long usage the term *acid* is frequently used to refer to a proton donor, whereas the term *Lewis acid* usually refers to charged electrophiles in general.

I. STABILITY OF CARBOCATIONS

Reactions in acid often involve the formation of carbocations—trivalent, positively charged carbon atoms—as intermediates. The order of stability of carbocations containing only alkyl substituents is $3° > 2° > 1° > CH_3$. Cation stability is influenced by several factors:

1. *Hyperconjugation.* An increase in the number of alkyl substituents increases the stability of the carbocation due to orbital overlap between the

adjacent σ bonds and the unoccupied p orbital of the carbocation. The resulting delocalization of charge, which can be represented by resonance structures, stabilizes the cation.

    2. *Inductive effects.* Neighboring alkyl groups stabilize a cation because electrons from an alkyl group, which is relatively large and polarizable compared to hydrogen, can shift toward a neighboring positive charge more easily than can electrons from an attached hydrogen.

    3. *Resonance effects.* Conjugation with a double bond increases the stability of a carbocation. Thus, allylic and benzylic cations are more stable than their saturated counterparts. (For example, see Problem 1.4.c.) Heteroatoms with unshared electron pairs, e.g., oxygen, nitrogen, or halogen, can also provide resonance stabilization for cationic centers, as in the following examples:

$$
\begin{array}{ccc}
\underset{|}{\overset{R}{\underset{+}{R-C}-\ddot{O}-CH_3}} & \longleftrightarrow & \underset{|}{\overset{R}{R-C=\underset{+}{\ddot{O}}-CH_3}}
\end{array}
$$

$$
\underset{+}{R-C}=\ddot{O} \longleftrightarrow R-C\equiv\underset{\cdot\cdot}{O}{}^{+}
$$

$$
\underset{+}{\overset{:\ddot{C}l:}{\underset{|}{R-C}-R}} \longleftrightarrow \overset{\overset{+}{:}Cl:}{\underset{\|}{R-C}-R}
$$

    Cations at sp^2- or sp-hybridized carbons are especially unstable. In general, the more s character in the orbitals, the less stable the cation. An approximate order of carbocation stability is CH_3CO^+ (acetyl cation) $\sim$ $(CH_3)_3C^+ \gg PhCH_2^+ > (CH_3)_2CH^+ > H_2C{=}CH{-}CH_2^+ \gg CH_3CH_2^+ > H_2C{=}CH^+ > Ph^+ > CH_3^+$. The stabilities of various carbocations can be determined by reference to the order of stability for alkyl carbocations, $3° > 2° > 1° > CH_3$. The acetyl cation has a stability similar to that of the *t*-butyl cation. Secondary carbocations, primary benzylic cations, and primary allylic cations are all more stable than primary alkyl cations. Vinyl, phenyl, and methyl carbocations are less stable than primary alkyl cations.

2. FORMATION OF CARBOCATIONS

A. Ionization

    A compound can undergo unimolecular ionization to a carbocation and a leaving group. If the final product formed is due to substitution, the process is called S_N1. If it is due to elimination, the process is called E1. In both cases, the rate-determining step is the ionization, not the product-forming step.

Example 4.1. *Acid-catalyzed loss of water from a protonated alcohol.*

In this process, protonation of the alcohol group is the first step. This occurs much faster than the rate-determining step, loss of water from the protonated alcohol.

Example 4.2. *Spontaneous ionization of a triflate.*

In this example, ionization is favored by several factors. First, the benzyl ion formed is resonance-stabilized and bears a methoxy group in the *para* position that can further stabilize the cation by resonance. In addition, the leaving group is triflate, an exceptionally good leaving group. Finally, the ionization takes place in water, a polar solvent that can stabilize the two charged species formed.

B. Addition of an Electrophile to a π Bond

Intermediate cations are often produced by addition of a proton or a Lewis acid to a π bond.

Example 4.3. *Protonation of an olefin.*

Example 4.4.   *Protonation of a carbonyl group.*

In acid, carbonyl compounds are in equilibrium with their protonated counterparts. Protonation is often the first step in nucleophilic addition or substitution of carbonyl groups. For aldehydes and ketones, the protonated carbonyl group is a resonance hybrid of two forms: one with positive charge on the carbonyl oxygen and one with positive charge on the carbonyl carbon.

When esters are protonated at the carbonyl group, there are three resonance forms: two corresponding to the ones that form with aldehydes and ketones and a third with positive charge on the alkylated oxygen.

Example 4.5.   *Reaction of a carbonyl compound with a Lewis acid.*

Carbonyl groups form complexes or intermediates with Lewis acids like $AlCl_3$, BF_3, and $SnCl_4$. For example, in the Friedel–Crafts acylation reaction in nonpolar solvents, an aluminum chloride complex of an acid chloride is often the acylating agent. Because of the basicity of ketones, the products of the acylation reaction are also complexes. For more detail on electrophilic aromatic substitution, see Section 7.

C. Reaction of an Alkyl Halide with a Lewis Acid

Example 4.6. *Reaction of 2-chlorobutane and aluminum trichloride.*

The Lewis acid removes the halide ion to give a carbocation.

3. THE FATE OF CARBOCATIONS

Once formed, carbocations have several options for further reaction. Among these are substitution, elimination, addition, and rearrangement.

1. Substitution (S_N1) occurs when the carbocation reacts with a nucleophile.

2. Elimination (E1) usually occurs with loss of a proton, as in the formation of 2-methyl-1-propene from the *t*-butyl cation.

3. The carbocation can react with an electron-rich reagent.

Example 4.7.   *Cationic olefinic polymerization.*

$$CH_3CH\overset{\overset{\displaystyle BF_3}{\curvearrowright}}{=\!=}CH_2 \longrightarrow CH_3\overset{+}{C}HCH_2\overset{\ominus}{B}F_3$$

$$CH_3CH\overset{\uparrow}{=\!=}CH_2 \longrightarrow CH_3\underset{\underset{\displaystyle CH_3\overset{+}{C}HCH_2}{|}}{CHCH_2\overset{\ominus}{B}F_3} \longrightarrow \quad \longrightarrow$$

4. The carbocation can undergo rearrangement (see the following section).

(The negative sign on the boron indicates charge only; there is not an unshared pair of electrons on boron.)

4. REARRANGEMENT OF CARBOCATIONS

Carbocations tend to rearrange much more easily than carbanions. Under common reaction conditions, a carbocation rearranges to another carbocation of equal or greater stability. For example, a secondary carbocation will rearrange to a tertiary carbocation or a different secondary carbocation, but ordinarily it will not rearrange to a less stable primary carbocation. This generalization is not absolute, and because there is not a high energy barrier to the rearrangement of carbocations, rearrangement to a less stable cation can occur if it offers the chance to form a more stable product.

Hint 4.1   Rearrangement of a carbocation frequently involves an alkyl, phenyl, or hydride shift to the carbocation from an adjacent carbon (a 1,2-shift).

In many cases there are several different pathways by which rearrangement may take place. In these situations, the question of which group will migrate (migratory aptitude) is a complex one. In general, aryl and branched alkyl chains migrate in preference to unbranched chains, but the selectivity is not high. Similarly, the tendency of hydrogen to migrate is unpredictable: sometimes hydrogen moves in preference to an aryl group, at other times it migrates less readily than an alkyl. Very often, other factors such as stereochemistry, relief of strain, and reaction conditions are as important as the structure of the individual migrating group. *Frequently it is difficult to predict the product of a reaction in which a carbocation is formed; it is much easier to identify a reasonable pathway by which an experimentally obtained product is derived from starting material.*

Example 4.8.   *A hydride shift in the rearrangement of a carbocation.*

Treatment of isobutyl alcohol with HBr and H_2SO_4 at elevated temperatures leads to *t*-butyl bromide. In the first step, the hydroxyl group is protonated by the sulfuric acid to convert it into a better leaving group, the water molecule. Water then leaves, giving the primary isobutyl carbocation, **4-1**.

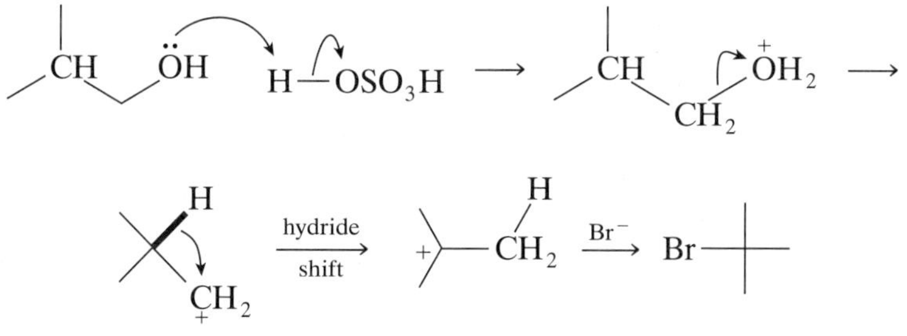

The hydrogen and the electrons in its bond to carbon, highlighted in **4-1**, move to the adjacent carbon. Now the carbon from which the hydride left is deficient by one electron and is, thus, a carbocation. Because the new carbocation, **4-2**, is tertiary, the molecule has gone from a relatively unstable primary carbocation to the much more stable tertiary carbocation. In the final step, the nucleophilic bromide ion reacts with the positive electrophilic tertiary carbocation to give the alkyl halide product.

In rearrangement reactions, the method of numbering both the starting material and the product, introduced in Chapter 2, can be very helpful.

Hint 4.2

Example 4.9.   *An alkyl shift in the rearrangement of a carbocation.*

Consider the following reaction:

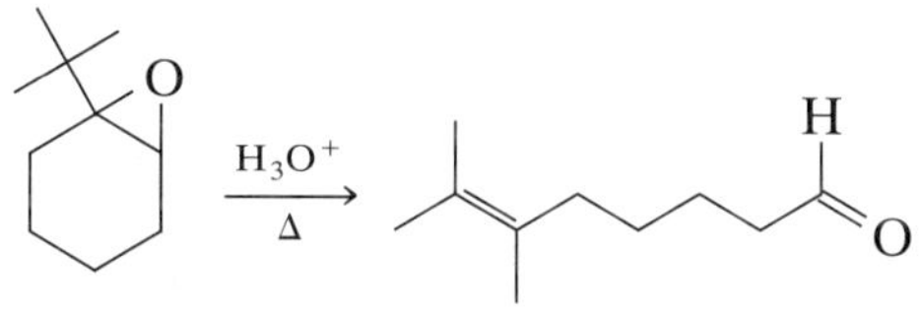

(This example is derived from Corona, T.; Crotti, P.; Ferretti, M.; Macchia, F. *J. Chem. Soc., Perkin Trans.* 1 **1985**, 1607–1616.)

Because the product has neither a *t*-butyl group nor a six-membered ring, a rearrangement must have taken place. Numbering of both the starting material and the product helps to visualize what takes place during the course of the reaction.

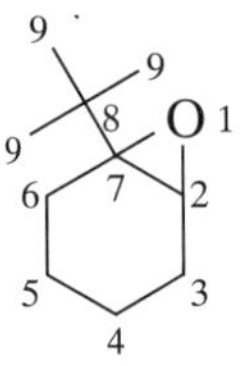

4-3 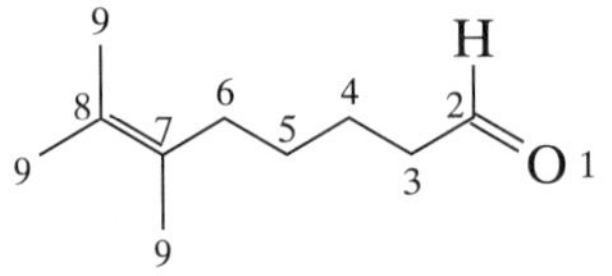   **4-4**

The three methyls in **4-3** are given the same number because they are chemically equivalent. The product **4-4** has been numbered so that the system conforms as closely as possible to that of the starting material, i.e., with the least possible rearrangement. Comparison of the numbering in **4-3** to that in **4-4** shows that one of the methyl groups shifts from carbon-8 to carbon-7 and that the bond between carbon-2 and carbon-7 is broken. Because a rearrangement to carbon-7 takes place, that carbon must have a positive charge during the course of the reaction.

The oxygen is the only basic atom in the molecule, so protonation of the oxygen must be the first step.

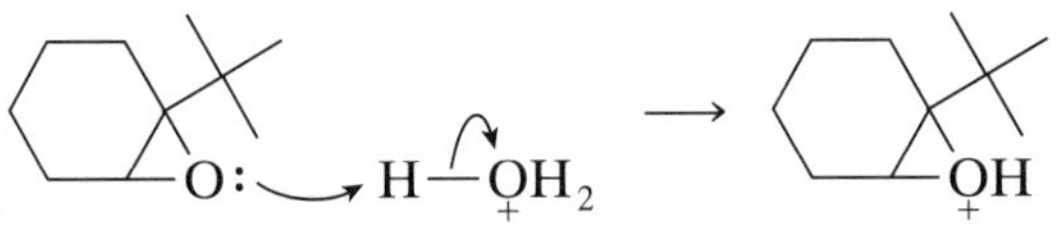

The protonated epoxide is unstable because of the high strain energy of the three-membered ring and opens readily. There are two possible modes of ring opening:

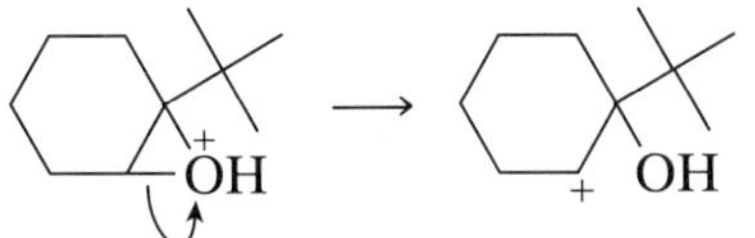

4-5

or

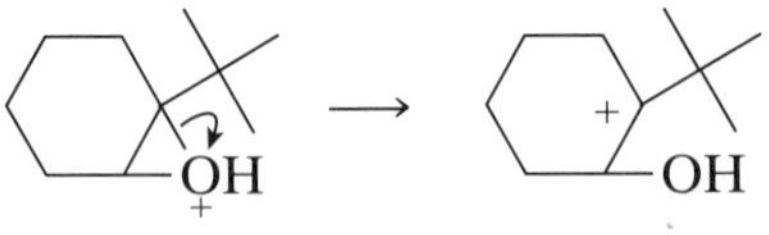

4-6

Because **4-6**, a tertiary carbocation, is more stable than **4-5**, a secondary cation, **4-6** would be expected to be formed preferentially. (However, if the tertiary carbocation did not lead to the product, we would go back to consider the secondary cation.) In addition, the formation of **4-6** appears to lead toward the product, because the carbon bearing the positive charge is number 7 in **4-3**. The tertiary carbocation can undergo rearrangement by a methyl shift to give another tertiary carbocation.

Now, one of the lone pairs of electrons on oxygen facilitates breaking the bond between C-2 and C-7. The formation of two new π bonds compensates for the energy required to break the bond between C-2 and C-7.

The only remaining step is deprotonation of the protonated aldehyde, **4-7**, to give the neutral product. This would occur during the workup, probably in a mild base like sodium bicarbonate.

A. The Dienone – Phenol Rearrangement

The dienone–phenol rearrangement is so named because the starting material is a dienone and the product is a phenol.

Example 4.10.   *Rearrangement of a bicyclic dienone to a tetrahydronaphthol system.*

Inspection shows that a skeletal change occurs in the following transformation, that is, rearrangement occurs. (See the answer to Problem 2.4.b for an application of Hint 2.13 to analyzing the bonding changes involved.)

The first step in the mechanism for this reaction is protonation of the most basic atom in the molecule, the oxygen of the carbonyl.

The intermediate carbocation, **4-8**, undergoes an alkyl shift to give another resonance-stabilized carbocation, **4-9**.

Finally, **4-9** undergoes another alkyl shift, followed by loss of a proton, to give the product. Looking at the resonance structures that can be drawn for all the cations involved in the mechanism suggests that their energy should be comparable to that of the initial cation, **4-8-2**. The driving force for the reaction comes from the formation of the aromatic ring.

What initially appears to be a complicated reaction is the result of a series of simple steps. For other examples of this reaction, see Miller, B. *Acc. Chem. Res.* **1975**, *8*, 245–256.

Write step-by-step mechanisms for the following transformations.   **PROBLEM 4.1**

a.

b.

B. The Pinacol Rearrangement

Many common rearrangement reactions are related to the rearrangement of 1,2-dihydroxy compounds to carbonyl compounds. Often these reactions are called pinacol rearrangements, because one of the first examples was the transformation of pinacol to pinacolone:

OH
$\xrightarrow{H_2SO_4}$
OH

pinacol   pinacolone

Example 4.11. *The rearrangement of 1,2-diphenyl-1,2-ethanediol to 2,2-diphenyl-ethanal.*

OH   OH $\xrightarrow{H_2SO_4}$ O   Ph
Ph   Ph   H   Ph

The mechanism of this reaction involves formation of an intermediate carbocation, a 1,2-phenyl shift, and loss of a proton to form the product:

4-10

In reactions of this type, it is possible to form more than one initial carbocation if the starting material is not symmetrical. In this situation, the more stable carbocation is usually formed in step 1. Once this initial carbocation has been formed, the course of step 2 is more difficult to predict because it depends on the propensity of one group to migrate in preference to another (*migratory aptitude*), which often depends on reaction conditions.

It has been suggested that in pinacols containing an aryl group, the initial carbocation formed by loss of hydroxyl can be stabilized by neighboring group participation of the phenyl group to give a bridged phenonium ion:

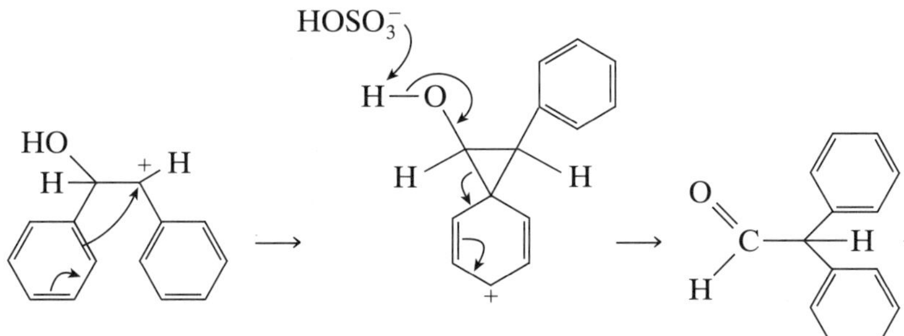

4-11

Although these types of bridged phenonium ions are accepted intermediates in a number of reactions, they do not appear to be involved in the pinacol rearrangement (see Schubert, W. M.; LeFevre, P. H. *J. Am. Chem. Soc.* **1972**, *94*, 1639). Stabilization of the carbocation **4-10** by resonance with the oxygen substituent may be a factor in determining the preference for phenyl migration over phenonium ion formation in the pinacol rearrangement.

Write a step-by-step mechanism for the transformation of pinacol to pinacolone in the presence of sulfuric acid.   **PROBLEM 4.2**

For the following reaction write a step-by-step mechanism that accounts for the observed stereospecificity.   **PROBLEM 4.3**

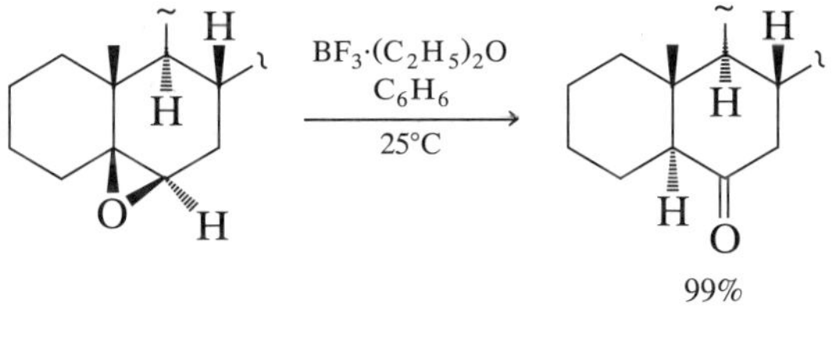

Heubest, H. B.; Wrigley, T. I. *J. Chem. Soc.* **1957**, 4596–4765.

PROBLEM 4.4   **Write a step-by-step mechanism for the following reaction.**

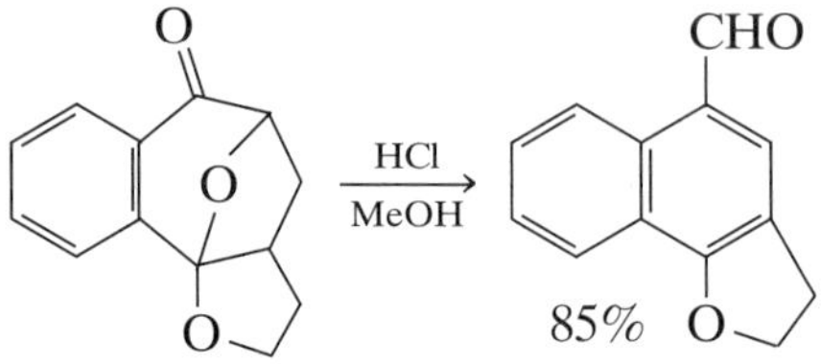

Padwa, A.; Carter, S. P.; Nimmesgern, H.; Stull, P. D. *J. Am. Chem. Soc.* **1988**, *110*, 2894–2900.

---

5. ELECTROPHILIC ADDITION

Addition of electrophiles is a reaction typical of aliphatic π bonds (see Example 4.3). Such additions involve two major steps: (1) addition of the electrophile to the nucleophilic π bond to give an intermediate carbocation, and (2) reaction of the carbocation with a nucleophile. Typical electrophiles are bromine, chlorine, a proton supplied by HCl, HBr, HI, H_2SO_4, or H_3PO_4, Lewis acids, and carbocations. The nucleophile in step 2 is often the anion associated with the electrophile, e.g., bromide, chloride, iodide, etc., or a nucleophilic solvent like water or acetic acid.

A. Regiospecificity

Because the more stable of the two possible carbocations is formed predominantly as the intermediate in addition of the electrophile (step 1), electrophilic additions are often regiospecific.

Example 4.12.   *Regiospecificity in electrophilic additions.*

When HI adds to a double bond, the proton acts as an electrophile, giving an intermediate carbocation that then reacts with the nucleophilic iodide ion to give the product. In the reaction of HI with 1-methylcyclohexene, there is only one product, 1-iodo-1-methylcyclohexane; no 1-iodo-2-methyl-cyclohexane is formed.

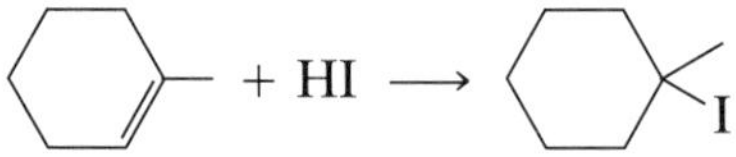

This reaction is said to be regiospecific because the iodide might occupy the ring position at either end of the original double bond, but only one of these products is actually formed. The reaction is regiospecific because the proton adds to form the more stable tertiary carbocation, **4-12**, and not the secondary carbocation, **4-13**.

4-12   **4-13**

B. Stereochemistry

anti Addition

In some electrophilic additions an unusual three-membered-ring intermediate is formed. When this intermediate is stable under the reaction conditions, an *anti* addition of electrophile and nucleophile takes place.

Example 4.13. *Stereospecific* **anti** *addition of bromine to* **cis-** *and* **trans-2-butene.**

The bromination of *cis-* and *trans*-2-butene occurs stereospecifically with each isomer.

4-14   **4-15**

4-14   **4-16**   **4-14**   **4-17**

From *cis*-2-butene, the products are the enantiomeric dibromides **4-16** and **4-17**, which are formed in equal amounts. The enantiomers are formed in equal amounts because bromine adds to the top and bottom faces of the alkene to give the intermediate bromonium ions, **4-14** and **4-15**, in equal amounts. Either carbon of each of these bromonium ions can then react with the nucleophile on the side opposite the bromine to give the product dibromides.

Note that **4-16** and **4-17** are mirror images and nonsuperimposable. In like manner, the reaction of **4-15** with bromide ion also gives **4-16** and **4-17**.

From *trans*-2-butene, *anti* reaction of bromide at either carbon of the bromonium ion, **4-18**, gives only **4-19-1**, which is a *meso* compound because it has a mirror plane. This is most easily recognized in the eclipsed conformation, **4-19-2**, rather than the staggered conformation, **4-19-1**.

4-18	**4-19-1**	**4-19-2**

In the addition of bromine to *cis*- and *trans*-2-butene, the stereochemistry of each product occurs because the two bromines were introduced into the molecule on opposite faces of the original double bond.

Under some experimental conditions, electrophilic addition of either Cl^+ or a proton may form stable three-membered-ring intermediates. Thus, when a double bond undergoes stereospecific *anti* addition, formation of a three-membered-ring intermediate analogous to the bromonium ion is often part of the mechanism.

syn Addition

Sometimes *syn* addition to a double bond may occur. These reactions usually occur in very nonpolar media.

Example 4.14. *The chlorination of indene to give cis-1,2-dichloroindane.*

There are several factors that influence the course of this halogen addition: (1) the reaction takes place with chlorine rather than bromine; (2) the double bond of the starting material is conjugated with an aromatic ring; and (3) the reaction takes place in a nonpolar solvent.

4-20

Chlorine is smaller and less polarizable than bromine and so has less of a tendency to form a bridged halonium ion than bromine. Also, the position of the double bond means that electrophilic addition of chlorine gives a stabilized benzylic cation, which is expected to be planar. In the nonpolar solvent, the planar cation is strongly attracted to the negative chloride ion to form the ion pair, **4-20**, because the carbocation and the chloride ion are not as strongly stabilized by solvation as they are in more polar solvents. Thus, the chloride ion remains in the position at which it was originally formed. Whereas a bridged chloronium ion would react most easily by backside reaction with the planar benzylic cation, the chloride ion can recombine without having to move to the other face of the cation. This step gives *syn* addition.

Nonstereospecific addition

Electrophilic addition is not always stereospecific. Some substrates and reaction conditions lead to products from both *syn* and *anti* additions.

Example 4.15. *The nonstereospecific bromination of* **cis-*stilbene in acetic acid.***

Buckles, R. E.; Bader, J. M.; Theimaier, R. J. *J. Org. Chem.* **1962**, *27*, 4523.

In the highly polar solvent, acetic acid, the reaction is completely nonstereospecific. The product distribution is consistent with reaction of *cis*-stilbene with bromine to form an intermediate carbocation, followed by reaction of bromide on either face of the planar intermediate to give **4-21** and **4-22**. In the nonpolar solvent, carbon tetrachloride, the product is exclusively the *dl* product **4-21** that would result from *anti* addition of bromide to a bridged bromonium ion. In the polar solvent, the localized charge on an intermediate planar carbocation would be more stabilized than the bromonium ion by solvent interactions, because the charge on the bridged bromonium ion is more dispersed.

The side product of the reaction is most likely a mixture of bromoacetoxy compounds (unspecified stereochemistry is indicated by the wavy bond lines). Electrophilic additions in nucleophilic solvents often give a mixture of products because the nucleophile derived from the electrophilic reagent (e.g., Br$^-$) and the solvent compete for the intermediate carbocation.

PROBLEM 4.5   **Write step-by-step mechanisms for the following transformations.**

At −50°C, **4-23** is the only product; at 0°C, **4-23** is still the major product, but **4-24** and **4-25** are also produced. Note that, as the wavy bond lines indicate, both the *exo* and *endo* isomers of the 2-bromo compound **4-24** are produced.

Harmandar, M.; Balci, M. *Tetrahedron Lett.* **1985**, *26*, 5465–5468.

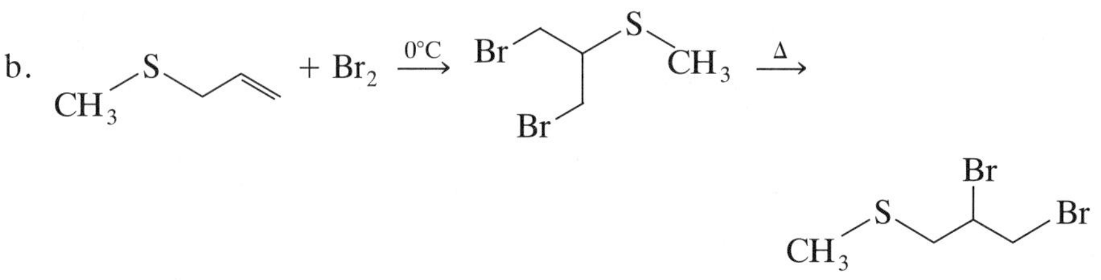

b.

$$CH_3\text{---}S\text{---}CH_2\text{---}CH\text{=}CH_2 + Br_2 \xrightarrow{0°C} $$

(structures shown)

Bland, J. M.; Stammer, C. H. *J. Org. Chem.* **1983**, *48*, 4393–4394.

6. ACID-CATALYZED REACTIONS OF CARBONYL COMPOUNDS

Several examples of the importance of acid catalysis have already been given: Example 2.5 gives one of the steps in the acid-catalyzed formation of an ester, and Example 3.16 shows the acid-catalyzed mechanism for the formation of a hydrazone.

A. Hydrolysis of Carboxylic Acid Derivatives

Acidic hydrolysis of all derivatives of carboxylic acids (e.g., esters, amides, acid anhydrides, acid chlorides) gives the corresponding carboxylic acid as the product. These hydrolyses can be broken down into the following steps.

(1) Protonation of the oxygen of the carbonyl group. This enhances the electrophilicity of the carbonyl carbon, increasing its reactivity with nucleophiles.

(2) The oxygen of water acts as a nucleophile and adds to the carbonyl carbon.

(3) The oxygen of the water, which has added and which is positively charged, loses a proton.

(4) A leaving group leaves. In the case of acid halides, the leaving group leaves directly; in the case of esters or amides, the leaving group leaves after prior protonation.

(5) A proton is lost from the protonated carboxylic acid.

Example 4.16.   *Hydrolysis of an amide.*

The overall reaction is as follows:

$$CH_3CONH_2 + H_3O^+ \longrightarrow CH_3CO_2H + {}^+NH_4$$

The mechanism of this reaction is as follows:

(1) The initial protonation of the carbonyl oxygen gives a cation that is a

resonance hybrid with positive character on carbon and nitrogen, as well as an oxygen.

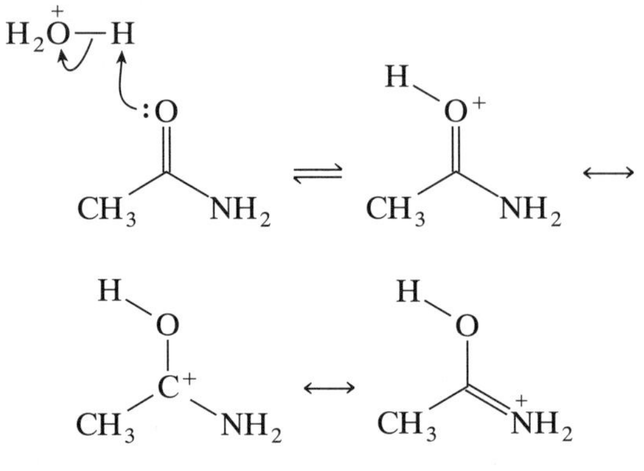

(2) The electrophilic cation reacts with the nucleophilic oxygen of water.

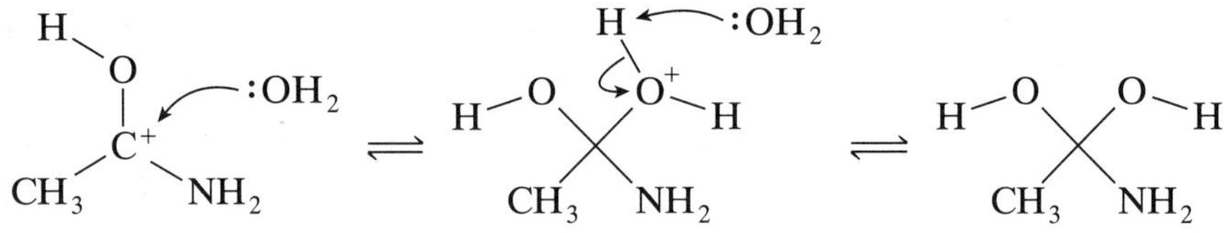

4-26

(3) Loss of a proton gives **4-26**:

(4) The neutral intermediate, **4-26**, can be protonated on either oxygen or nitrogen, but only protonation on nitrogen leads to product formation. Notice that the NH$_2$ group is now an amine and is much more basic than the NH$_2$ group of the starting amide.

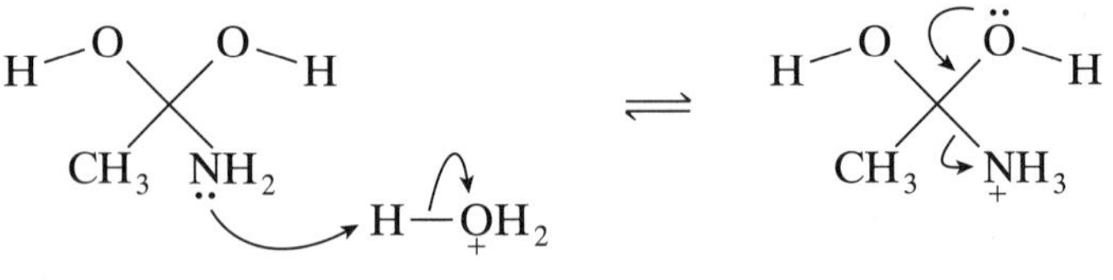

4-26

Note that the leaving group is ammonia rather than $^-NH_2$, which would be a very poor leaving group. Whereas the loss of ammonia is a potentially reversible process, the protonation of ammonia to give the ammonium ion occurs much more rapidly in the acidic medium. Thus, the loss of ammonia is irreversible, not because the addition of ammonia in the reverse process is energetically unfavorable, but because there is no ammonia present.

(5) Finally, a proton is removed from the protonated carboxylic acid to give the carboxylic acid product.

Example 4.17. *Hydrolysis of an ester.*

For esters, all of the steps of the hydrolysis reaction are reversible, and the mechanism of ester formation is the reverse of ester hydrolysis. The course of the reaction is controlled by adjusting the reaction conditions, chiefly the choice of solvent and the concentration of water, to drive the equilibrium in the desired direction. For hydrolysis, the reaction is carried out in an excess of water; for ester formation, the reaction is carried out with an excess of the alcohol component under anhydrous conditions. Frequently, an experimental set-up is designed to remove water as it is formed in order to favor ester formation.

Write step-by-step mechanisms for the following transformations.   **PROBLEM 4.6**

a.

Hauser, F. M.; Hewawasam, P.; Baghdanov, V. M. *J. Org. Chem.* **1988**, *53*, 223–224.

b.

78%

Serafin, B.; Konopski, L. *Pol. J. Chem.* **1978**, *52*, 51–62.

B. Hydrolysis and Formation of Acetals and Orthoesters

Acetals, ketals, and orthoesters are polyethers, represented by the following structural formulas:

ketal   orthoester
If R′ = H, the compound
is an acetal.

The formation and hydrolysis of these groups are acid-catalyzed processes. Other derivatives of carbonyl groups, e.g., enamines, are also formed and hydrolyzed under acidic conditions by very similar mechanisms.

Example 4.18.   *Hydrolysis of ethyl orthoformate to ethyl formate.*

As in the case of other acid-catalyzed hydrolyses, the first step involves protonation of the most basic atom in the molecule. (In the case of ethyl orthoformate, all three oxygen atoms are equally basic.)

Protonation creates a better leaving group, that is, ethanol is a better leaving group than ethoxide ion.

4-27

The electrophilic carbocation, **4-27**, reacts with nucleophilic water. Because water is present in large excess over ethanol, this reaction occurs preferentially and shifts the equilibrium toward the hydrolysis product. The protonated intermediate loses a proton to give **4-28**.

4-27   **4-28**

The neutral intermediate, **4-28**, can be protonated on either a hydroxyl or ethoxy oxygen. Protonation on the hydroxyl oxygen is simply the reverse of the deprotonation step. Although this is a reasonable step, it leads to starting material. However, protonation on the oxygen of the ethoxy group leads to product.

Loss of ethanol, followed by removal of a proton by water, gives the product ester.

$$
\text{EtOH} + \quad
\underset{\text{EtO}}{\overset{\text{H}}{\underset{\;}{\overset{\text{O}}{\underset{\;}{\text{C}^{+}}}}}} \text{H} \;\;\; \cdot\cdot\text{OH}_2
\qquad \rightleftharpoons \qquad
\underset{\text{EtO}}{\overset{\text{O}}{\text{C}}}\text{H}
$$

$$
\underset{\text{EtO}}{\overset{+\text{O}}{\text{C}}}\text{H} \quad \overset{\text{H}}{\underset{\ddot{\text{O}}\text{H}_2}{}}
$$

All of the steps in this reaction are reversible. Why, then, do the hydrolysis conditions yield the formate ester and not the starting material? The key, as we saw in the preceding section with ester formation and hydrolysis, lies in the overall reaction. Water is present on the left-hand side of the equation and ethanol on the right-hand side. Thus, an excess of water would shift the equilibrium to the right, and an excess of ethanol would shift the equilibrium to the left. In fact, in order to get the reaction to go to the left, water must be removed as it is produced. Depending on the reaction conditions, the hydrolysis may proceed further to the corresponding carboxylic acid.

In Problem 3.14.a, we saw that the reaction of an amine with a carbonyl compound in the presence of an acid catalyst can be driven toward the enamine product by removing water from the reaction mixture as it is formed. The reverse of this reaction is an example of the acid hydrolysis of an enamine, a mechanism that is very similar to that of the orthoacetate hydrolysis shown in Example 4.18.

C. 1,4-Addition

Electrophilic addition to α,β-unsaturated carbonyl compounds is analogous to electrophilic addition to isolated double bonds, except that the electrophile adds to the carbonyl oxygen, the most basic atom in the molecule. After that, the nucleophile adds to the β carbon, and the resulting intermediate enol tautomerizes to the more stable carbonyl compound. These reactions may also be considered as the electrophilic counterparts of the nucleophilic Michael and 1,4-addition reactions discussed in Chapter 3, Section 3.C.

Example 4.19. *Electrophilic addition of HCl to acrolein.*

The overall reaction is as follows:

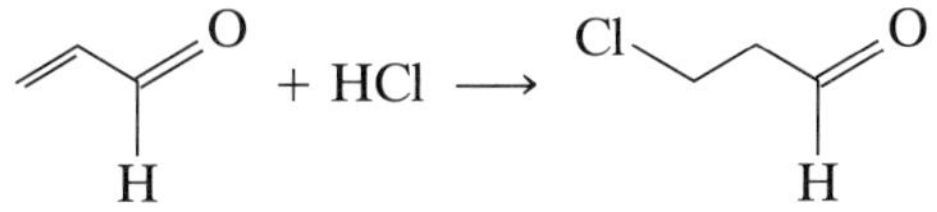

The first step in a mechanism is reaction of the nucleophilic oxygen of the carbonyl group with the positive end of the HCl molecule.

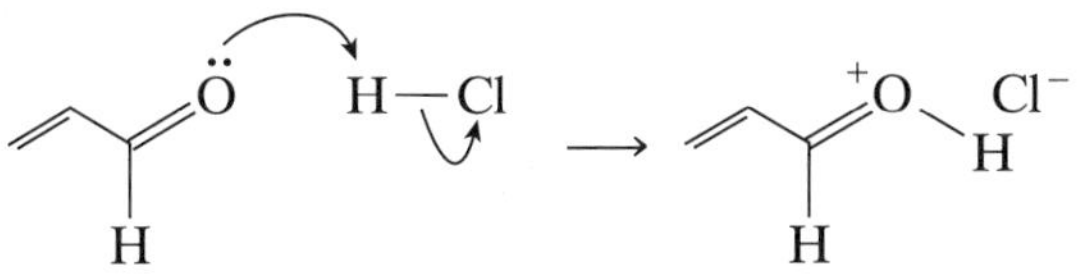

The resulting cation is a resonance hybrid with a partial positive charge on carbon as well as on oxygen:

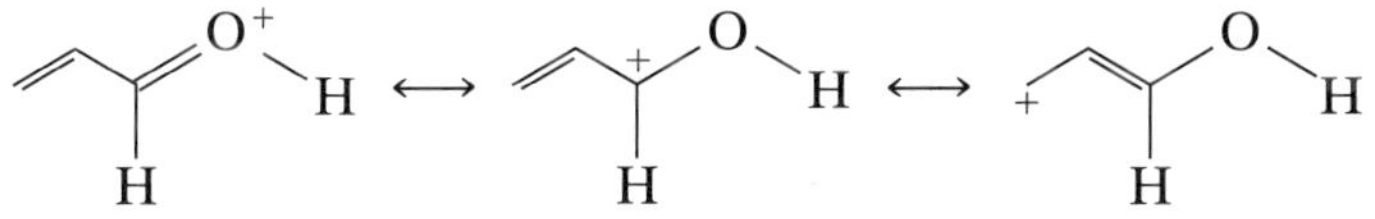

The electrophilic β position now reacts with the nucleophilic chloride ion to give an enol, which then tautomerizes to the keto form.

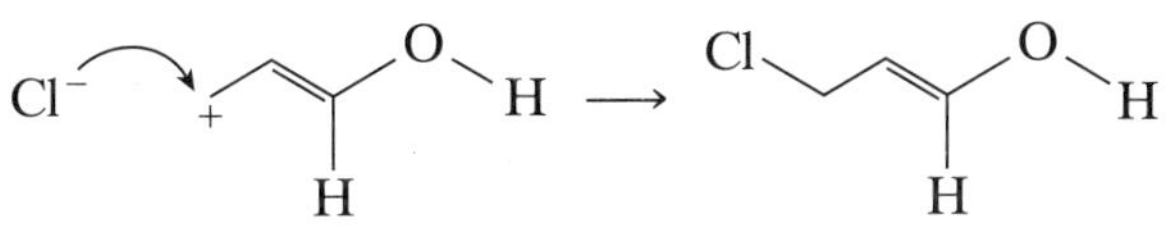

This reaction is the acid-catalyzed counterpart of a 1,4-addition reaction to an α,β-unsaturated carbonyl compound. Chloride ion, without an acid present, will not add to acrolein. That is, chloride ion is not a strong enough nucleophile to drive the reaction to the right. However, if the carbonyl is protonated, the intermediate cation is a stronger electrophile and will react with chloride ion.

Rationalize the regiochemistry of the protonation shown in Example 4.19 by comparing it to protonation at other sites in the molecule.   **PROBLEM 4.7**

PROBLEM 4.8   **Write a mechanism for the following tautomerization in the presence of anhydrous HCl.**

PROBLEM 4.9   **Write a step-by-step mechanism for the following transformation.**

7. ELECTROPHILIC AROMATIC SUBSTITUTION

The interaction of certain electrophiles with an aromatic ring leads to substitution. These electrophilic reactions involve a carbocation intermediate that gives up a stable, positively charged species (usually a proton) to a base to regenerate the aromatic ring. Typical electrophiles include chlorine and bromine (activated by interaction with a Lewis acid for all but highly reactive aromatic compounds), nitronium ion, SO_3, the complexes of acid halides and anhydrides with Lewis acids (see Example 4.5) or the cations formed when such complexes decompose ($R\overset{+}{-}C\overset{}{=}O$ or $Ar\overset{+}{C}\overset{}{=}O$), and carbocations.

Example 4.20. *Electrophilic substitution of toluene by sulfur trioxide.*

In this reaction, the aromatic ring is a nucleophile and the sulfur of sulfur trioxide is an electrophile.

The positive charge on the ring is stabilized by resonance.

4-29-1   **4-29-2**   **4-29-3**

A sulfonate anion, acting as a base, can remove a proton from the intermediate to give the product:

Because the aromatic ring acts as a nucleophile, the reaction rate will be enhanced by electron-donating substituents and slowed by electron-withdrawing substituents. Furthermore, the intermediate cation is especially stabilized by an adjacent electron-donating group, as in resonance structure **4-29-2**. Because of this, electrophiles react at positions *ortho* or *para* to electron-donating groups. Such groups are said to be *ortho-* and *para*-directing substituents. Conversely, because electron-withdrawing groups destabilize a directly adjacent positive charge, the electrophile will react at the *meta* position in order to avoid this stabilization. A review of directing and activating–deactivating effects of various substituents is given in Table 4.1.

The effect of fluorine, chlorine, or bromine as a substituent is unique in that the ring is deactivated, but the entering electrophile is directed to the *ortho* and *para* positions. This can be explained by an unusual competition between resonance and inductive effects. In the starting material, halogen-substituted benzenes are deactivated more strongly by the inductive effect than they are activated by the resonance effect. However, in the intermediate carbocation, halogens stabilize the positive charge by resonance more than they destabilize it by the inductive effect.

TABLE 4.1 Influence of Substituents in Electrophilic Aromatic Substitution

Strongly activating and *ortho-* or *para*-directing
 $-NR_2$, $-NRH$, $-NH_2$, $-O^-$, $-OH$
Moderately activating and *ortho* or *para*-directing
 $-OR$, $-NHCOR$
Weakly activating and *ortho-* or *para*-directing
 $-R$, $-Ph$
Weakly deactivating and *ortho-* or *para*-directing
 $-F$, $-Cl$, $-Br$
Strongly deactivating and *meta*-directing
 $-\overset{+}{S}R_2$, $-\overset{+}{N}R_3$, $-NO_2$, $-SO_3H$, $-CO_2H$, $-CO_2R$,
 $-CHO$, $-COR$, $-CONH_2$, $-CONHR$, $-CONR_2$, $-CN$

PROBLEM 4.10 **For the following reactions, explain the orientations in the product by drawing resonance forms for possible intermediate carbocations and rationalize their relative stabilities.**

a. The reaction in Example 4.20.

b.

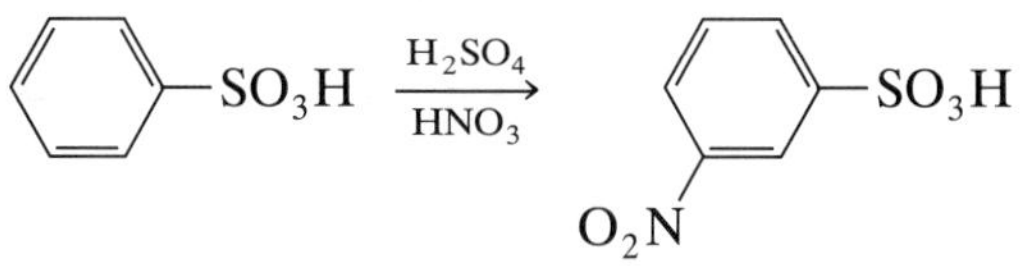

c.

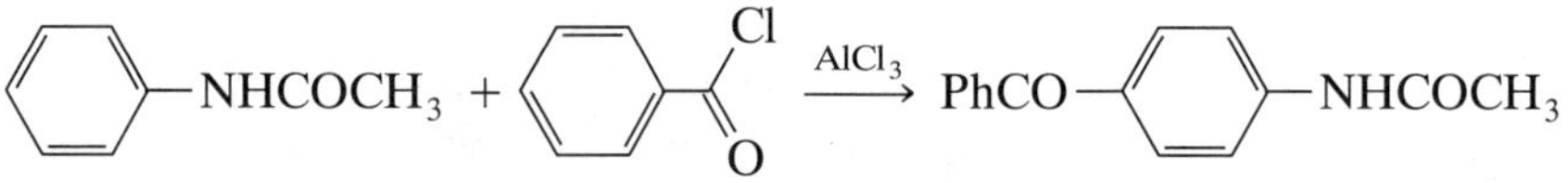

Example 4.21. *A metal-catalyzed, intramolecular, electrophilic aromatic substitution.*

Write a mechanism for the following transformation.

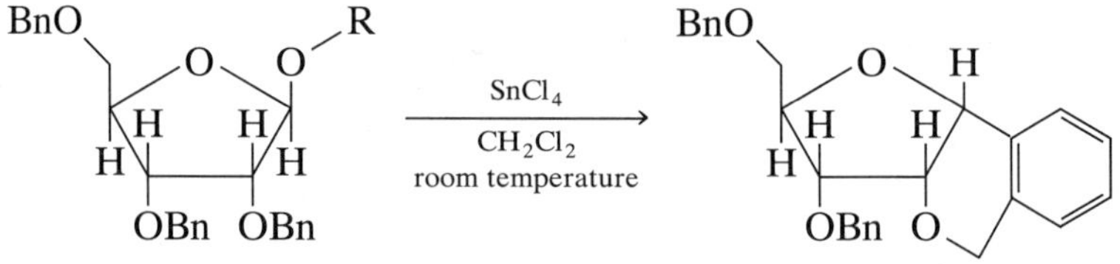

Bn = benzyl ($PhCH_2-$)

Tin(IV) chloride can undergo nucleophilic substitution,

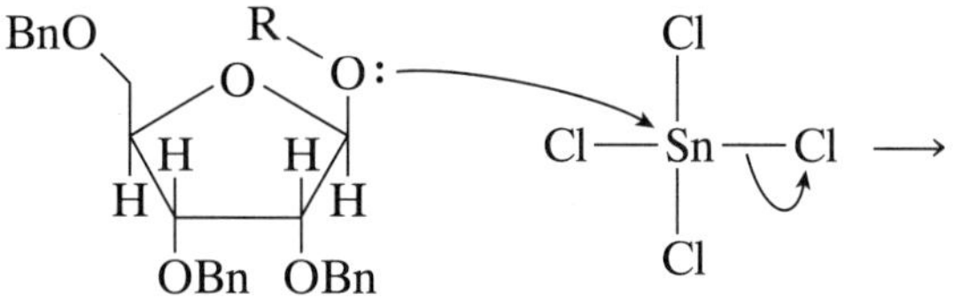

which converts the acetal OR group into a better leaving group. (The positively charged oxygen of the OR group of an acetal is a much better leaving group than the oxygen of an ether because the resulting cation is stabilized by resonance interaction with the remaining oxygen of the original acetal.) After the leaving group leaves, the aromatic ring of a benzyl group at position 2 in **4-30** (suitably situated geometrically for this interaction) acts as a nucleophile toward the positive center. The resulting carbocation then loses a proton to give the product:

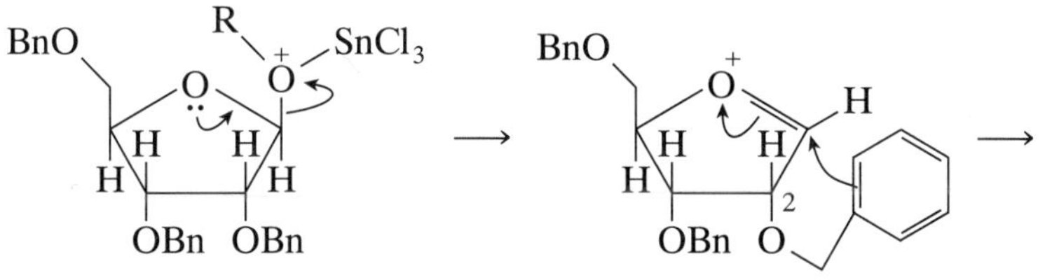

4-30

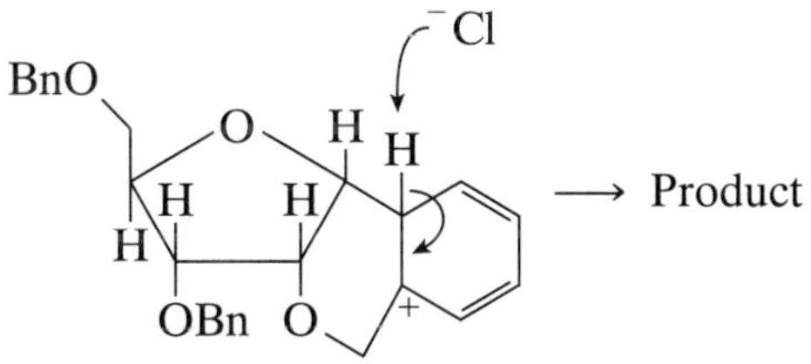

Similar intramolecular electrophilic aromatic substitution reactions are common, especially when five- or six-membered rings are formed.

Martin, O. R. *Tetrahedron Lett.* **1985**, *26*, 2055–2058.

PROBLEM 4.11   **Write step-by-step mechanisms for the following transformations.**

a.

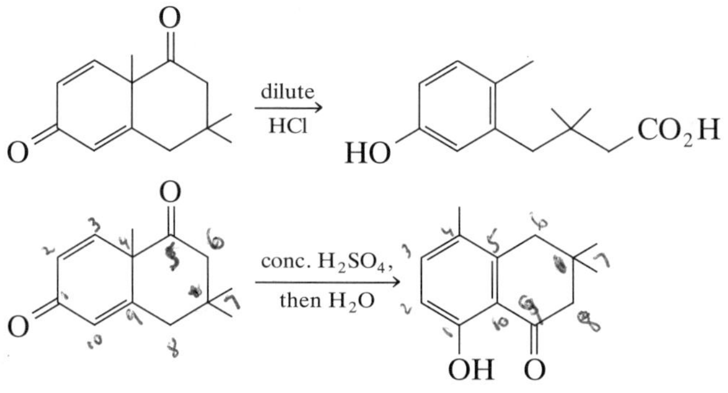

b.

(conc. = concentrated)

The reaction in dilute HCl was discussed in another context in Example 2.11. Modify the mechanism given there to account for the result shown here for concentrated sulfuric acid.

Waring, A. J.; Zaidi, J. H. *J. Chem. Soc., Perkin Trans. 1* **1985**, 631–639.

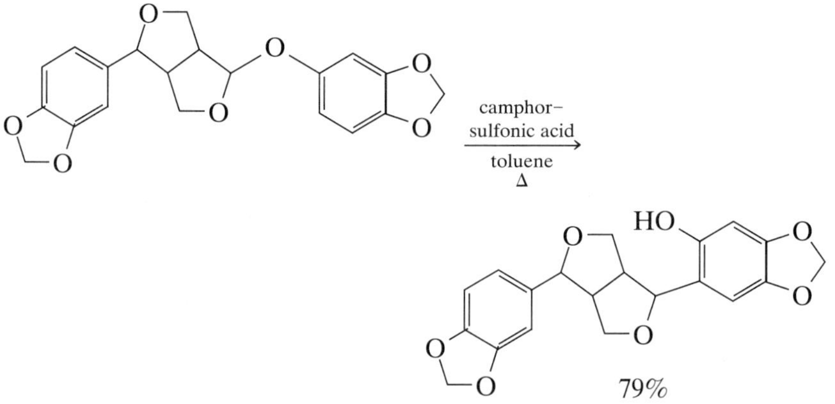

Fukuda, Y.; Isobe, M.; Nagata, M.; Osawa, T.; Namiki, M. *Heterocycles* **1986**, *24*, 923–926.

8. CARBENES

Carbenes are very reactive intermediates. Some have been isolated in matrices at low temperatures, but with few exceptions (see *Chem. Eng. News* **1991**, *28*, 19–20) they are very short-lived at ambient temperatures. Although carbenes are often generated in basic media, they usually act as electrophiles.

A. Singlet and Triplet Carbenes

A carbene is a neutral divalent carbon species containing two electrons that are not shared with other atoms. When these two electrons have *opposite* spins, the carbene is designated a *singlet* carbene; when they have *parallel* spins, the carbene is a *triplet*. In the ground state, a singlet carbene has a pair of electrons in a single orbital, whereas the triplet has two unpaired electrons, each occupying a separate orbital. The designations singlet and triplet originate in spectroscopy.

Singlet Carbene   Triplet Carbene   Carbocation

Radical   Carbanion   Carbanion

Note that the formal charge on the carbon atom in either a singlet or a triplet carbene is zero; the singlet carbene is *not* a carbanion.

The term *carbenoid* is used to refer to a carbene when the exact nature of the carbene species is uncertain and especially when referring to neutral electron-deficient species that are coordinated with a metal.

B. Formation of Carbenes

Example 4.22. *Generation of carbenes from alkyl halides and base.*

(1) Dichlorocarbene can be formed from aqueous KOH and chloroform:

$$HO^- \quad H{-}CCl_3 \longrightarrow \quad \longrightarrow \quad Cl{-}\ddot{C}{-}Cl + Cl^-$$

The base removes the acidic proton from chloroform. The resulting anion then loses chloride ion, giving a divalent carbon with two unshared electrons. In this case, the unshared electrons are paired (occupy the same orbital), i.e., dichlorocarbene is a singlet. The carbon of the carbene has no charge.

(2) A carbenoid is formed when potassium *t*-butoxide reacts with benzal bromide in benzene. The reactivity of the carbenoid, **4-31**, is similar to that of phenylbromocarbene. The exact nature of the interaction between the carbenoid carbon and the metal halide, in this case potassium bromide, is not known.

$$\text{C}_6\text{H}_5{-}\overset{H}{\underset{Br}{C}}{-}Br + KOC(CH_3)_2 \longrightarrow \text{C}_6\text{H}_5{-}\overset{K}{\underset{Br}{C}}{-}Br + t\text{-BuOH}$$

4-31

(3) Some alkyl halides react with alkyllithium reagents under aprotic conditions to give carbenoids by a process called "halogen–metal exchange."

$$\overset{OH}{\underset{CHBr_2}{C}} + 2BuLi \xrightarrow{\text{THF}} \overset{OLi}{\underset{CHBrLi}{C}}$$

4-32

In some cases, proton removal from the carbon may predominate over exchange with the halogen. For details, see Kobrich, G. *Angew. Chem., Int. Ed. Engl.* **1972**, *11*, 473–485.

Example 4.23. *Generation of ICH$_2$ZnI (the Simmons–Smith reagent) from methylene iodide and zinc–copper couple.*

When diiodomethane is treated with zinc–copper couple, a carbenoid is formed.

$$CH_2I_2 + ZnCu \longrightarrow ICH_2ZnI$$

4-33

The Simmons–Smith reagent (**4-33**), other carbenoids, and carbenes are very useful in the synthesis of cyclopropanes (see Example 4.25).

Example 4.24. *Generation of carbenes from diazo compounds.*

Loss of nitrogen from a diazo compound can be effected by heat, light, or a copper catalyst. This gives either a carbene (heat or light) or a carbenoid (copper).

C. Reactions of Carbenes

Once generated by the methods outlined in the preceding section, the highly reactive carbene intermediates can react in a number of different ways, including addition, substitution, insertion, rearrangement, and hydrogen or halogen abstraction. Two important reactions involving carbenes are addition to carbon–carbon double and triple bonds to generate cyclopropanes and cyclopropenes, respectively, and electrophilic aromatic substitution (the Reimer–Tiemann reaction), in which electrophilic addition of a carbene is the first step.

Addition

Electrophilic addition of carbenes to carbon–carbon double and triple bonds has been extremely useful synthetically. In many cases, the reaction goes with 100% stereospecificity, so that the stereochemistry about a double bond in the starting material is maintained in the product. Cases in which addition is not 100% stereospecific are rationalized on the basis of a triplet or diradical intermediate. If the triplet carbene is relatively unreactive, the formation of the two new carbon–carbon bonds may be a stepwise process that allows for rotation and, therefore, loss of stereochemistry in the intermediate.

Example 4.25. *Addition of singlet dichlorocarbene to cis-2-butene and trans-2-butene.*

The addition of dichlorocarbene to the double bond of *cis*-2-butene goes with 100% stereospecificity, that is, the only product is *cis*-1,2-dimethyl-3,3-dichlorocyclopropane. The addition of dichlorocarbene to *trans*-2-butene gives only the corresponding *trans* isomer. The stereospecificity of the reaction has been interpreted to mean that dichlorocarbene is a singlet and that both ends of the double bond react simultaneously, or nearly so, with the carbene.

Example 4.26. *A nonstereospecific addition of a carbene.*

Because the reaction is nonstereospecific, the mechanism would be written as a stepwise addition to the double bond by a diradical carbene.

Note that the arrows used to show the flow of unpaired electrons (radicals) have only a half head. Also, the intermediate in the reaction is a diradical. (Radicals are discussed in more detail in Chapter 5.) Rotation about the highlighted single bond takes place fast enough that the stereochemistry of the starting olefin is lost.

Substitution

Carbenes also add to other nucleophiles, such as hydroxide, thiolate, and phenoxide. The reaction with phenoxide is the classic Reimer–Tiemann reaction.

Example 4.27. *The Reimer–Tiemann reaction.*

Phenols react with chloroform in the presence of hydroxide ion in water to give *o*- and *p*-hydroxybenzaldehydes. The steps of the reaction are (1) the formation of dichlorocarbene, as shown in Example 4.22; (2) nucleophilic reaction of the phenoxide with the electrophilic carbene; and (3) hydrolysis.

Complete the mechanism for Example 4.27 showing the steps leading from the intermediate carbanion to the product *o*-salicylaldehyde. **PROBLEM 4.12**

PROBLEM 4.13   **Write a step-by-step mechanism for the following transformation.**

Wenkert, E.; Arrhenius, T. S.; Bookser, B.; Guo, M.; Mancini, P. *J. Org. Chem.* **1990**, *55*, 1185–1193.

Insertion

In insertion, as the name suggests, the carbene inserts itself between two atoms. Insertions have been observed into C—H, C—C, C—X, N—H, O—H, S—S, S—H, and M—C bonds, among others. The mechanism of the process is often concerted. A three-centered transition state is usually written for the concerted mechanism:

Example 4.28.   *A synthetically useful insertion reaction.*

Carbenoid **4-32** can be generated as shown in Example 4.22. The insertion of **4-32** into the C1—C2 bond leads to the formation of cycloheptanone in 70% yield. Other homologues give even higher yields of ring-expanded products.

The intermediate enolate, **4-34**, forms the corresponding ketone in acid. From Taguchi, H.; Yamamoto, H.; Nozaki, H. *J. Am. Chem. Soc.* **1974**, *96*, 6510–6511.

Rearrangement

Because of their reactivity, it is frequently difficult to decide whether a free carbene has been generated or whether an electrophilic carbon undergoes a synchronous reaction. It is generally accepted that a free carbene is an intermediate in the Wolff rearrangement, in which a diazoketone rearranges to a highly reactive ketene. The versatile ketene intermediate then reacts to give the carboxylic acid, ester, or amide by reaction with water, alcohol, or amine, respectively.

Example 4.29. *The Wolff rearrangement of diazoacetophenone to methyl phenylacetate.*

$$\underset{\text{80\%}}{PhCCH=\overset{+}{N}=\overset{-}{N}:} \xrightarrow[\text{CH}_3\text{OH--CH}_3\text{CN}]{\text{CuI}} PhCH_2COCH_3 + N_2$$

The mechanism involves decomposition of the diazoketone **4-35** to a carbene **4-36**. The decomposition can be brought about by activation with metal salts, particularly copper and silver, as well as by heat or light.

$$Ph-\overset{O}{\overset{\|}{C}}CH=\overset{+}{N}=\overset{-}{N}: \longleftrightarrow Ph-\overset{O}{\overset{\|}{C}}CH-\overset{+}{N}\equiv N: \longrightarrow Ph-\overset{O}{\overset{\|}{C}}CH + N_2$$

$$\textbf{4-35} \qquad\qquad\qquad\qquad \textbf{4-36}$$

$$Ph-\overset{O}{\overset{\|}{C}}\overset{..}{C}H \longrightarrow Ph-CH=C=O$$

$$\textbf{4-37}$$

The carbene then rearranges to the ketene **4-37**. An important feature of the rearrangement is that the migrating group moves with *retention* of configuration. In the example shown, the ester is formed by reaction of the ketene with methanol in the solvent.

9. ELECTROPHILIC HETEROATOMS

A number of reactions involve electrophilic heteroatom species that are analogous to their carbon counterparts. In this section, we will consider electrophilic nitrogen and oxygen species.

A. Electron-Deficient Nitrogen

Nitrenes

A nitrene is the nitrogen analogue of a carbene. In other words, it is a neutral, univalent nitrogen that contains two lone pairs of electrons not shared with other atoms.

phenyl nitrene

Nitrenes are generated and react very similarly to carbenes, but because of their greater reactivity, the presence of free nitrenes is difficult to demonstrate experimentally.

Example 4.30. *Generation of nitrenes.*

Common methods for generating nitrene intermediates are the photolysis and thermolysis of azides. Generation of nitrenes from acyl azides can only be effected photochemically; thermolysis of an acyl azide gives the corresponding isocyanate.

The reactivity of nitrenes is similar to that of carbenes. They readily add to double bonds to give the corresponding three-membered heterocycles, aziridines. Insertion reactions are also common.

Nitrenium Ions

A nitrenium ion is a positively charged divalent nitrogen with one lone pair of electrons. Thus, it is a cation that is isoelectronic with a carbene, a neutral species.

Example 4.31. *Generation and reaction of a nitrenium ion.*

Silver-ion-assisted loss of chloride ion from the starting material gives a nitrenium ion, which acts as an electrophile toward the aromatic ring to give **4-38**.

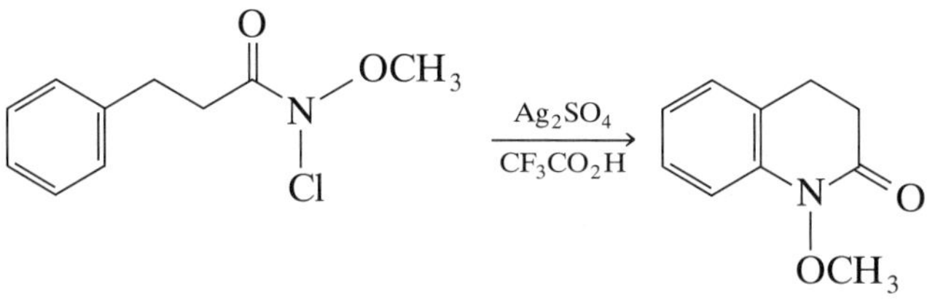

4-38

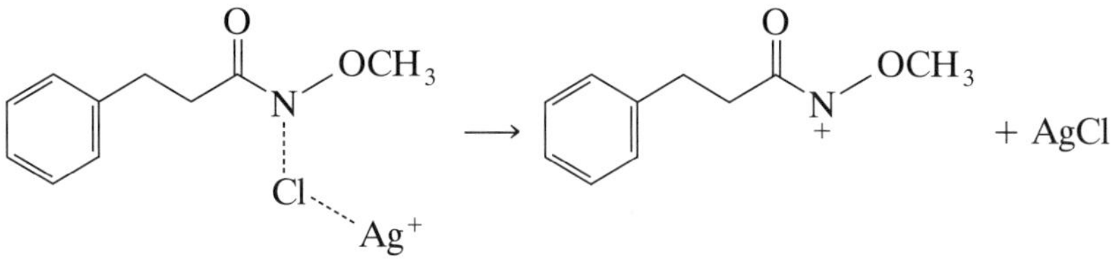

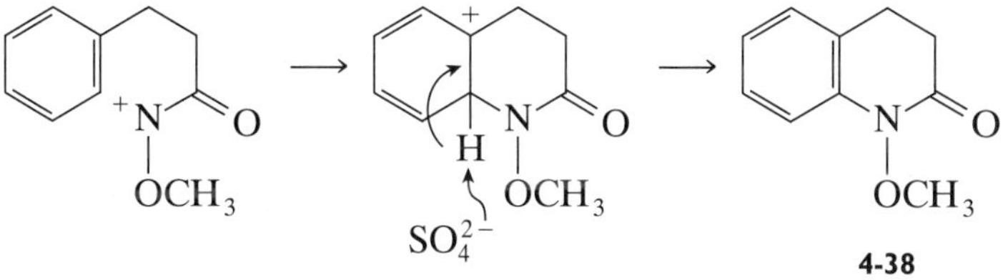

+ AgCl

4-38

B. Rearrangements Involving Electrophilic Nitrogen

The Beckmann Rearrangement

In strong acids, or when treated with reagents such as thionyl chloride or phosphorus pentachloride, an oxime will react to give a rearranged amide. This is known as the Beckmann rearrangement. When the reaction gives products other than amides, these products are referred to as abnormal products. One such abnormal pathway is illustrated in Problem 4.14.

Example 4.32. *Beckmann rearrangement of benzophenone oxime.*

The overall reaction is as follows:

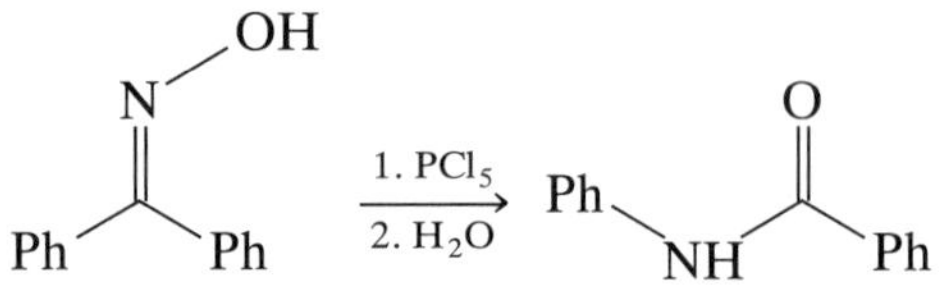

The first two steps of the reaction mechanism convert the original oxime to a derivative, **4-39**. This process converts the hydroxyl group of the oxime into a better leaving group.

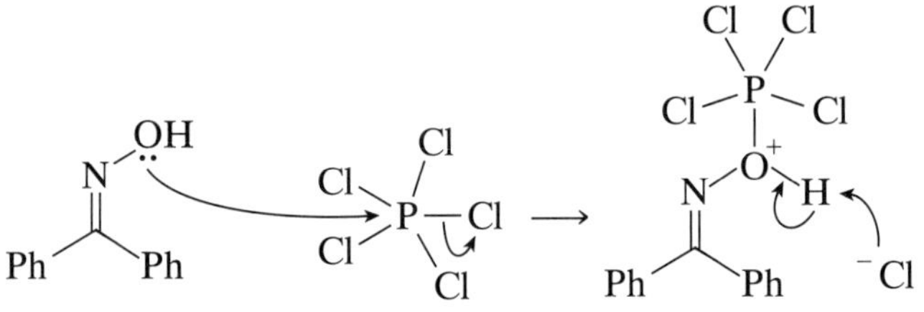

The strongly electron-withdrawing leaving group creates a substantial partial positive charge on nitrogen. The phenyl group, on the side opposite the leaving group, moves with its pair of electrons to the electron-deficient nitrogen as the leaving group leaves. The resulting ions then collapse to form **4-40**. During workup, **4-40** will undergo hydrolysis and tautomerization to the final product, the amide.

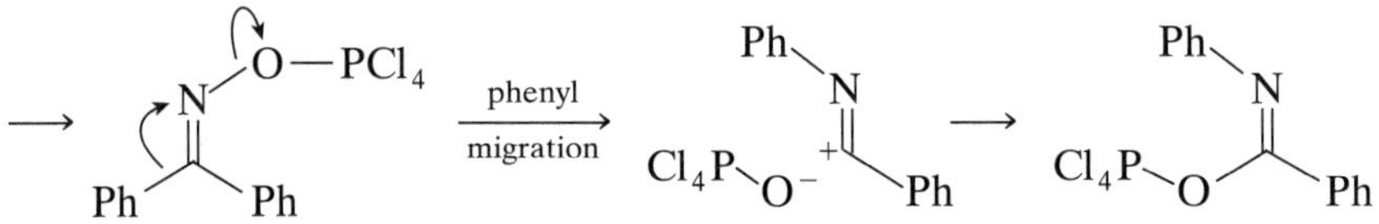

Write a step-by-step mechanism for the following reaction.　　　　**PROBLEM 4.14**

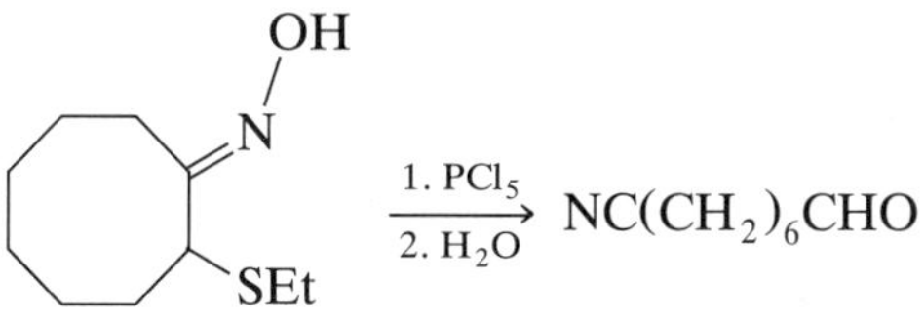

$$\xrightarrow[\text{2. H}_2\text{O}]{\text{1. PCl}_5} \text{NC(CH}_2)_6\text{CHO}$$

Ohno, M.; Naruse, N.; Torimitsu, S.; Teresawa, I. *J. Am. Chem. Soc.* **1966**, *88*, 3168–3169.

Nitrogen Analogues of the Wolff Rearrangement

Important rearrangements involving electrophilic nitrogen are the Hofmann, Curtius, and Schmidt rearrangements, which are nitrogen analogues of the Wolff rearrangement discussed in Section 8. In all of these rearrangements, it is possible to write a discrete nitrene intermediate; however, it is generally considered more likely that the reactions proceed through a less reactive intermediate in which rearrangement accompanies loss of the leaving group. The electron deficiency of the nitrogen is due to withdrawal of electron density by a good leaving group. As with the Wolff rearrangement, all of these reactions proceed with *retention* of configuration by the group migrating to nitrogen. Table 4.2 compares rearrangements involving various electron-deficient heteroatom species.

Example 4.33. *The Hofmann rearrangement of hexanamide to pentylamine.*

In the Hofmann rearrangement, a halide leaving group confers electrophilic character on the nitrogen atom.

$$\text{CH}_3(\text{CH}_2)_4\overset{\overset{\text{O}}{\|}}{\text{C}}\text{NH}_2 \xrightarrow[\text{2. HCl}]{\text{1. NaOCl, H}_2\text{O}} \text{CH}_3(\text{CH}_2)_4\text{NH}_2$$

95%

Magnieri, E. *J. Org. Chem.* **1958**, *23*, 2029.

TABLE 4.2 Parallels in Electrophilic Rearrangements

Name reaction	Starting material	Reaction conditions	Reactive intermediate	Intermediate product	Final product
Wolff	$R-\overset{O^-}{\underset{}{C}}-\overset{+}{C}H-\overset{-}{N}\equiv N:$	$Ag^+, Cu^+, \Delta, h\nu$		$N_2 + O=C=CHR \xrightarrow{H_2O} RCH_2CO_2H$	
Hofmann	$R-\overset{O^-}{\underset{}{C}}-NH-Br$	OH^-, H_2O	$R-\overset{O}{\underset{}{C}}-\ddot{N}-Br$	$Br^- + O=C=N-R \xrightarrow{H_2O} RNH_2 + CO_2$	
Curtius	$R-\overset{O^-}{\underset{}{C}}-\overset{+}{N}\equiv N:$	Δ	$R-\overset{O^-}{\underset{}{C}}=\ddot{N}-\overset{+}{N}\equiv N:$	$N_2 + O=C=N-R \xrightarrow{H_2O} RNH_2 + CO_2$	
Schmidt	$R-CR'$	$H^+ + NaN_3$	$R-C=\ddot{N}-\overset{+}{N}\equiv N:$, R'	$N_2 + R-\overset{+}{C}=N-R' \xrightarrow{H_2} RCNHR'$	
Beckman	$\underset{R}{\overset{R'}{>}}C=\ddot{N}-OH$	H^+	$\underset{R}{\overset{R'}{>}}C=\ddot{N}-\overset{+}{O}H_2$	$H_2O + R-\overset{+}{C}=N-R \xrightarrow{H_2O} R-CNHR$	
Baeyer–Villager	$\underset{R'}{\overset{R}{>}}C=O$	$H^+, R''CO_3H$	$\underset{R}{\overset{R'}{\underset{}{}}}C\overset{OH}{\underset{O-OCR''}{<}}$	$\underset{R-OR'}{\overset{OH}{\underset{+}{C}}} \xrightarrow{H_2O} R-COR'$; $\underset{R'-OR}{\overset{OH}{\underset{+}{C}}} \longrightarrow R'-COR$	

The first step in the mechanism is formation of the *N*-chloroamide.

$$CH_3(CH_2)_4C-N \quad \longrightarrow \quad CH_3(CH_2)_4C-N-H \quad \longrightarrow$$

$$CH_3(CH_2)_4C-N-H + OH^-$$
$$\overset{|}{Cl}$$

$$CH_3(CH_2)_4C-N-H \quad \longrightarrow \quad CH_3(CH_2)_4-C-N: \quad \longrightarrow$$
$$\overset{|}{Cl} \qquad\qquad\qquad\qquad \overset{|}{Cl}$$

$$CH_3(CH_2)_4-N{=}C{=}O + Cl^-$$

4-41

Because of the electron-withdrawing effect of the chloro group, the remaining hydrogen attached to the amide nitrogen is acidic and easily removed by aqueous base. The tendency of the chloro group to leave with the N—Cl bonding electrons confers electrophilic character on the amide nitrogen, so that as the chloro group departs, the alkyl group moves to compensate for the developing electron deficit on nitrogen, yielding the isocynate **4-41**.

$$CH_3(CH_2)_4-N{=}C{=}O \quad \longrightarrow \quad CH_3(CH_2)_4-N{=}C-O^-$$
$$\qquad\qquad\qquad \overset{\cdot\cdot}{:}OH \qquad\qquad\qquad\qquad \overset{|}{OH}$$

4-41

$$CH_3(CH_2)_4NH-C{=}O \quad \longrightarrow \quad CH_3(CH_2)_4\overset{\cdot\cdot}{N}H + CO_2 \longrightarrow CH_3(CH_2)_4NH_2$$
$$\qquad\quad \overset{\cdot\cdot}{:}O{:}^- \qquad\qquad\qquad\qquad\qquad H-OH$$

4-42

Basic hydrolysis of the reactive isocyanate **4-41** leads to an intermediate that tautomerizes under the reaction conditions to give **4-42**, which spontaneously decarboxylates. The irreversible decarboxylation yields the amine, which contains one less carbon (lost as CO_2) than the starting material.

PROBLEM 4.15   Propose a reasonable mechanism for the transformation shown.

$$\xrightarrow[\text{H}_2\text{SO}_4]{\text{NaN}_3}$$

46%

Greco, C. V. *Tetrahedron* **1970**, *26*, 4329.

C. Rearrangement Involving Electron-Deficient Oxygen

The Baeyer–Villager reaction involves rearrangement of an oxygen-deficient species. Because of the high electronegativity of oxygen, we would not expect a positively charged electron-deficient oxygen, and the reaction is considered to proceed by a concerted mechanism. As in the rearrangement involving electron-deficient nitrogen, the configuration at the migrating carbon is maintained.

Example 4.33.   *Baeyer–Villager oxidation of cyclopentanone to δ-valerolactone.*

$$\xrightarrow[\text{10–15°C}]{\text{CF}_3\text{CO}_3\text{H–CF}_3\text{CO}_2\text{H}}$$

81%

Protonation of the ketone carbonyl is the first step. The generally accepted mechanism is as follows:

Deprotonation yields the lactone product.

Baeyer – Villager oxidation of the heavily functionalized cyclopentanone yields a mixture of products. Use a step-by-step mechanism to decide whether this result is unexpected.   **PROBLEM 4.16**

TBDMS = *t*-butyldimethylsilyl
MCPBA = *m*-chloroperbenzoic acid

Clissold, C.; Kelly, C. L.; Lawrie, K. W. M.; Willis, C. L. *Tetrahedron Lett.* **1997**, 8105.

PROBLEM 4.17   **Write step-by-step mechanisms for the following transformations:**

a.

$+\ CF_3SO_3H\ \xrightarrow{\Delta}$

Capozzi, G.; Chimirri, A.; Grasso, S.; Romeo, G. *Heterocycles* **1984**, *22*, 1759–1762.

b.

$\xrightarrow{HCO_2H}$

Ent, H.; de Koning, H.; Speckamp, W. N. *J. Org. Chem.* **1986**, *51*, 1687–1691.

c.

$-\!\!\!-\!\!\!\!\bigcirc\!\!\!-OH\ +\ (CH_3OCONH)_2CHCOCH_3\ \xrightarrow[CH_2Cl_2]{CH_3SO_3H}$

Ben-Ishai, D.; Denenmark, D. *Heterocycles* **1985**, *23*, 1353–1356.

d.

$\xrightarrow[CH_3NO_2]{SnCl_4}\ \xrightarrow{H_2O}$

Hantawong, K.; Murphy, W. S.; Boyd, D. R.; Ferguson, G.; Parvex, M. *J. Chem. Soc., Perkin Trans. 2* **1985**, 1577–1582.

e.

$$\xrightarrow[\text{CF}_3\text{CO}_2\text{H}]{\text{Ag}_2\text{CO}_3}$$

Kikugawa, Y.; Kawase, M. *J. Am. Chem. Soc.* **1984**, *106*, 5728–5729.

f.

$$\xrightarrow[\text{2. H}_2\text{O}]{\text{1. SnCl}_4, \text{CH}_2\text{Cl}_2}$$

White, J. D.; Skeean, R. W.; Trammell, G. L. *J. Org. Chem.* **1985**, *50*, 1939–1948.

ANSWERS TO PROBLEMS

a. This is an ionization, followed by an alkyl shift and nucleophilic reaction of solvent with the electrophilic carbocation intermediate. In other words, it is an example of an S_N1 reaction with rearrangement. As noted in Table 3.1, the tosylate anion is an excellent leaving group.

Problem 4.1

Problem 4.1
continued

b. Because the hydroxyl group is lost and an alkene is formed, the reaction appears to be a dehydration involving a carbocation rearrangement. Numbering of the carbons in starting material and product is helpful. Numbering of the first five carbons in the product is straightforward because of the methyl carbons.

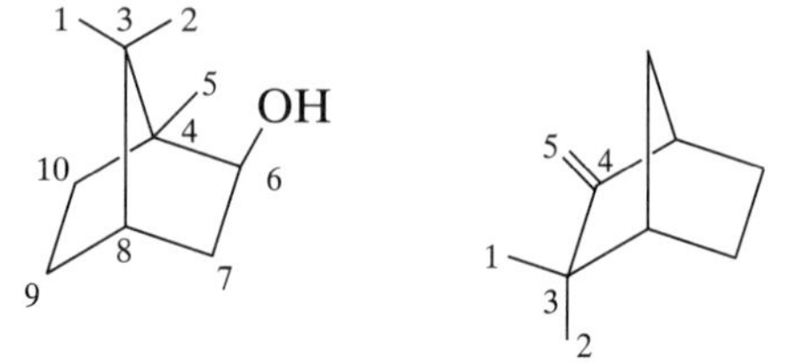

If we minimize bond reorganization there are only two ways to number the remaining carbons of the product:

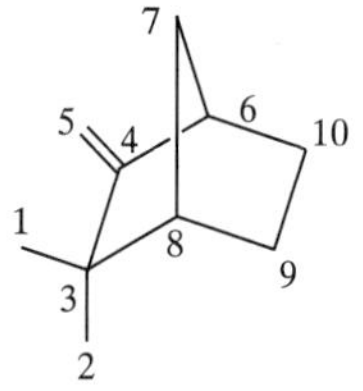

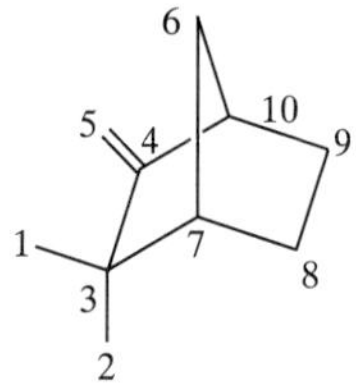

A major difference between **4-43** and the starting material is that, at carbon-10 (C-10), a bond to C-4 has been replaced by a bond to C-6. This is equivalent to an alkyl shift of C-10 from C-4 to C-6. In **4-44** at C-6, the bond to C-4 has been replaced by a bond to C-10. This is equivalent to a alkyl shift of C-6 from C-4 to C-10. However, the carbocation will be formed at C-6, and thus this carbon cannot be the one that shifts. Thus, **4-43** must be the product.

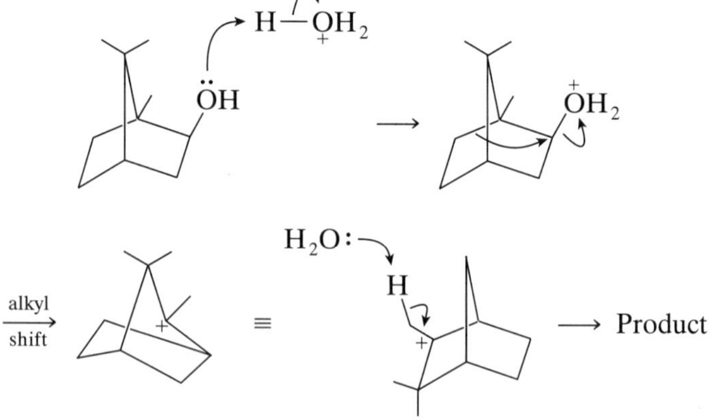

An alternative approach to solving this problem is to start by forming the secondary carbocation at C-6 and then to assess possible 1,2 alkyl shifts that could occur. If we recognize that the relationship of atoms C-1 through C-5 is unchanged in the product, two 1,2 alkyl shifts are possible:

Problem 4.1
continued

The first leads to formation of a primary carbocation. The instability of the primary carbocation is not so great that it is grounds for eliminating this step from consideration; however, the intermediate ion also contains a four-membered ring, a structural feature not found in the product. The second alkyl shift leads to the carbocation encountered in the mechanism shown previously.

For some reactions, it will be necessary to consider both alkyl and hydride shifts in order to account for the products formed.

The mechanism of this reaction involves protonation of oxygen and loss of water to form a carbocation. Because the molecule is symmetrical, the hydroxyls are equivalent; thus, protonation of either oxygen leads to the product.

Problem 4.2

The subsequent alkyl shift gives a carbocation, **4-45**, which is resonance-stabilized by interaction with the adjacent oxygen. Loss of a proton then leads to the product.

Problem 4.2
continued

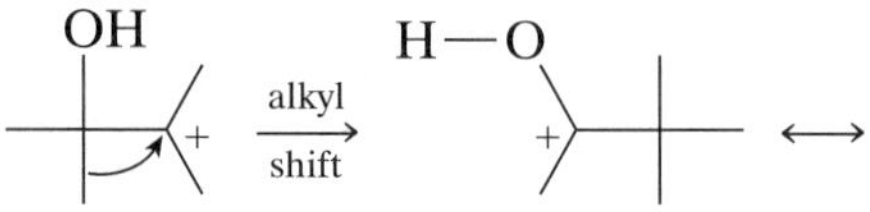

4-45-1

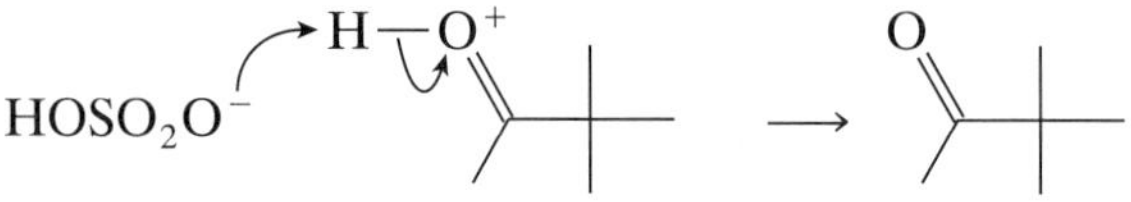

4-45-2

Problem 4.3   The Lewis acid BF_3 can complex with the epoxide oxygen to induce ring opening.

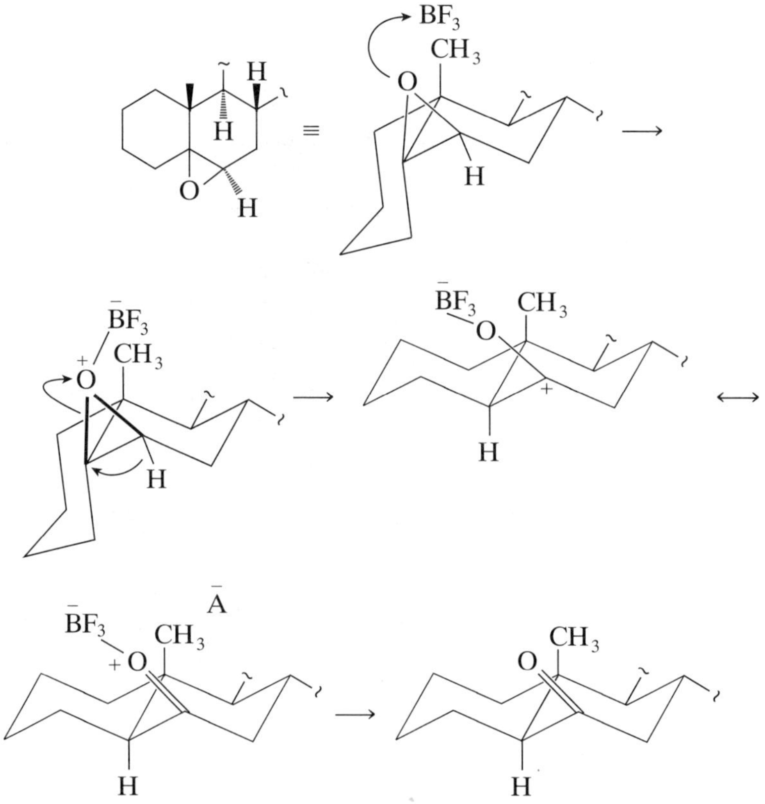

If migration of the hydride and ring opening of the epoxide are concerted, stereospecificity of the reaction is assured. Nucleophilic displacement by ethyl ether (derived from boron trifluoride etherate) removes boron trifluoride to generate the product ketone.

Problem 4.3
continued

There are several possible mechanistic varitions for this reaction. Because the starting material contains several basic oxygens and acid is one of the reagents, protonation is a reasonable first step. Dimethyl ether more basic than acetophenone by only 0.5 pK_a units (see Appendix C). An acetal would be expected to be less basic than an ether because of the electron-withdrawing effect of the second oxygen. Thus, protonation at both the acetal oxygen of the center ring and the carbonyl oxygen will be examined. (Protonation of the other acetal oxygen is unlikely to lead to product, because the ring containing this oxygen is not rearranged in the final product.)

Problem 4.4

Mechanism 1

In this sequence, initial protonation of the acetal oxygen in the center ring is considered.

4-46

Ring opening of the protonated intermediate could produce either cation **4-46** or cation **4-47** depending upon which bond is broken. Carbocation **4-46** is stabilized by resonance with both the aromatic ring and the adjacent ether oxygen. Carbocation **4-47** is a hybrid of only two resonance forms, and **4-47-2** is quite unstable because the positive oxygen does not have an octet of electrons. Thus, the ring-opening reaction to give **4-46** is preferred.

4-47-1   **4-47-2**

Problem 4.4
continued

Reaction of **4-46** continues with elimination of a proton to form a double bond, followed by protonation of the carbonyl group.

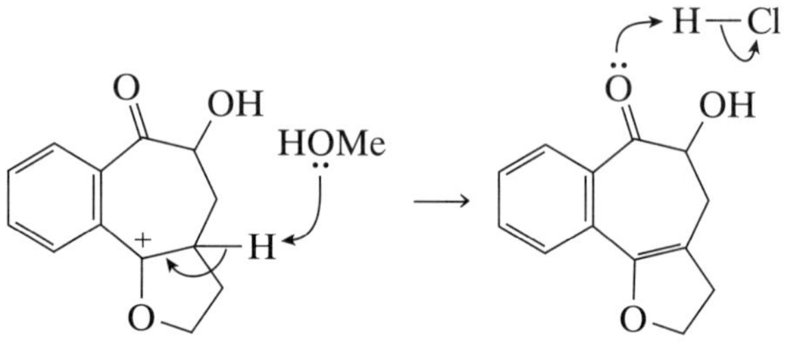

4-46

Then an alkyl shift occurs, followed by a deprotonation.

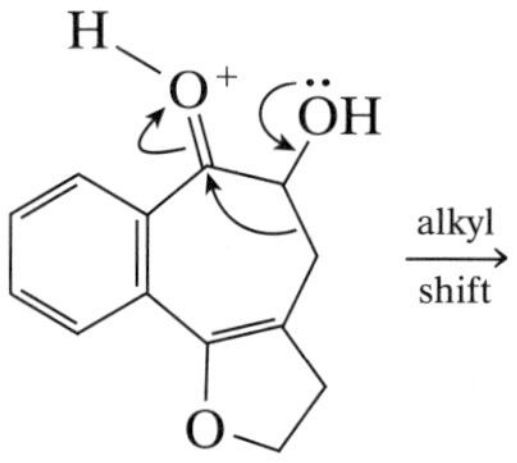

Finally, a dehydration takes place, forming the product.

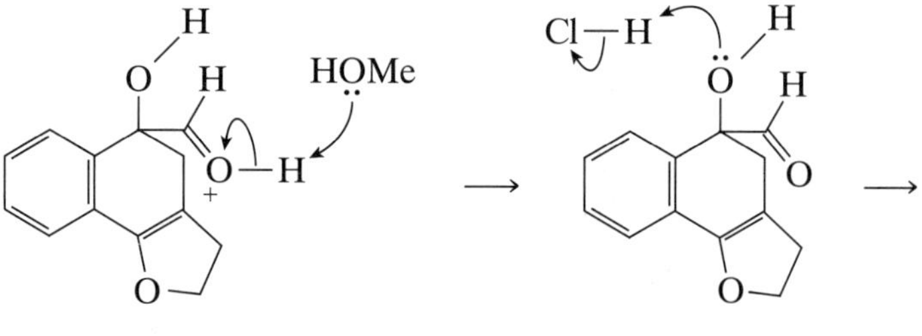

4-48

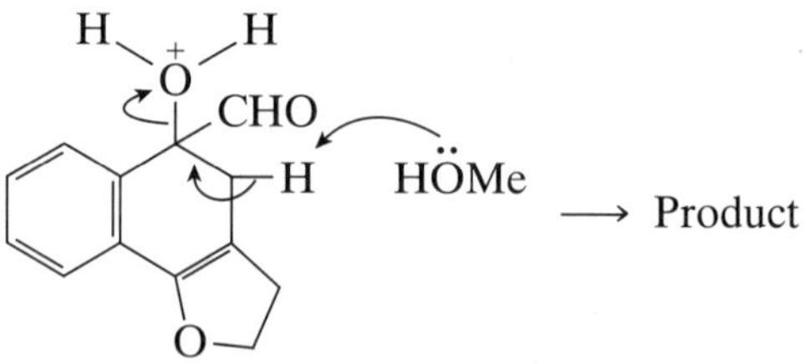

In this sequence (mechanism 1), methanol was consistently used as the base, because data in Appendix C show that it is a stronger base than chloride ion.

Problem 4.4
continued

Mechanism 2

Mechanism 2 starts with protonation of the carbonyl oxygen to produce cation **4-49**.

An alkyl shift gives cation **4-50**, which is resonance-stabilized. This cation

then undergoes bond cleavage to form **4-51**, which is stabilized in the same manner as **4-46** in mechanism 1. Loss of a proton gives **4-48**, which is also an intermediate in mechanism 1.

Problem 4.4
continued

Mechanism 3

Intermediate **4-49** of mechanism 2 undergoes a hydride shift, instead of an alkyl shift, to give the following:

4-49 **4-52**

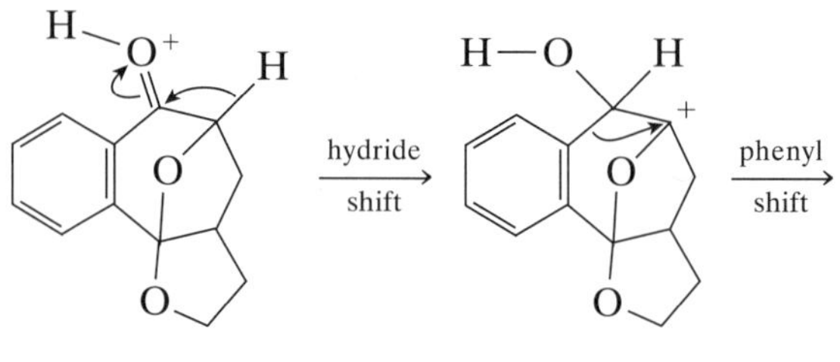

A phenyl shift gives the cation, **4-53**, which can lose a proton to give the aldehyde functional group. The acetal oxygen in the center ring can be protonated and open to give cation, **4-51**, which behaves as shown previously in mechanism 2.

4-53

4-51

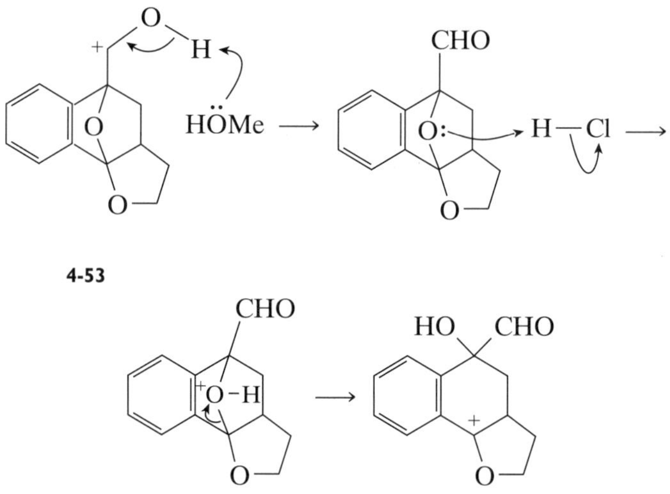

The problem with this mechanism is that the likelihood of forming a carbocation at the position shown in intermediate **4-52** is vanishingly small. Several factors combine to make this cation very unstable. First, although it is tertiary, the carbocation is at a bridgehead position and is constrained from assuming the planar geometry of an sp^2 hybridized carbocation. The bridgehead location also precludes resonance stabilization by the adjacent bridging oxygen, which nevertheless has a destabilizing inductive effect. Finally, there is a hydroxyl group on the carbon adjacent to the cationic center, and its inductive effect would further destabilize the carbocation in **4-52**. Because of these factors this would not be considered a viable mechanism.

Problem 4.4
continued

a. The formation of **4-23** could involve formation of a bromonium ion, **4-52**, similar to **4-14**, **4-15**, and **4-18**. This ion undergoes an aryl shift and then reacts with bromide to give the product. The bromonium ion must form on the *exo* side for the rearrangement to take place because the migrating aryl group must enter from the side opposite the leaving bromo group.

Problem 4.5

4-54-1

4-23

4-23

In the last step of the preceding sequence, the bromide ion enters on the side opposite the phenonium ion, in analogy to the stereochemistry of reaction of bromide ion with a bromonium ion.

There are several possible explanations for the different result at higher temperatures. One is that the bromonium ion forms on both the *exo* and *endo* sites. Reaction of the *exo* bromonium ion gives **4-23**, as before, and **4-25** is formed in an elimination reaction.

4-54-1

Reaction of the *endo* bromonium ion gives the isomers, **4-24**.

Another explanation is that at the higher temperature, only the *exo* bromonium ion **4-54-1** forms, but that under these conditions **4-54-1** can rearrange to give **4-23**, undergo elimination to form **4-25**, or react with bromide ion approaching from the *endo* side to give the two isomers **4-24**.

b. The displacement of Br by sulfur in the intermediate bromonium ion, **4-55**, is similar to the ring closure in Problem 3.4a.

Problem 4.5
continued

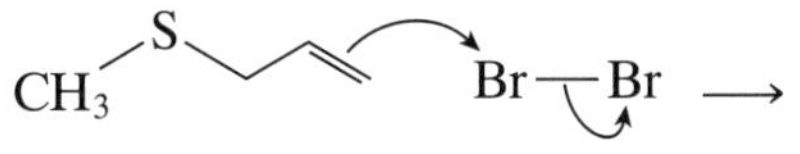

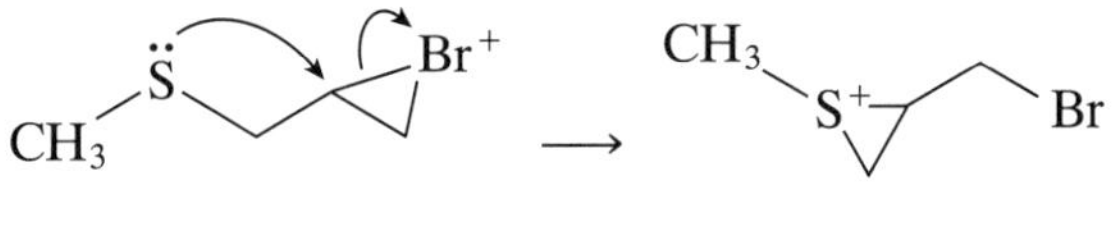

4-55   **4-56**

Representing the reaction as follows would be incorrect. The doubly charged structure **4-57** is not a resonance form of the starting material because the positions of the atoms and the bond angles of the sigma bonds are different. Also, because it is not stabilized by resonance, **4-57** would be very unstable.

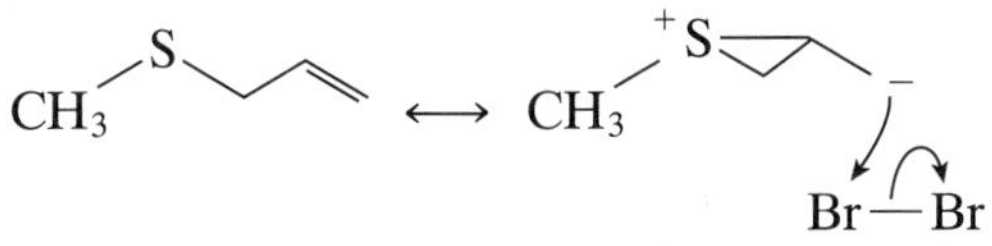

4-57

At low temperatures, the three-membered-ring episulfonium ion, **4-56**, which resembles a bromonium ion, reacts with bromide only at the primary carbon. The product of this S_N2 reaction is determined by the reaction rate, which is faster at the primary carbon than at the secondary carbon. This is a reaction in which product formation is rate-controlled.

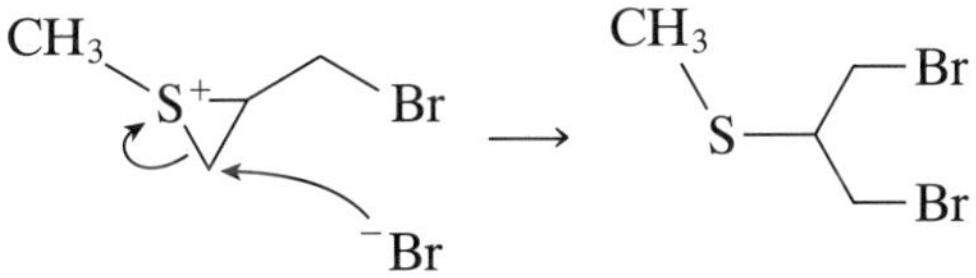

4-56   **4-58**

Problem 4.5 *continued*

At elevated temperatures, **4-58** is in equilibrium with **4-56**. Under these conditions, **4-56** also reacts at the secondary carbon to give **4-59**. Because **4-59** is more stable than **4-58**, its reverse reaction to **4-56** is slower, and **4-59** accumulates. Thus, at higher temperatures, product formation is equilibrium-controlled.

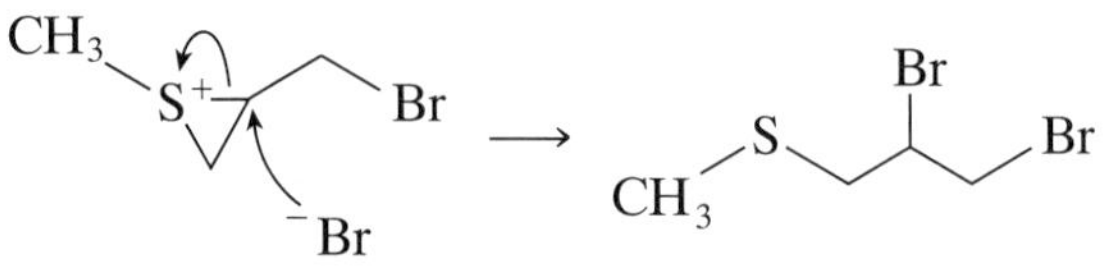

4-56 **4-59**

Problem 4.6

a. This reaction involves loss of the diethylamino group of the amide and ring closure with the aldehyde to form the hydroxylactone product. Because the timing of the ring closure is open to question, two possible mechanistic sequences are given.

Mechanism 1

In this mechanism, the second ring is formed early. The carbonyl oxygen of the amide is protonated, and then the oxygen of the aldehyde adds to this protonated functional group to form a new ring.

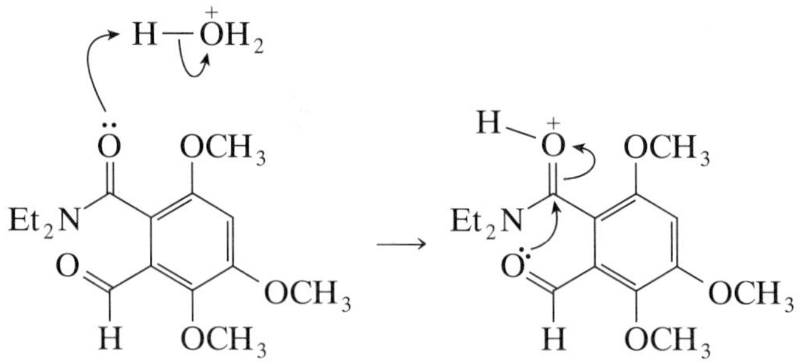

The four steps, after participation of the aldehyde oxygen, are (1) nucleophilic reaction of water with the most electrophilic carbon; (2) loss of a
proton; (3) protonation of the nitrogen; and (4) loss of diethylamine.
Writing the steps in this order avoids formation of more than one positive
charge in any intermediate.

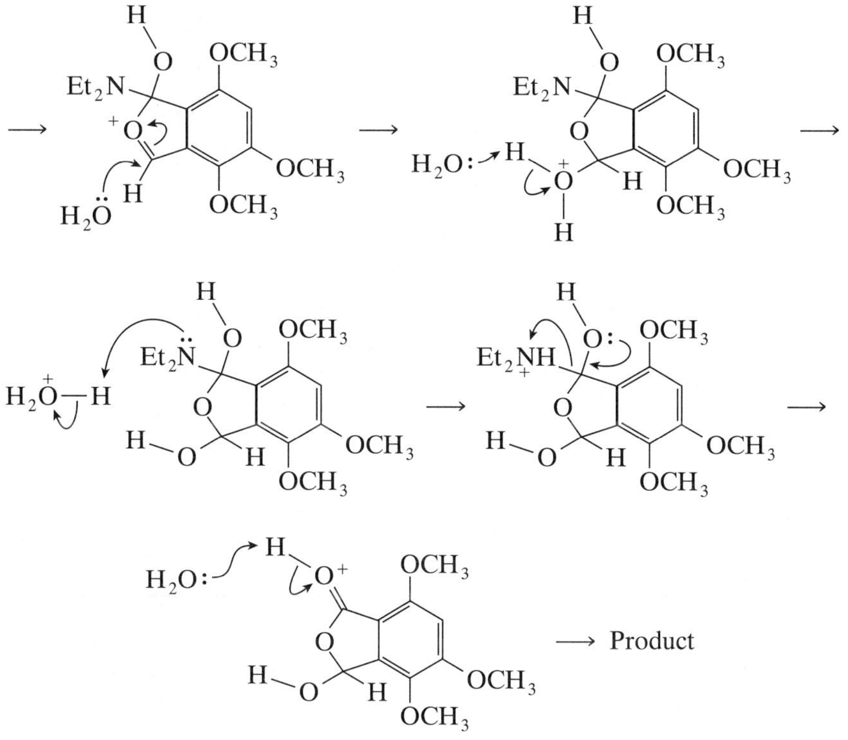

Problem 4.6
continued

4-60

Mechanism 2

Another possibility is hydrolysis of the amide to a carboxylic acid followed
by closure with the aldehyde. The hydrolysis of the amide mimics the steps
in Example 4.16.

Problem 4.6
continued

Because it is easier to protonate an aldehyde than a carboxylic acid
(compare benzoic acid and benzaldehyde in Appendix C), the ring closure
would be written best as protonation of the aldehyde oxygen followed by
nucleophilic reaction of the carboxylic acid carbonyl oxygen. The carbonyl
oxygen acts as the nucelophile because a resonance-stabilized cation is
produced. If the hydroxyl oxygen acts as a nucleophile, the cation is not
resonance-stabilized.

Problem 4.6
continued

The last step is the same as that for mechanism 1. Mechanism 1 seems better than mechanism 2 because no tetrahedral intermediate is formed at the amide carbon prior to cyclization. The amide position is sterically congested, because it is flanked by *ortho* substituents. A tetrahedral intermediate (sp^3-hybridized carbon) adds to the congestion and might be of such high energy that its formation would be unlikely.

b. Both the ester and nitrile undergo hydrolysis. (There are many acids present. The strongest acids, H_3O^+ and H_2SO_4, can be used interchangeably.) Normally, esters are hydrolyzed more rapidly than nitriles, so that the ester will be hydrolyzed first.

Hydrolysis of the nitrile produces the imino form of the amide, which readily tautomerizes by the usual mechanism to form the amide.

Problem 4.6
continued

If reaction conditions are controlled, the hydrolysis of nitriles can be stopped at the amide stage. In this case, once the amide is produced, the nitrogen will react as a nucelophile with the protonated carbonyl of the acid. Because a six-membered-ring transition state is involved, this intramolecular reaction is very favorable.

Problem 4.6
continued

⟶ Product

Reaction of the nitrile, acting as a nucleophile, with the electrophilic protonated ester is unlikely as a ring-forming reaction. Because of the linearity of the nitrile group, it is difficult for the lone pair of electrons on nitrogen to reach close enough to overlap the *p* orbital on the carbon of the carbonyl group. (See the boxed groups on the last structure in the sequence.) Also, because the lone pair of electrons on the nitrile occupies an *sp* orbital, these electrons would be much less nucleophilic than the lone pair occupying a *p* orbital on the amide nitrogen.

⟶

Problem 4.7

None of the cations produced by protonation at the carbons of acrolein is as stable as the cation produced by protonation at oxygen. We will consider the possibilities.

Protonation of the aldehyde carbonyl carbon gives cation **4-61**. This is a very unstable intermediate. It is not stabilized by resonance, and the positive oxygen lacks an octet of electrons. Notice that the double bond and the positive oxygen are not conjugated because they are separated by an sp^3-hybridized carbon.

4-61   **4-62**

Problem 4.7
continued

The intermediate produced by protonation at the α carbon also gives a very unstable primary cation, **4-62**, which is not stabilized by resonance. Protonation at the β carbon gives a delocalized cation, **4-63**, but the resonance form (**4-63-2**) is especially unstable because the oxygen has a positive charge and does not have an octet of electrons.

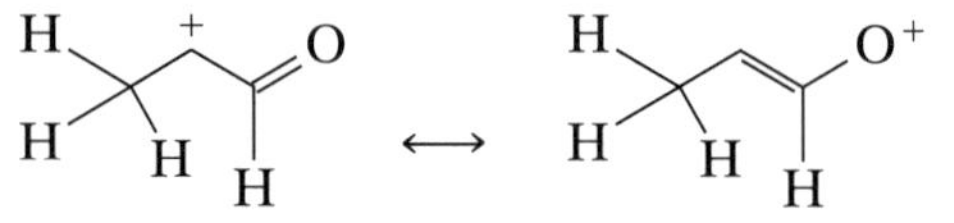

          4-63-1   **4-63-2**

Problem 4.8

HCl is a catalyst for the transformation, so it must be regenerated at some point in the mechanism.

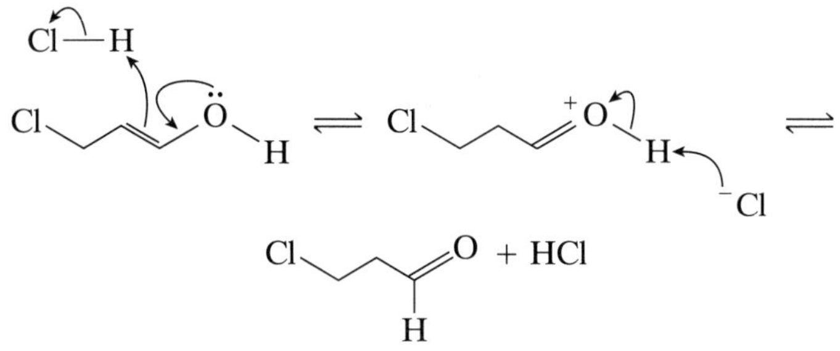

Problem 4.9

By analogy to Example 4.19 and Problem 4.7, protonation at the carbonyl group, rather than at the double bond, of the starting material will give a more stable cation. (That is not to say that all reactions occur via a mechanism involving formation of the most stable carbocation. But if the most stable cation leads to product, that is the one to use.)

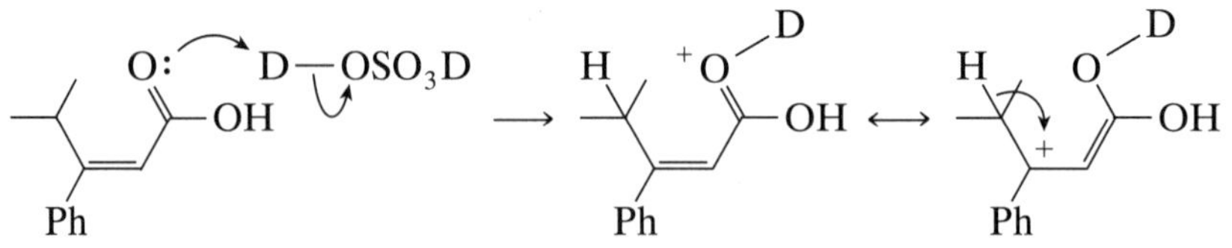

                           4-64-1   **4-64-2**

The carbocation formed initially, **4-64**, can undergo a hydride shift to give a tertiary carbocation. The oxygen of the OH or OD group reacts as a nucleophile with the electrophilic carbocation. In order for the ring to form, the oxygen that reacts must be the one on the same side of the double bond as the carbocation. In **4-65**, this happens to be the OD oxygen, but because these two oxygens can equilibrate rapidly under the reaction conditions, the oxygen *cis* to the alkyl group could just as easily be protonated as deuterated.

Problem 4.9
continued

Intermediate **4-66** undergoes acid-catalyzed tautomerization to give the product.

There is another cation that could undergo the same rearrangement as this one. This cation could be formed by direct protonation of the double bond with Markovnikoff regiospecificity:

Problem 4.9
continued

The remaining steps in this mechanism would be very similar to those of the first. However, the first mechanism is better because the carbocation formed initially is more highly stabilized by resonance and, thus, should form faster.

Problem 4.10

a. The intermediate, formed by reaction of the electrophile at the *para* position, has three resonance forms. Form **4-67** is especially stable because the positive charge is next to an alkyl group, which stabilizes positive charge by the inductive effect and by its polarizability.

4-67

Reaction of the electrophile at the *ortho* position would give an intermediate of essentially the same stability. However, reaction of the electrophile at the *meta* position gives an intermediate, **4-68**, in which the positive character cannot be located at the alkyl group position. Thus, this intermediate is not as stable and is not formed as rapidly as the intermediates from reaction at either the *ortho* or *para* position.

4-68

b. The intermediate, formed by reaction of a nitronium ion at the *meta* position, does not have positive charge on the carbon bearing the positively charged sulfur of the sulfonic acid group.

The intermediates, formed by electrophilic reaction at the *ortho* or *para* positions, are not as stable as the intermediate formed by reaction at the *meta* position because, for *ortho* and *para* attack, one of the three resonance forms has positive charge on the carbon bearing the positively charged sulfur of the sulfonic acid group. For example, for the intermediate resulting from reaction at the *para* position we can draw three resonance forms. Of these **4-69-2** is particularly unstable because of the destabilizing effect of two centers of positive charge in close proximity. Because the intermediate involved in *meta* substitution lacks this unfavorable interaction, it is more stable and *meta* substitution is favored.

Problem 4.10
continued

4-69-1   **4-69-2**

4-69-3

For reaction at the *ortho* position, the resonance forms of the resulting intermediate are destabilized in the same way as those that result from reaction at the *para* position.

c. The intermediate, formed by reaction of the electrophile at the *para* position, has a resonance form, **4-70-2**, with positive character at the substituent position. This means that the unshared pair of electrons on the nitrogen of the acetamido group can overlap with the adjacent positive charge, as shown in **4-70-4**.

4-70-1   **4-70-2**

4-70-3   **4-70-4**

This extra delocalization of the positive charge adds to the stability of the

Problem 4.10
continued

intermediate. On the other hand, if the electrophile reacts at the *meta* position, the positive charge cannot be placed on the nitrogen.

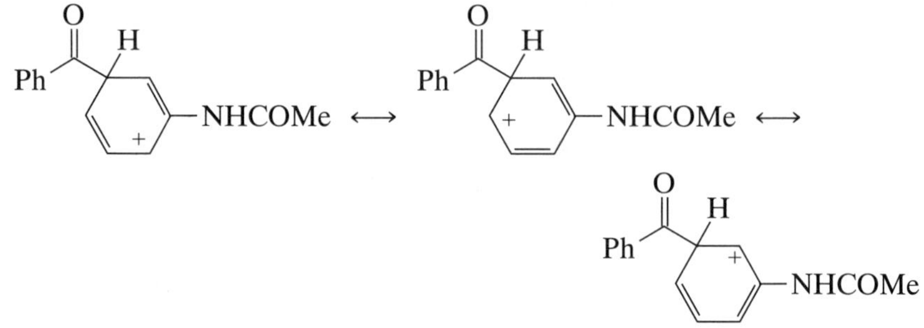

Problem 4.11

a. The mechanism in sulfuric acid might occur by the steps typical for a dienone–phenol rearrangement. Protonation of the A-ring carbonyl gives a more highly resonance-stabilized intermediate than protonation of the B-ring carbonyl.

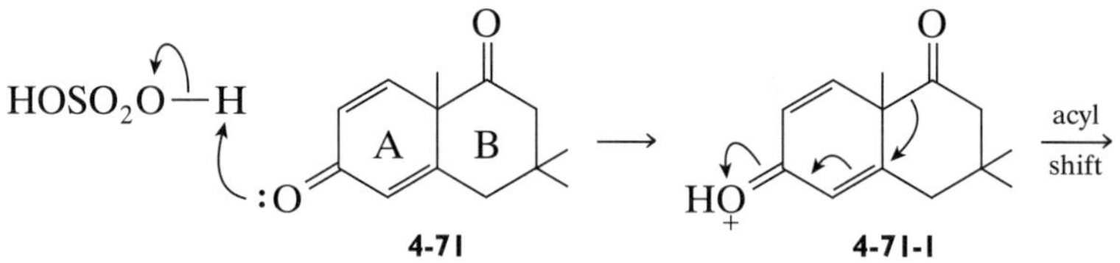

4-71 **4-71-1**

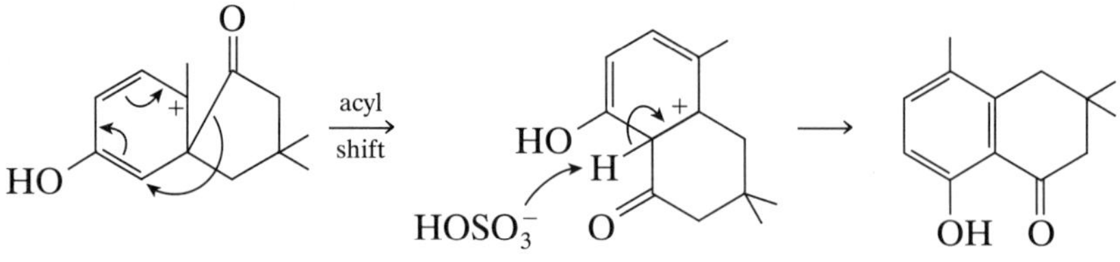

4-72

Formation of the product, **4-73**, formed in dilute HCl requires cleavage of the B ring. A mechanism involving the formation of the hydrate of the keto group in the B ring, followed by ring opening, is shown next. (Example 2.11 shows a slightly different possibility.)

Problem 4.11
continued

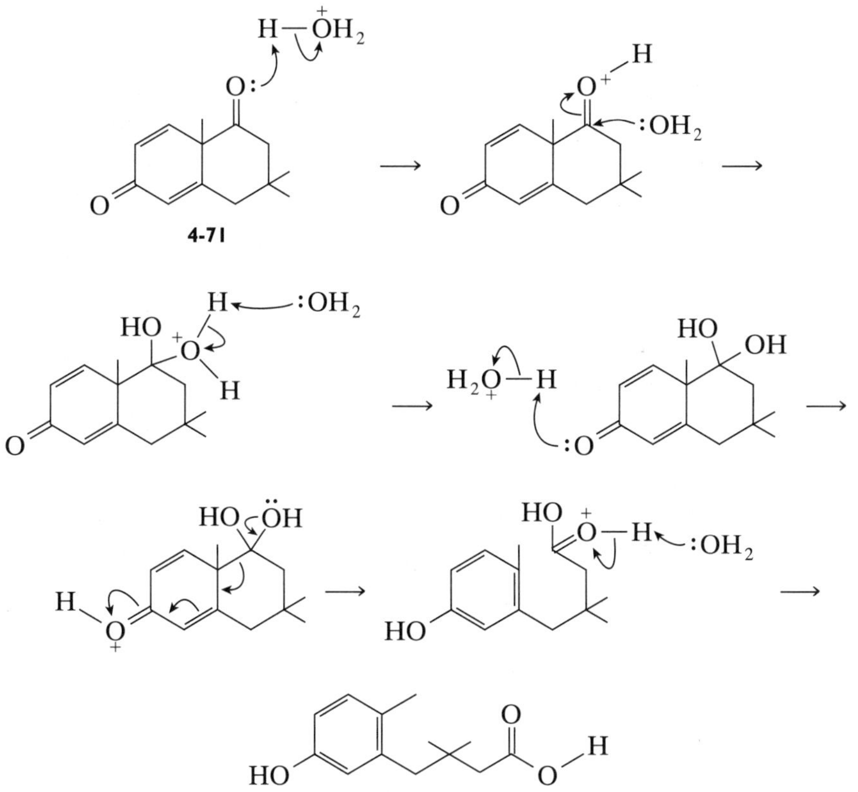

4-71

4-73

The mechanism in dilute acid suggests an alternative route to **4-72** in the stronger acid, sulfuric acid, in which initial cleavage of the B ring occurs. In this pathway, the ring of the intermediate cation **4-71-1** can open to give the acylium ion, **4-74**.

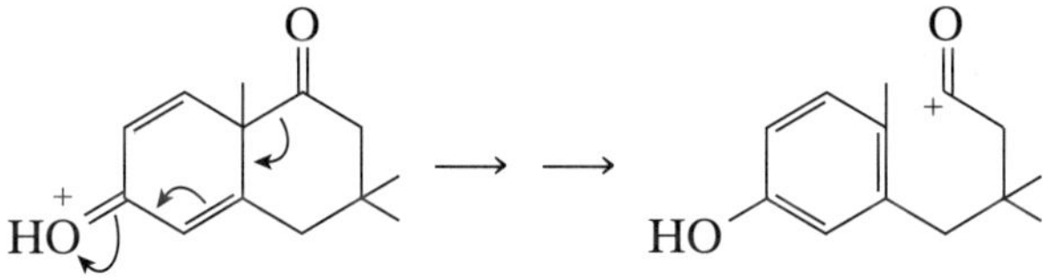

4-71-1   **4-74**

Rewriting **4-74**, with the acylium ion in proximity to the position *ortho* to the hydroxyl group clarifies the intramolecular acylation reaction that forms a six-membered ring.

Problem 4.11
continued

4-74

The acylium ion would have a longer lifetime in concentrated sulfuric acid than in aqueous hydrochloric acid because the concentration of water in the former acid is extremely low. In aqueous hydrochloric acid, the starting ketones are more likely to be in equilibrium with their hydrates, and the intermediate acylium ion might never form. If it does form, it will react so rapidly with the nucleophilic oxygen of water that the electrophilic aromatic substitution cannot occur.

Experimental data in the cited paper support the opening of the B ring in concentrated as well as dilute acid. It was found that on treatment with sulfuric acid, **4-73** and **4-71** both react to give **4-72** at the same rate. However, the alternative mechanism involving successive acyl shifts appears to occur in trifluoroacetic acid. In this medium, **4-71** rearranges to **4-72**, but **4-73** does not give **4-72**. This means that **4-73** cannot be an intermediate in the reaction. Consequently, the mechanism involving opening of ring B is ruled out, and the acyl shift mechanism is a reasonable alternative.

b. The reaction proceeds by a cleavage–recombination reaction.

Support for the formation of **4-75-1** and **4-75-2** comes from a crossover experiment reported in the paper. In the presence of *m*-cresol, **4-76** was obtained.

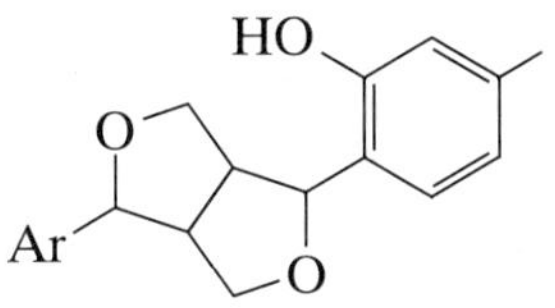

Problem 4.11
continued

4-76

This product is formed by capture of the intermediate carbocation **4-75-1** by the added *m*-cresol. Loss of a proton gives **4-76**. Furthermore, as the concentration of *m*-cresol increases, the amount of **4-76** increases. This supports the idea that the *m*-cresol is competing with **4-75-2** for **4-75-1**. The term, crossover experiment, refers to the fact that **4-75-1** reacts with an external reagent rather than **4-75-2**, its co-cleavage product.

The occurrence of crossover rules out the following mechanism:

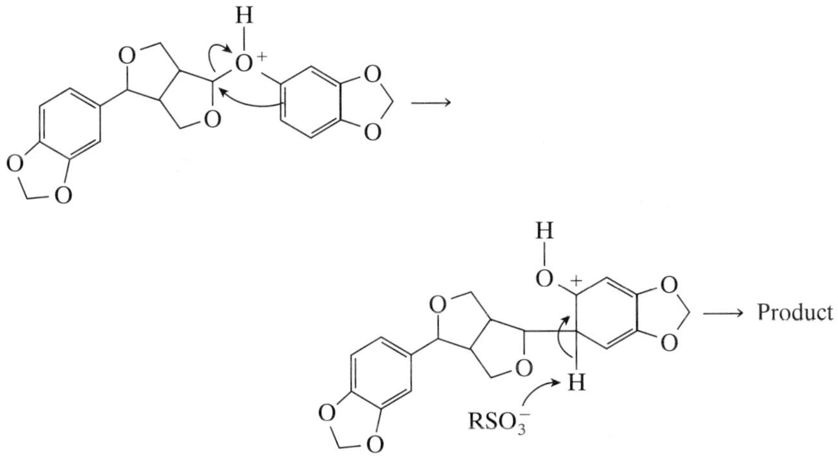

The intermediate anion picks up a proton: removal of the proton on the *sp*³-hybridized carbon *α* to the carbonyl group gives the phenolate ion. The phenolate loses chloride ion to give **4-77**, which undergoes addition of hydroxide at the carbon at the *ortho* position. Loss of the remaining chloride and removal of a proton gives the product phenolate.

Problem 4.12

Problem 4.12
continued

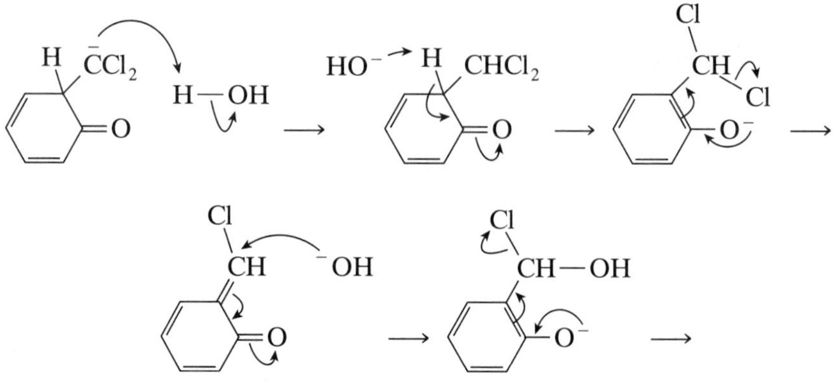

4-77

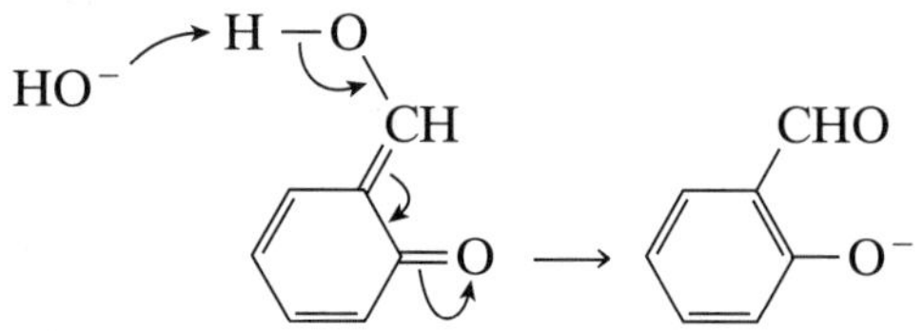

Acidic workup will give the phenol itself.

Problem 4.13   Dibromocarbene can be produced from bromoform and base:

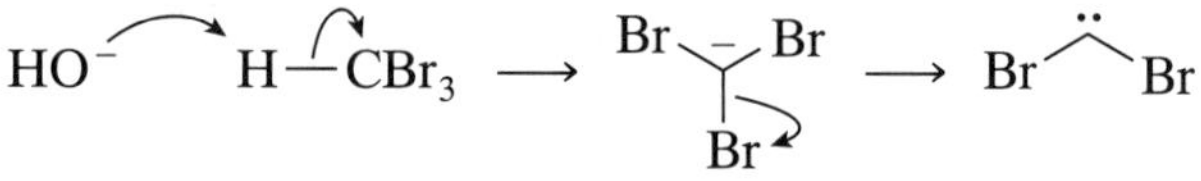

Phenolate, formed in the basic medium, reacts as a nucleophile with the electrophilic carbene. The resulting anion is protonated by water to give the product.

Problem 4.13
continued

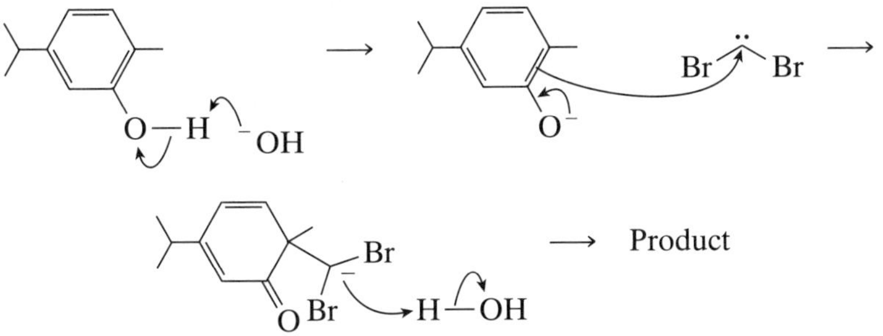

Problem 3.4b gives a subsequent reaction of this product and its literature reference.

An alternative mechanism with the ring acting as an electrophile and the carbene as a nucleophile would be incorrect. The dihalocarbenes are quite electrophilic. Also, because the phenol readily forms a salt with sodium hydroxide, the aromatic ring would not be electrophilic.

This reaction is a fragmentation, which often accompanies a Beckmann rearrangement. Reaction of the hydroxyl group with phosphorus pentachloride converts it into a better leaving group. In **4-78**, instead of an alkyl group migration, the ring bond on the side opposite the leaving group cleaves, producing a ring-opened cation, **4-79**. This mimics the stereochemical requirement of the Beckmann rearrangement.

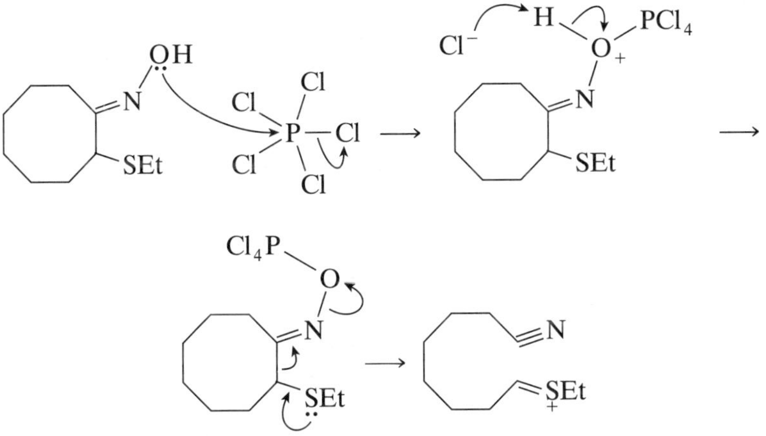

Problem 4.14

Problem 4.14
continued

The electrophilic cation, **4-79**, reacts with a nucleophile at carbon to give the neutral derivative, **4-80**.

The product, **4-80**, will undergo hydrolysis readily when the reaction is worked up in water. This hydrolysis is acid-catalyzed because the phosphorus compounds in the reaction mixture are acidic.

One student wrote an alternative mechanism that involved ring expansion of **4-78** rather than fragmentation.

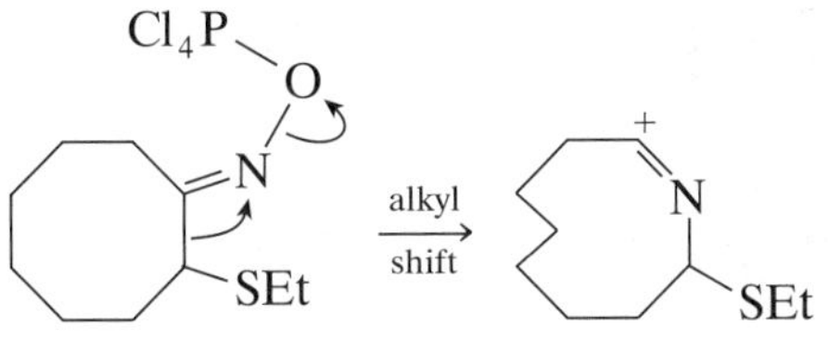

Problem 4.14
continued

4-78

This is a good step because it is the normal Beckmann rearrangement. However, to get from this intermediate to the open-chain product took a fairly large number of "creative" steps. Application of Occam's razor suggests that the initial fragmentation occurs early in the mechanism.

The initial step is a protonation, followed by nucleophilic reaction of azide ion.

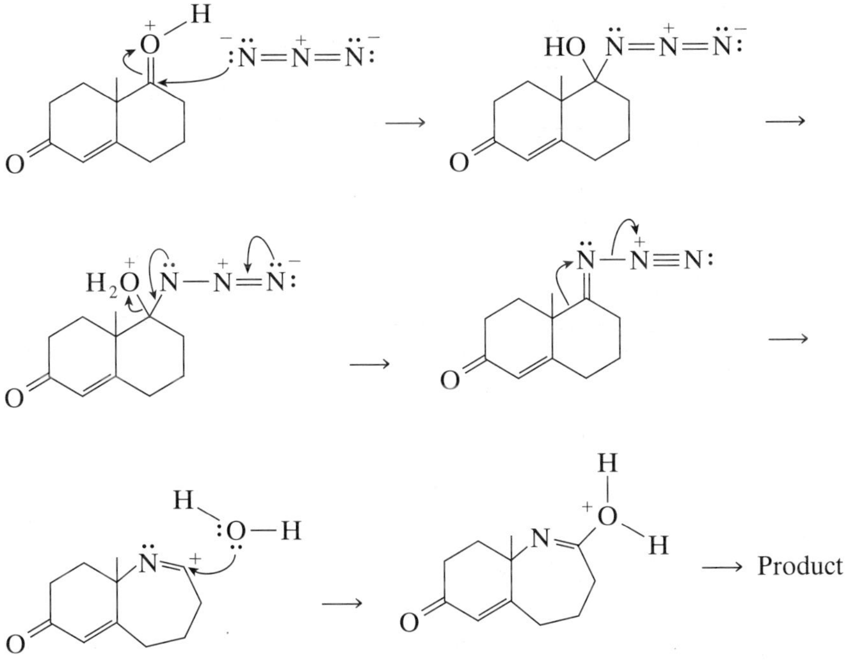

Problem 4.15

→ Product

Problem 4.15
continued

The reaction is a Schmidt reaction and its mechanism closely resembles that of the Beckman and Wolff rearrangements. The final step is deprotonation, followed by tautomerization to the lactam. Note that the alternate lactam formed by migration of the secondary carbon was not found.

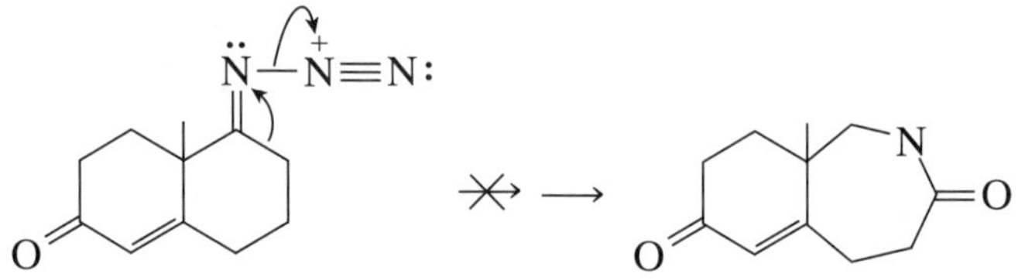

One might expect that reaction could also occur at the α-β-unsaturated keto group, and, depending on the reaction conditions, several different products were formed by reaction at this site. The course of reaction when there are two similar functional groups present in a molecule frequently is difficult to predict. The authors of the article cited offer rationalizations for the formation of the various products of the reaction.

Problem 4.16

In the presence of $NaHCO_3$, deprotonation of the peracid is the first step, followed by nucleophilic attack on the carbonyl group:

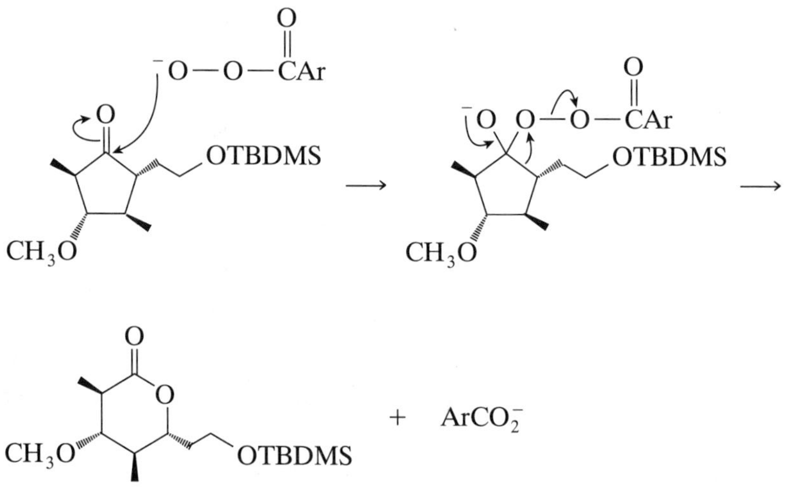

The minor product is formed by migration of the other alkyl substituent. Note that both products are formed with retention of configuration at the migrating carbon.

$+ \; ArCO_2^{-}$

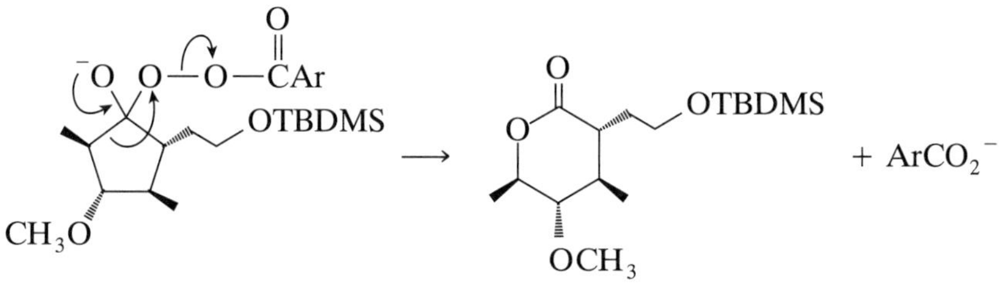

There is no readily apparent reason why one group should migrate in preference to the other. The rearrangement of a closely related structure that differs only in the side-chain substitution is regiospecific:

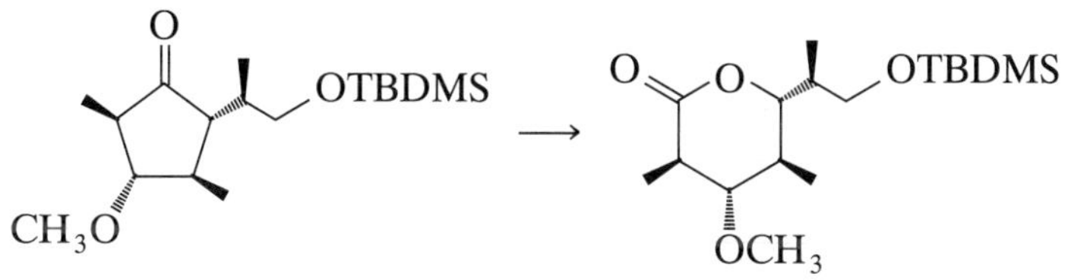

sole product

The reaction is buffered by $NaHCO_3$ to prevent hydrolysis of the TBDMS protecting group by the acid produced in the course of the reaction.

a. Trifluoromethanesulfonic acid is a very strong acid, and the most basic atom in the amide is the carbonyl oxygen. Protonation of the carbonyl group is a likely process, but it is not one that leads to product. That is, protonation of the hydroxyl oxygen is on the pathway to product, whereas protonation of the carbonyl oxygen is not. The nitrogen bearing the protonated hydroxyl group can act as an electrophile, and the phenyl ring, situated to form a favorable six-membered ring, can act as a nucleophile.

Problem 4.17
continued

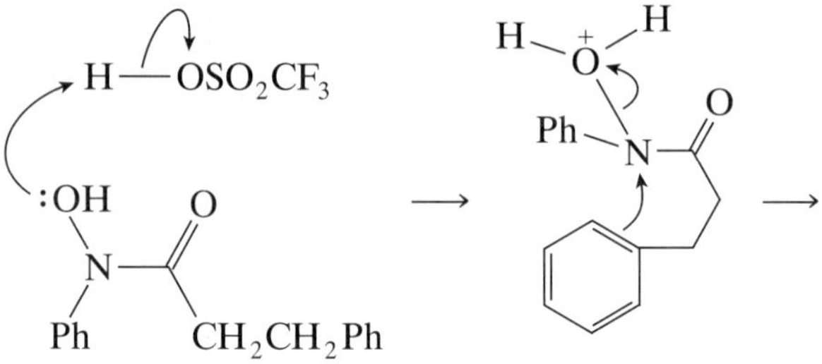

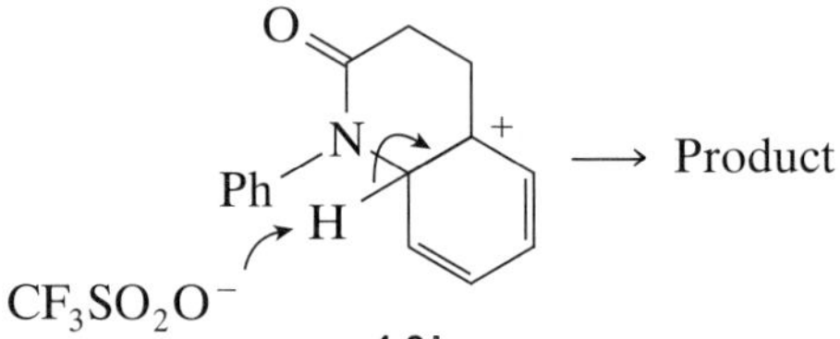

In an alternate mechanism, water leaves the protonated starting material prior to involvement of the nucleophile (the aromatic ring). This gives a nitrenium ion, which acts as the electrophile. Cyclization leads to the same intermediate carbocation, **4-81**, as before.

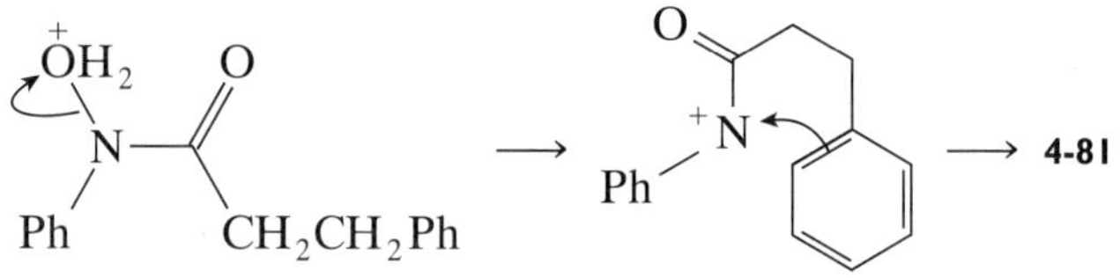

The following reaction in trifluoromethanesulfonic acid provides support for initial formation of positive nitrogen (Endo, Y.; Ohta, T.; Shudo, K.; Okamoto, T. *Heterocycles* **1977**, *8*, 367–370).

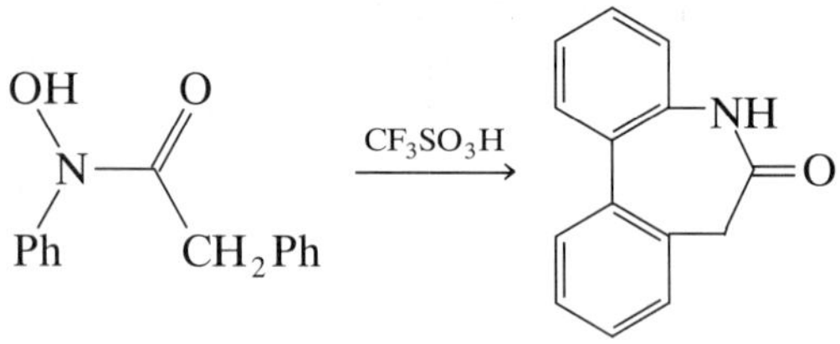

76%

Problem 4.17
continued

This reaction is an electrophilic aromatic substitution in which one benzene ring acts as the electrophile and the other benzene ring acts as the nucleophile. In order to develop substantial positive charge in the electrophilic ring (a driving force for the reaction), resonance must occur. This requires prior loss of water. The intermediate nitrenium ion and the direction of the substitution reaction are shown:

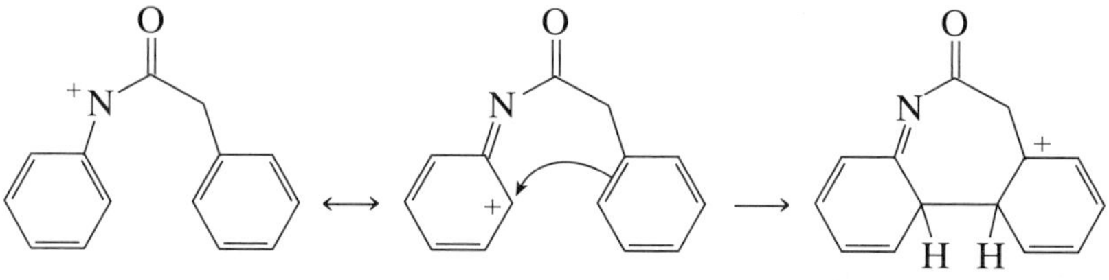

Formation of the product requires loss of a proton to regenerate a neutral compound and tautomerization to regenerate the other benzene ring.

b.

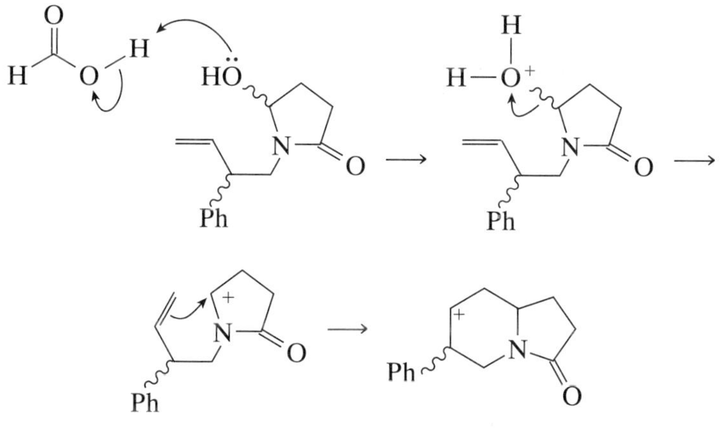

Nucleophilic reaction of formic acid at the electrophilic carbocation can be either *cis* or *trans* to the phenyl group in the bicyclic intermediate and leads to the two products.

c. There are several different mechanistic sequences that might lead to the product.

Mechanism 1

The two equivalent amide carbonyl oxygens are the most basic atoms in the starting materials. Protonation of one of these oxygens gives a carbocation, **4-82**, which is stabilized by delocalization onto oxygen and nitrogen as well as onto carbon.

The intermediate, **4-82**, can decompose to give a new electrophile, with which the nucleophilic phenol then reacts.

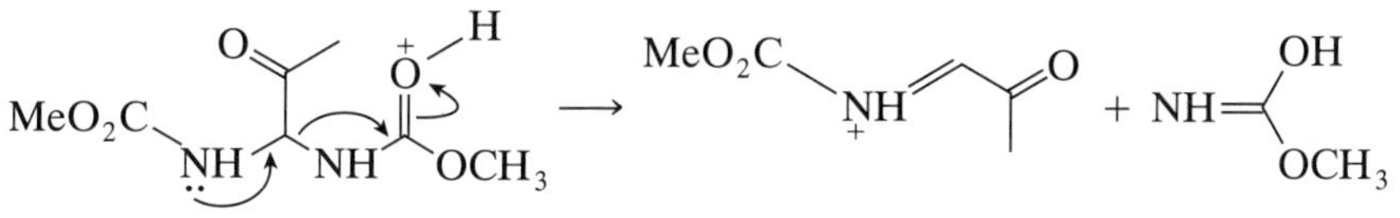

Problem 4.17
continued

4-82

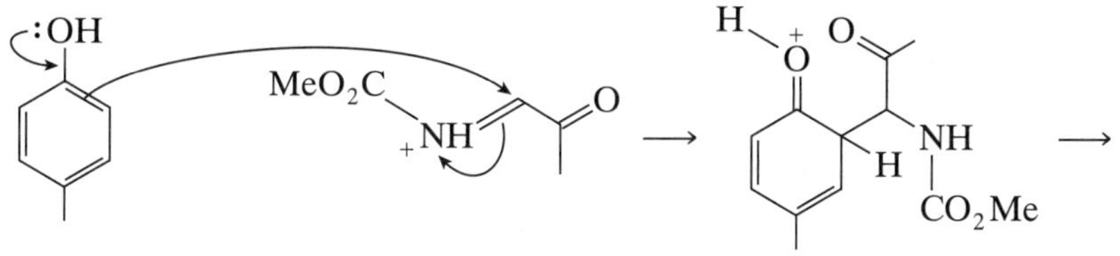

4-83

An alternative mechanism to **4-83** might be protonation of one of the nitrogens in the starting material, followed by nucleophilic reaction of the phenol with the carbon bearing the protonated leaving group (S$_N$2 reaction). However, this carbon is quite sterically congested, so that this is not an attractive option.

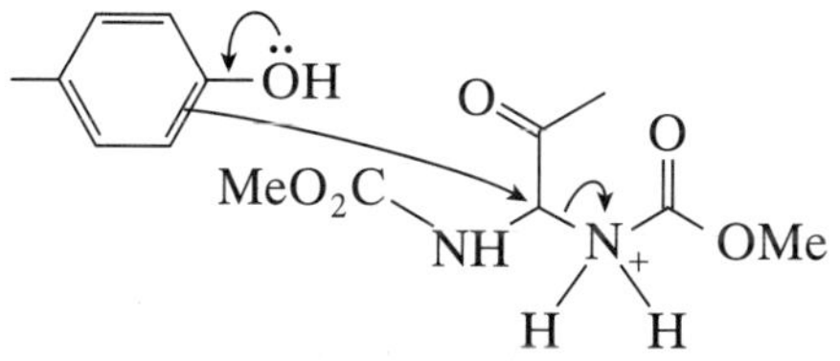

After aromatization of **4-83** and protonation of the keto carbonyl, cycliza-

Problem 4.17
continued

tion can occur:

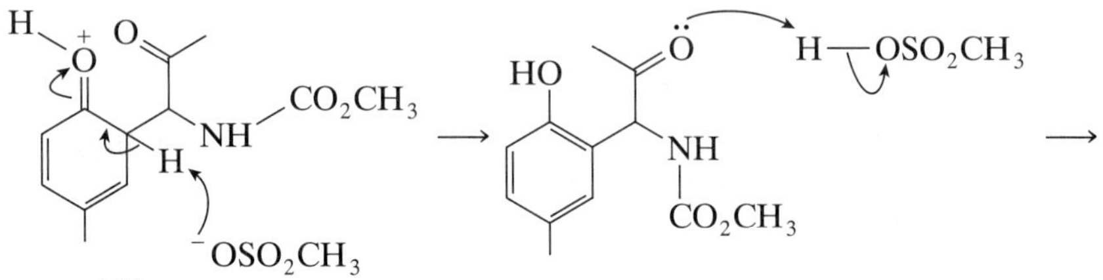

4-83

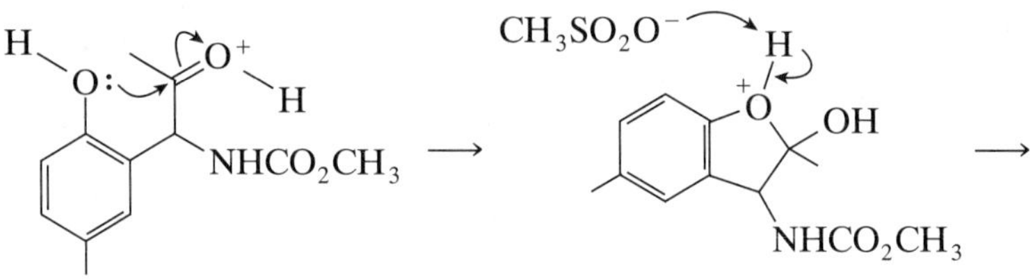

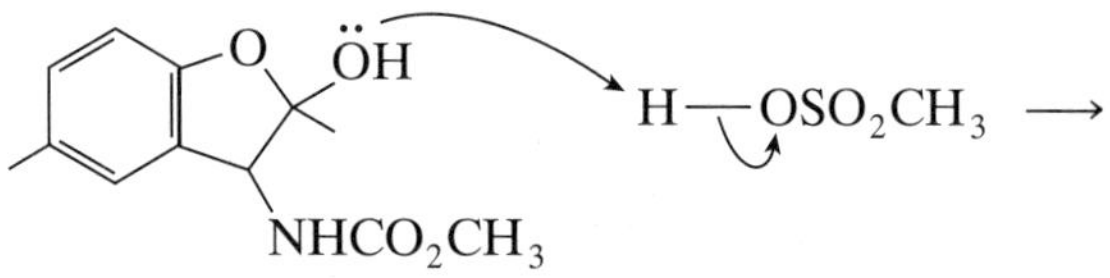

4-84

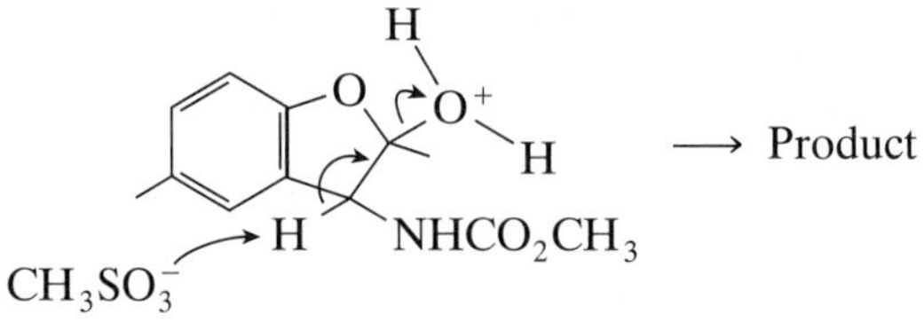

→ Product

Mechanism 2

If the keto group in the starting material were protonated, the phenolic oxygen could react as a nucleophile at the carbonyl carbon.

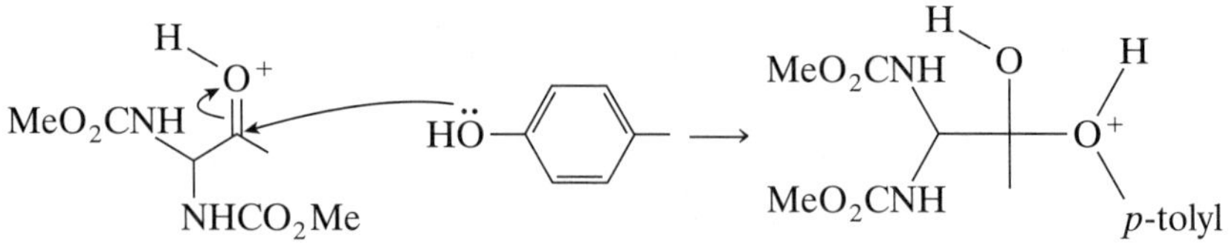

Following proton loss from the positive oxygen and protonation of one of the amide nitrogens, intermediate **4-85** is formed. This intermediate can lose methyl carbamate to form **4-86**, in which the ring acts as a nucleophile.

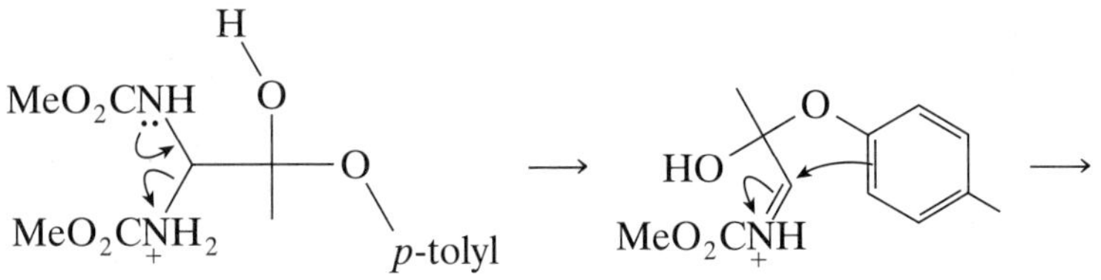

Problem 4.17
continued

Loss of a proton gives **4-84**, which undergoes dehydration as in mechanism 1.

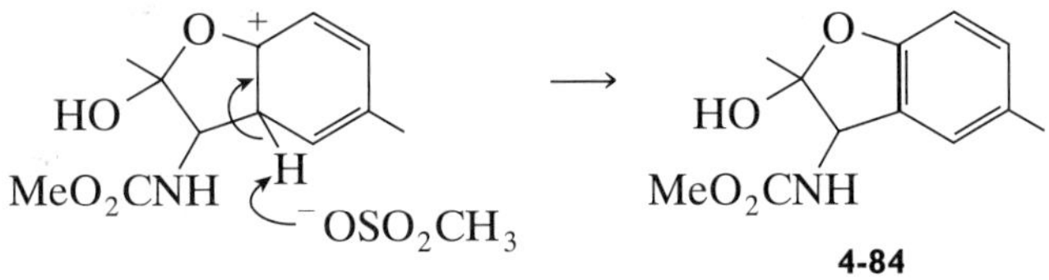

d. The tin intermediate can be written as either the product of a nucleophilic displacement on tin, or as an expanded orbital on tin, as in **4-87**.

Mechanism 1

Ring opening occurs readily in the cyclopropyl-substituted cation, **4-87**, to give **4-88**. The driving forces are release of the strain energy of the three-membered ring and stabilization of the resulting cation by the *p*-methoxyphenyl group.

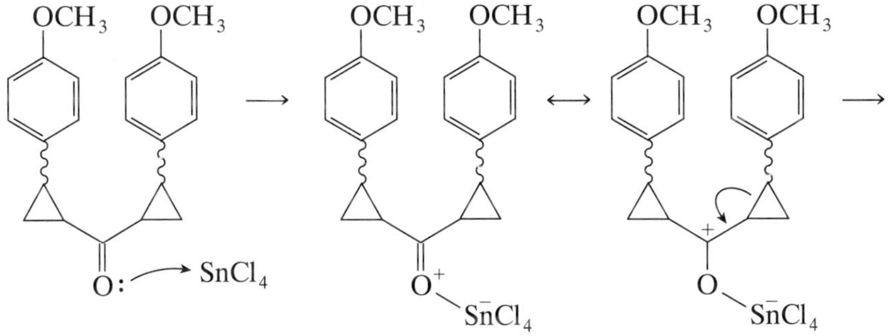

Problem 4.17
continued

Intermediate **4-88** must equilibrate with the *cis* isomer, **4-89**, which under-goes further ring opening and cyclization to **4-90**, with the same driving forces as the reaction of **4-87** to give **4-88** (Ar = aryl).

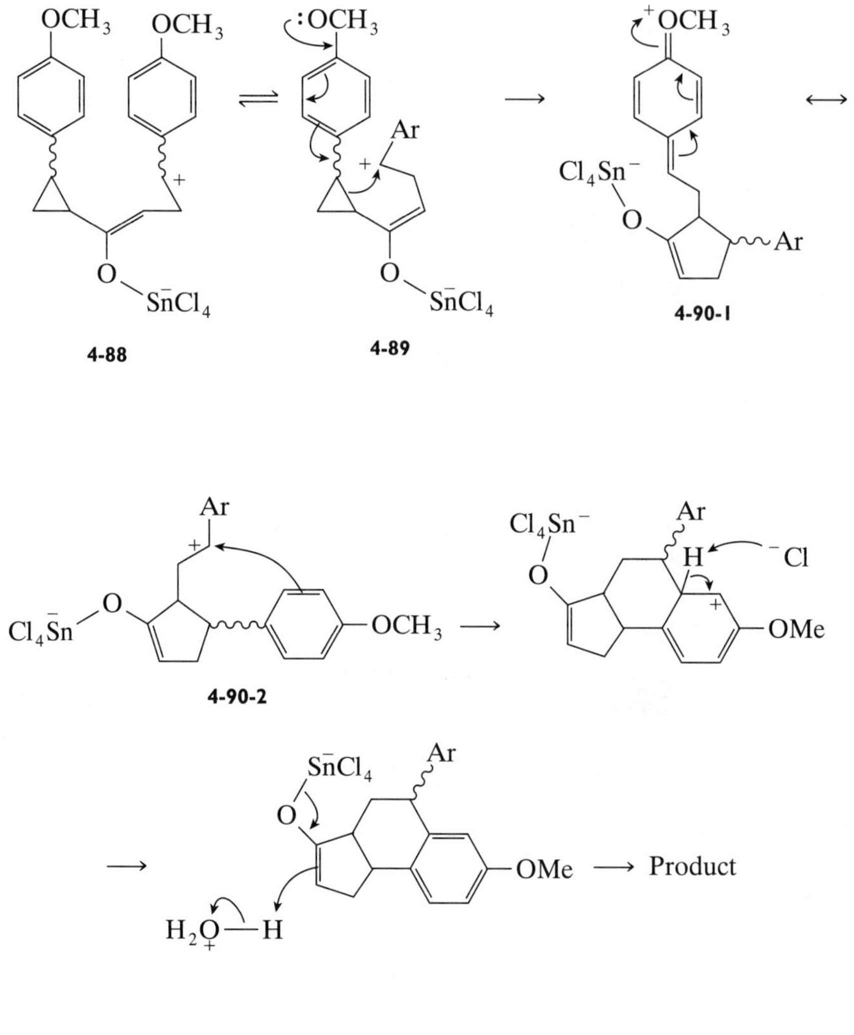

Problem 4.17
continued

Mechanism 2

An alternative mechanism involves formation of an eight-membered-ring intermediate from **4-89**:

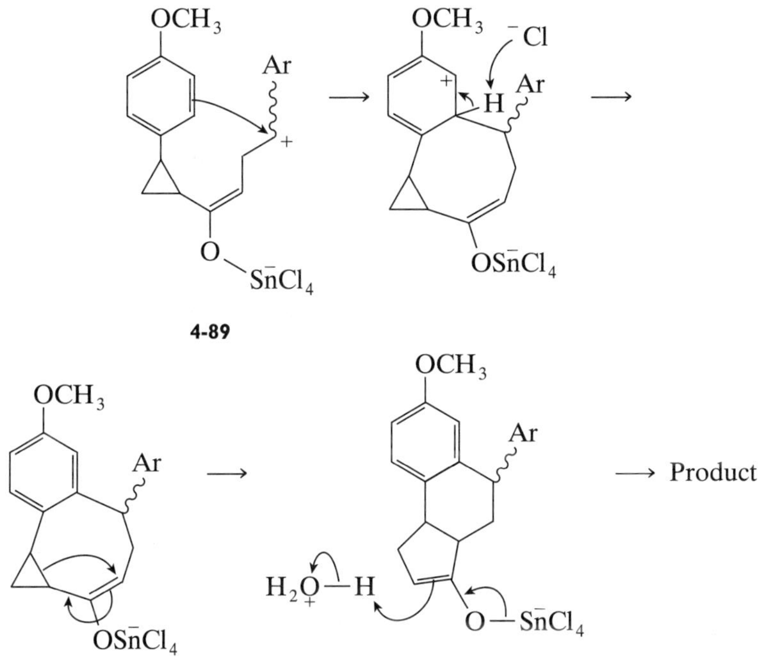

4-89

→ Product

Formation of the eight-membered ring is entropically less favorable than mechanism 1. Another advantage of mechanism 1 is that the methoxy group participates in resonance stabilization of the positive charge produced when the five-membered ring is formed. Direct participation of the methoxy group is not possible in the intermediates formed in mechanism 2.

The source of chloride ion, used as a base in these mechanisms, would be the following equilibrium:

$$ROS\bar{n}Cl_4 \rightleftharpoons ROSnCl_3 + Cl^-$$

Nitromethane is the sole solvent for the reaction.

e. Silver ion coordinates with the chlorine and increases its ability to act as a leaving group. After the positive center (a nitrenium ion) has been created, an intramolecular electrophilic aromatic substitution takes place.

f. The nucleophilic ester carbonyl oxygen reacts with the electrophilic tin(IV) chloride. Loss of a proton from the intermediate formed initially gives **4-91**.

The HCl that is produced can add in a regiospecific manner to the isolated double bond, giving a tertiary carbocation. An intramolecular nucleophilic reaction with this cation generates the product.

Problem 4.17
continued

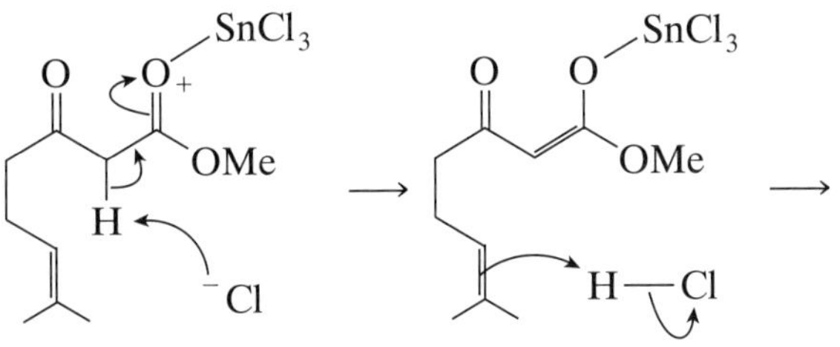

4-91

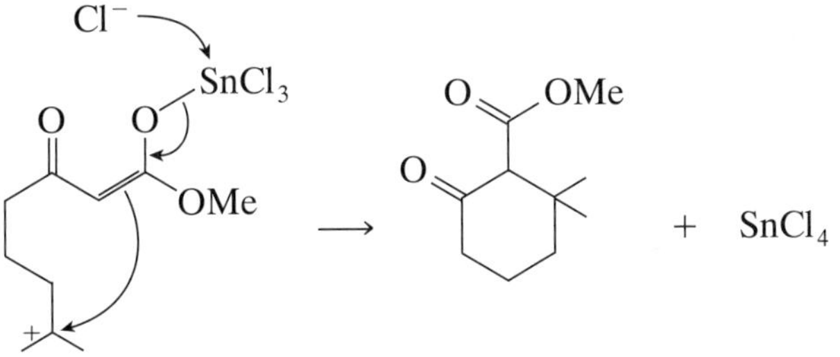

The following would *not* be a good mechanistic step because the anion that is generated is so unstable. Also, we would not expect such a strong base to be formed in a strongly acidic medium.

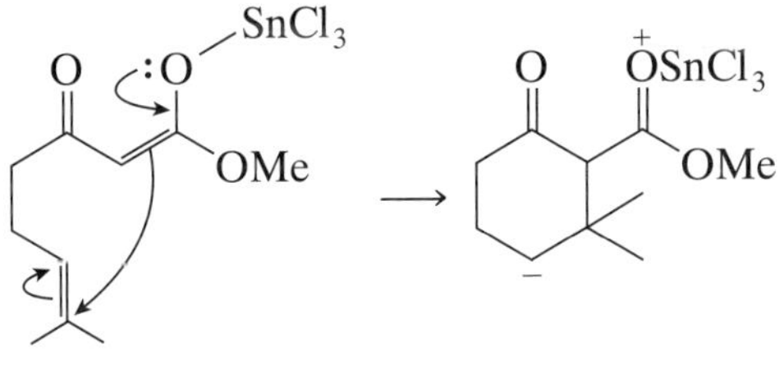

4-91

5

Radicals and
Radical Anions

I. INTRODUCTION

Radicals are species that contain one or more unpaired electrons. They are encountered in many reactions used in chemical industry (e.g., the production of polyethylene), in the processes responsible for the spoiling of foods (e.g., autoxidation by molecular oxygen), and in many biological systems (e.g., signaling by nitrogen oxide, NO). Radical anions, as their name implies, are radical species that have an unpaired electron and a negative charge.

In chemical structures, the unpaired electron of a radical is represented by a dot. Radical mechanisms are depicted in one of two ways. Most commonly, each individual step of the mechanism is written without the use of arrows to show electron movement. The resulting series of equations shows the order of events, and it is assumed that one-electron transfers are taking place throughout. A second method uses curved, half-headed arrows ($\rightarrow$) to show electron movement. The half-headed arrow is used to denote movement of a single electron, whereas the normal arrowhead is used to denote movement of a pair of electrons.

Various studies indicate that although radicals tend to be pyramidal, the pyramids are shallow and the barriers to inversion are low. This means that the stereochemical results of radical reactions are very similar to those of carbocations, in other words, stereochemistry is usually lost at a reactive center.

Whereas most carbon free radicals are highly reactive species, there are notable exceptions.

2. FORMATION OF RADICALS

Many radicals are produced by homolytic cleavage of bonds. The energy for this kind of bond breaking comes from thermal or photochemical energy or from electron-transfer reactions effected by either inorganic compounds or electrochemistry. These kinds of processes initiate reactions that proceed by a radical mechanism. Compounds that readily produce radicals are called initiators or free radical initiators.

A. Homolytic Bond Cleavage

Radicals produced from chlorine and bromine can be generated photochemically and/or thermally. Because bromine atoms are less reactive than chlorine atoms, brominations are often done in the presence of both heat (Δ) and light ($h\nu$).

$$Cl_2 \xrightarrow{h\nu} 2Cl^{\bullet}$$

$$Br_2 \xrightarrow[\Delta]{h\nu} 2Br^{\bullet}$$

Many peroxides and azo compounds can be heated to generate radicals. Peroxides decompose readily because of the weak $O-O$ bond, whereas azo compounds cleave readily because of the driving force provided by the formation of the stable nitrogen molecule. Common examples of these decompositions follow. (The $t_{1/2}$ is the time it takes one-half of the material to decompose.)

benzoyl peroxide

$$(CH_3)_3C—O—O—C(CH_3)_3 \quad \xrightarrow[t_{1/2}=1\,h]{150°C} \quad 2(CH_3)_3C—O\cdot$$

di-*t*-butyl peroxide

$$(CH_3)_2CN{=}NC(CH_3)_2 \quad \xrightarrow[t_{1/2}=1\,h]{85°C} \quad 2(CH_3)_2\dot{C}—CN \; + \; N_2$$

with CN groups below each carbon.

5-1

Example 5.1. *Decomposition of AIBN [AIBN = azobisisobutyronitrile].*

Half-headed arrows show the movement of a single electron:

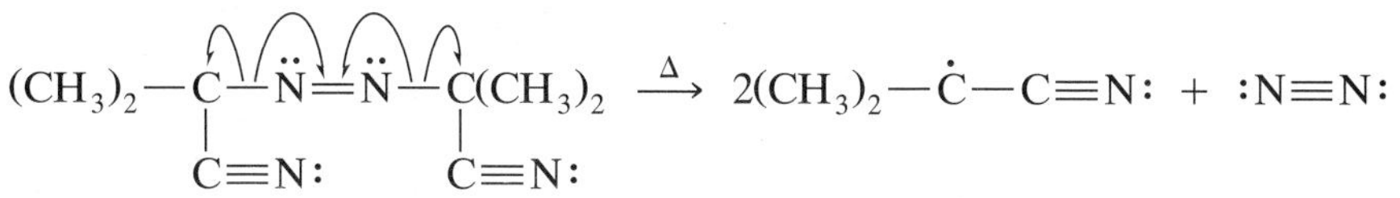

$$(CH_3)_2—C\cdots\ddot{N}{=}\ddot{N}\cdots C(CH_3)_2 \quad \xrightarrow{\Delta} \quad 2(CH_3)_2—\dot{C}—C{\equiv}N{:} \; + \; {:}N{\equiv}N{:}$$

with C≡N groups below each carbon.

(handwritten margin notes)

Why is the reaction shown in Example 5.1 a highly favorable process? **PROBLEM 5.1**

B. Hydrogen Abstraction from Organic Molecules

The carbon–hydrogen bond is relatively stable, so that direct formation of an organic radical by homolytic cleavage of a C—H bond is rarely observed. However, many radicals can remove hydrogen atoms from organic molecules to form carbon radicals. This process is known as hydrogen abstraction. For example,

$$F\cdot \; + \; CH_4 \quad \longrightarrow \quad HF \; + \; \cdot CH_3$$

Some radicals, like fluorine, are so reactive that they form radicals with almost any organic compound, whereas others are so stable that they can abstract hydrogen from very few organic compounds. An example of such a stable radical is **5-1**, formed from AIBN. When selectivity is desired in a radical initiator, a relatively stable radical like the one formed from AIBN is a good choice.

Rates of Hydrogen Abstraction and Relative Stability of Radicals

Because most radicals are electrophilic, the effects of structure on their rate of formation are very similar to those for the formation of carbocations. Thus, the rates of hydrogen abstraction are $1° < 2° < 3°$, corresponding to the order of stability of the resulting radicals: $1° < 2° < 3°$. Also, it is relatively easy to abstract a hydrogen from an allylic or benzylic position because the resulting radicals are stabilized by delocalization. In contrast, it is quite difficult to abstract vinyl and aromatic hydrogens because the electrophilicity of the resulting radicals is higher due to the higher s character of an sp^2-hybridized orbital relative to an sp^3-hybridized orbital. In these cases, the sp^2-hybridized orbital is perpendicular (orthogonal) to the π system; thus, the radical *cannot* be stabilized by resonance. Finally, it is difficult to abstract a hydrogen from the alcohol functional group. The resulting alkoxy radical is quite unstable due to the high electronegativity of oxygen.

The approximate relative stabilities of organic radicals are as follows:

$$Ph\cdot \ < \ CH_2=CH\cdot \ < \ RO\cdot \ < \ RCH_2\cdot \ < \ R_2CH\cdot$$
$$< \ R_3C\cdot \ < \ PhCH_2\cdot \ < \ RCH-CH_2\cdot$$

C. Organic Radicals Derived from Functional Groups

Reactive radicals are often produced by abstraction of a halogen atom from a substrate. A commonly used halogen-abstracting reagent is tri-*n*-butyltin radical, formed from tri-*n*-butyltin hydride using AIBN as an initiator. AIBN generates **5-1**, which abstracts a hydrogen atom from the tri-*n*-butyltin hydride, generating the tri-*n*-butyltin radical. This tin radical can abstract a halogen from a variety of substrates (e.g., alkyl, olefinic, or aryl chlorides, bromides, or iodides) to generate the corresponding radical and tri-*n*-butyltin halide.

$$(CH_3)_2\overset{\cdot}{C}-CN \ + \ (n\text{-Bu})_3SnH \ \longrightarrow \ (CH_3)_2CH-CN \ + \ (n\text{-Bu})_3Sn\cdot$$

5-1

$$(n\text{-Bu})_3Sn\cdot \ + \ RBr \ \longrightarrow \ (n\text{-Bu})_3SnBr \ + \ R\cdot$$

Tri-*n*-butyltin radicals can also be used to generate radicals from selenium compounds. An example is the formation of acyl radicals from seleno esters.

Boger, D. L.; Mathvink, R. J. *J. Org. Chem.* **1988**, *53*, 3377–3379.

Radicals can also be synthesized by the reduction of alkylmercury salts. For example, in the presence of sodium borohydride, compound **5-2** reacts to form the radical, **5-3**.

5-2   **5-3**

Giese, B.; Horler, H.; Zwick, W. *Tetrahedron Lett.* **1982**, *23*, 931–934.

Despite their toxicity, alkylmercury salts have been used widely in research due to their versatility as synthetic intermediates.

3. RADICAL CHAIN PROCESSES

Most synthetically useful radical reactions occur as chain processes. A radical chain process is one in which many moles of product are formed for every mole of radicals produced. These chain processes include the following steps:

1. *Initiation.* One or more steps produce a radical from starting material.
2. *Propagation.* The radical enters a series of steps that results in the formation of product and a new radical that can start the series of propagation steps over again.
3. *Termination.* Removal of radicals from the propagation steps, thus ending a chain.

Initiation has been discussed in the previous section. In order for propagation to occur, many product molecules need to be formed from just one radical produced in the initiation step, and this occurs *only if the propagation steps are exothermic*. Termination steps include coupling, disproportionation, and abstraction. Abstraction by a chain-transfer agent removes a radical from a propagation step, and a new radical is generated in its place. If this newly generated radical is sufficiently stable, the chain transfer results in termination.

The propagation steps of a chain process must add up to the overall equation for the reaction. The overall equation will not contain any radicals. This means that the radical produced in the last propagation step must be the same as a reactant in the first propagation step.

Hint 5.1

Hint 5.2   In radical reactions, the products are rarely generated by radical coupling.

Radical coupling reactions are those in which two radicals react to form a covalent bond. They are rarely a significant source of product, because both radicals are reactive intermediates and are present in extremely low concentrations. This means that the probability that they will react together is very small, and thus the rate of their reaction, which depends on their concentrations, is very low and other processes will compete effectively.

These various aspects of radical chain processes are illustrated in the following example.

Example 5.2. *Radical chain halogenation by* **t-butyl hypochlorite.**

The overall process is

$$t\text{-BuOCl} + \underset{}{\overset{}{\bigwedge}}\!\!H \xrightarrow[\text{or } h\nu]{\Delta} \underset{}{\overset{}{\bigwedge}}\!\!Cl + t\text{-BuOH}$$

This can be broken down into initiation, propagation, and termination steps.

Initiation Step

$$t\text{-BuOCl} \xrightarrow[\text{or } h\nu]{\Delta} t\text{-BuO}\cdot + \cdot Cl$$

In continuing with the mechanism, consider t-BuO· to be the chain-carrying radical. This means that t-BuO· will be used up in the first propagation step but regenerated in the last propagation step. In this case, the last propagation step is the second step.

Propagation Steps

(1) $\underset{}{\overset{}{\bigwedge}}\!\!H + t\text{-BuO}\cdot \longrightarrow t\text{-BuOH} + \underset{}{\overset{}{\bigwedge}}\!\!\cdot$

(2) $\underset{}{\overset{}{\bigwedge}}\!\!\cdot + t\text{-BuOCl} \longrightarrow \underset{}{\overset{}{\bigwedge}}\!\!Cl + t\text{-BuO}\cdot$

The radical formed in step 2 can start another reaction (step 1); that is why these processes are called chain processes. In this way, just one t-BuO· radical can initiate many sets of propagation steps. Addition of steps 1 and 2 gives the equation for the overall reaction; the radicals cancel when the addition is performed.

Is there another possible reaction pathway? Could the chlorine radical be the chain-carrying radical? If so, equation 3, an exothermic reaction, could represent one of the propagation steps.

(3)

$$\underset{}{\big\backslash}\!\!\!\!\big/\!\!-\!H + \cdot Cl \longrightarrow \underset{}{\big\backslash}\!\!\!\!\big/\!\!\cdot + HCl$$

Now a step must be found that forms the *t*-butyl chloride product *and* regenerates chlorine radical. This rules out equation 2 as a product-forming step because that reaction does not generate the same radical that is used in equation 3. Addition of equations 2 and 3 does not yield the overall reaction.

If we couple equation 3 with equation 4, we regenerate the chlorine radical but arrive at a different product for the reaction.

(4)

$$\underset{}{\big\backslash}\!\!\!\!\big/\!\!\cdot + \textit{t-BuOCl} \longrightarrow \underset{}{\big\backslash}\!\!\!\!\big/\!\!-\!\text{O-}\textit{t}\text{-Bu} + Cl \cdot$$

Another possible product-forming step would be reaction of the *t*-butyl radical with a chlorine atom. However, because this reaction removes radicals without forming new ones, it could not be one of the propagation steps of a chain process.

$$\underset{}{\big\backslash}\!\!\!\!\big/\!\!\cdot + \cdot Cl \longrightarrow \underset{}{\big\backslash}\!\!\!\!\big/\!\!-\!Cl$$

Termination Steps

(1) Disproportionation

$$\underset{}{\big\backslash}\!\!\!\!\big/\!\!\cdot + \underset{}{\big\backslash}\!\!\!\!\big/\!\!\cdot \longrightarrow \underset{}{\big\backslash}\!\!\!\!\big/ + \underset{}{\big\backslash}\!\!\!\!\big/$$

The disproportionation process involves the abstraction of hydrogen by one radical from another radical. Abstraction of a hydrogen atom from the carbon adjacent to the radical produces a double bond:

$$\underset{}{\big\backslash}\!\!\!\!\big/\!\!\cdot \quad H\!\!-\!\!\cdot\!\!\big/ \longrightarrow \underset{}{\big\backslash}\!\!\!\!\big/\!\!-\!H + \underset{}{\big\backslash}\!\!\!\!\big/$$

(2) Radical Coupling

Some possible coupling reactions follow:

$$\underset{}{\big\backslash}\!\!\!\!\big/\!\!\cdot + \cdot Cl \longrightarrow \underset{}{\big\backslash}\!\!\!\!\big/\!\!-\!Cl$$

$$\textit{t}\text{-BuO} \cdot + \textit{t}\text{-BuO} \cdot \longrightarrow \textit{t}\text{-BuOO-}\textit{t}\text{-Bu}$$

$$\underset{}{\big\backslash}\!\!\!\!\big/\!\!\cdot + \underset{}{\big\backslash}\!\!\!\!\big/\!\!\cdot \longrightarrow \underset{}{\big|}\!\!-\!\!\underset{}{\big|}$$

The following radical coupling does not affect the chain process because chlorine atoms are not involved in chain propagation.

$$\cdot Cl + \cdot Cl \longrightarrow Cl—Cl$$

PROBLEM 5.2   **What other radical coupling reactions are possible for the reaction of t-butyl hypochlorite in Example 5.2?**

4. RADICAL INHIBITORS

Radical reactions can be slowed or stopped by the presence of substances called inhibitors.

Hint 5.3   A reaction that is slowed by the addition of a free radical inhibitor can be assumed to proceed by a radical mechanism.

Common radical inhibitors include 2,6-di-*t*-butyl-4-methylphenol (BHT), 2-nitroso-2-propane, and oxygen.

The acronym BHT stands for butylated hydroxytoluene, the common name for 2,6-di-*t*-butyl-4-methylphenol. It is used widely as an antioxidant in foodstuffs and food packaging. The reaction of oxygen with unsaturated fats gives, after several steps, alkylperoxy radicals ($ROO\cdot$) that, upon further reaction, give smaller odiferous molecules that can ruin the palatability of foods. BHT acts as a scavenger for alkylperoxy radicals, $ROO\cdot$, because hydrogen abstraction by $ROO\cdot$ gives the stable free radical, **5-4**. The hydroperoxide ROOH, which also is formed in the reaction, is much less reactive than $ROO\cdot$ and consequently causes much less oxidation. The new radical, **5-4**, which is formed from BHT, is relatively unreactive for two reasons: (i) it is stabilized by resonance, and (ii) the groups attached to the ring sterically hinder further reaction.

5-4

Radicals also can be trapped by addition to the nitroso group to form nitroxide radicals. In the following equation, diacetyl peroxide is a source of methyl radicals, which are trapped by the nitroso compound. (For further discussion of the fragmentation of radicals, see Section 7.)

Oxygen can act as an inhibitor by reacting with radicals to produce less reactive hydroperoxyl radicals.

$$R\cdot + O_2 \longrightarrow ROO\cdot$$

Two compounds that effectively inhibit electron-transfer reactions are di-*t*-butyl nitroxide and 1,4-dinitrotoluene: *Benzene*

Transfer of an electron to 1,4-dinitrobenzene gives a radical anion so stable that it is unlikely to transfer an electron to anything else. The reaction of benzene radical anion with 1,4-dinitrobenzene proceeds to the right, because the radical anion of the 1,4-dinitrobenzene is much more stable than the radical anion of benzene itself.

Although 1,3-dinitrobenzene sometimes is used for the inhibition of electron-transfer reactions, it is not as effective as the 1,4-isomer, because its radical anion is not as stable (see Problem 1.5).

Nitroxides are also used to remove radical anions from a reaction sequence. For example, di-*t*-butyl nitroxide has often been used to inhibit the $S_{RN}1$ reaction (see Section 10). Apparently, nitroxides remove radicals via coupling reactions (Hoffmann, A. K.; Feldman, A. M.; Gelblum, E.; Hodgson, W. G. *J. Am. Chem. Soc.* **1964**, *86*, 639–646).

Hint 5.4   A radical reaction that is initiated only by light can be prevented by omitting light. In this case, any reaction that takes place in the dark must proceed by a nonradical pathway.

---

5. DETERMINING THE THERMODYNAMIC FEASIBILITY OF RADICAL REACTIONS

The bond dissociation energies in Table 5.1 can be used to determine the feasibility of radical reactions. In general, a radical process will give reasonable synthetic yields only if all propagation steps are exothermic, so that many propagation steps can occur before termination. The number of propagation steps per initiation step is the *chain length* of the reaction. Highly exothermic propagation steps involve highly reactive radicals, like alkyl radicals, and have a long chain length. More stable radicals, like aryl radicals, may fail to react until they encounter another radical. In this case, there is no chain reaction.

In Example 5.3, calculations show that the radical chlorination of 2-methylpropane by *t*-butyl hypochlorite is an energetically feasible process for the synthesis of *t*-butyl chloride.

Example 5.3.   *Determination of the enthalpy change in chlorination by* **t-*butyl hypochlorite.***

Consider the reactions shown in Example 5.2. If both propagation steps are exothermic, the chain length will be long enough that the thermochemistry of the initiation step(s) is unimportant relative to that of the propagation steps.

We will calculate the enthalpy change for the reactions in the propagation steps. In equation 1, the C—H bond broken has a bond dissociation energy (BDE) of 91.0 kcal/mol, and the O—H bond formed has a BDE of 103 kcal/mol. Therefore, this first reaction is exothermic by 91 − 103 or −12 kcal/mol.

$$(1) \quad \text{\ce{}}\overset{H}{\diagup} + t\text{-BuO}\cdot \longrightarrow t\text{-BuOH} + \diagup\cdot \qquad \Delta H = -12 \text{ kcal/mol}$$

TABLE 5.1 Bond Dissociation Energies (BDEs, kcal/mol) at 298 K[a]

		$Cl-Cl^b$	58	$Br-Br^b$	46	$I-I^b$ 36	$F-F^b$	37
		$H-Cl^b$	103	$H-Br^b$	87	$H-I^b$ 36	$H-F^b$	135
CH_3-H	104	CH_3-Cl	73	CH_3-Br	70			
C_2H_5-H	98	C_2H_5-Cl	81	C_2H_5-Br	69	C_2H_5-I 53	C_2H_5-F	106
$CH_3CH_2CH_2-H$	98	$n\text{-}C_3H_7-Cl$	82	$n\text{-}C_3H_7-Br$	69			
$(CH_3)_2CH-H$	94.5	$(CH_3)_2CH-Cl$	81	$(CH_3)_2CH-Br$	68			
$(CH_3)_3C-H$	91.0	$(CH_3)_3C-Cl$	79	$(CH_3)_3C-Br$	63			
$PhCH_2-H$	85	$PhCH_2-Cl$	68	$PhCH_2-Br$	51			
$CH_2=CHCH_2-H$	85							
CCl_3-H	95.7	CCl_3-Cl	73	CCl_3-Br	54			
$Ph-H$	104			$Ph-Br$	71			
$CH_2=CH-H$	104							
$HOCH_2-H$	92							
CH_3O-H	102							
C_2H_5O-H	102							
$i\text{-}C_3H_7O-H$	103							
$t\text{-}C_4H_9O-H$	103	$t\text{-}C_4H_9O-Cl$	44^c					
CH_3COO-H	112							
CH_3S-H	88							
$PhS-H$	75							
CH_3-CH_3	88							
$PhCO-H$	74	$PhCO-Cl$	74					
$HCO-H$	88							
CH_3CO-H	88							
CH_3COCH_2-H	92							

[handwritten:] $t\text{-}BuOCl \rightarrow t\text{-}BuO\bullet + Cl\bullet$ 4

[a]All values from Kerr, J. A. *Chem. Rev.* **1966**, *66*, 465–500, unless otherwise noted.
[b]*Handbook of Chemistry and Physics*, 63rd ed.; Weast, R. C.; Astle, M. J., Eds.; CRC Press: Boca Raton, FL, 1982–1983; F186ff.
[c]Walling, C.; Jacknow, B. B. *J. Am. Chem. Soc.* **1960**, *82*, 6108–6112.

For equation 2, the O—Cl bond broken has a BDE of 44 kcal/mol and the C—Cl bond formed has a BDE of 79 kcal/mol. Thus, the second reaction is exothermic by 44 − 79 kcal/mol or −35 kcal/mol. Because both reactions are substantially exothermic.

(2) $\quad$ [structure] $+ t\text{-BuOCl} \longrightarrow$ [structure with Cl] $+ t\text{-BuO}\cdot \qquad \Delta H = -25$ kcal/mol

this should be a highly favorable process with a long chain length. The enthalpy change for the overall reaction is $[(-12) + (-35)]$ or −47 kcal/mol.

What are the defects in the following mechanism for the chlorination of 2-methylpropane by *t*-butyl hypochlorite? You will need to consider the   **PROBLEM 5.3**

294

PROBLEM 5.3
continued

thermochemistry of the processes, as well as the fact that a chain process is not involved.

$$t\text{-BuOCl} \xrightarrow{\Delta} t\text{-BuO}\cdot + \cdot\text{Cl}$$

$$\text{+ }t\text{-BuO}\cdot \longrightarrow t\text{-BuOH +}$$

6. ADDITION OF RADICALS

Radical addition reactions are commonly used in organic synthesis. These additions range from the simple addition of halocarbons to π bonds to cyclization reactions with demanding stereoelectronic requirements.

A. Intermolecular Radical Addition

Hint 5.5

Common types of radicals that add to π bonds are those that can be generated from alkyl halides, mercaptans, thiophenols, thioacids, aldehydes, and ketones. Like the corresponding electrophilic additions to double bonds, many radical additions are either regiospecific or highly regioselective.

Example 5.4. *Photochemical addition of trifluoroiodomethane to allyl alcohol.*

In the initation step of this reaction, light induces homolytic cleavage of the weak C—I bond.

$$\text{CF}_3\text{I} \xrightarrow{h\nu} \cdot\text{CF}_3 + \cdot\text{I}$$

The trifluoromethyl radical adds to the double bond, giving the most stable intermediate radical:

$$\text{F}_3\text{C}\cdot \quad \text{CH}_2\!\!=\!\!\text{CHCH}_2\text{OH} \longrightarrow \text{F}_3\text{CCH}_2\dot{\text{C}}\text{HCH}_2\text{OH}$$

This radical then abstracts an iodide atom from trifluoroiodomethane to generate product and the chain-propagating radical, the trifluoromethyl radical.

$$F_3CCH_2\dot{C}HCH_2OH + ICF_3 \longrightarrow F_3CCH_2CHICH_2OH + \cdot CF_3$$

The product shown is the major regioisomer produced. For further details on this reaction, see Park, J. D.; Rogers, F. E.; Lacher, J. R. *J. Org. Chem.* **1961**, *26*, 2089–2095.

Examine the mechanistic steps for the reaction shown.   **PROBLEM 5.4**

a. For each step, use half-headed arrows to show the movement of electrons, and dots to indicate all the unpaired electrons.

$$CH_2{=}CHC_6H_{13} + CBr_4 \xrightarrow[\Delta]{(PhCO_2)_2} Br_3CCH_2CHBrC_6H_{13}$$

Step 1

Step 2

$$PhC{-}O{-}O{-}CPh \longrightarrow 2Ph\overset{O}{C}{-}O\bullet \qquad \text{init}$$

Step 3

$$Ph\overset{O}{C}{-}\dot{O} + Br{-}CBr_3 \longrightarrow Ph\overset{O}{C}{-}O{-}Br + \bullet CBr_3$$

Step 4

$$\bullet CBr_3 + CH_2{=}CHC_6H_{13} \longrightarrow Br_3CCH_2\dot{C}HC_6H_{13} \qquad Prop$$

$$Br_3CCH_2\dot{C}HC_6H_{13} + Br{-}CBr_3 \longrightarrow Br_3CCH_2CHBrC_6H_{13} + \dot{C}Br_3$$

b. Identify the initation and propagation steps.
c. Why is the addition in step 3 regiospecific? (→ most stble radical)

Example 5.5. *Addition of a radical formed by reduction of a C—Hg bond.*

Addition of radicals formed from mercury compounds to alkenes often produce good-to-excellent yields. The following mechanism illustrates a reaction with a yield of 64%.

Note: Cbz = $-CO_2CH_2$Ph.

Danishefsky, S.; Taniyama, E.; Webb, R. R., II *Tetrahedron Lett.* **1983**, *24*, 11–14.

B. Intramolecular Radical Addition: Radical Cyclization Reactions

Generation of a radical in a molecule that contains a site of unsaturation presents an opportunity for cyclization. These types of radical cyclizations have been developed into very useful synthetic reactions. The synthetic utility of these cyclizations is enhanced by the ability to predict the regiochemistry of cyclization by applying the Baldwin rules. The Baldwin rules cover the formation of three- to seven-membered rings by various reactions and are based on consideration of both kinetic and thermodynamic factors. (See Beckwith, A. L. J.; Easton, D. J.; Serelis, A. K. *J. Chem. Soc., Chem. Commun.* **1980**, 482–483, and Baldwin, J. E. *J. Chem. Soc., Chem. Commun.* **1976**, 734–736.)

The Baldwin rules distinguish two types of ring closure, *exo* and *endo*.

exo adduct   *endo* adduct

With regard to radical cyclization, the most important guidelines are as follows:

1. For unsubstituted ω-alkenyl radicals containing up to eight carbons, with the general structure previously shown, the preferred mode of cyclization is *exo*. (The symbol ω means that the alkene is at the terminus distant from the radical.) The reaction is controlled kinetically; the major product is the one that forms faster. The rates of formation of the two possible adducts are a result of the stereochemical requirements of their transition states. Often the *endo* product is more stable than the *exo* product. When this is the case, however, the more stable product is not the major product formed.

2. If the alkene is substituted at the nonterminal position, reaction to form the *exo* product is sterically hindered and the percentage of *endo* product increases. With substituted alkenes, the more stable product may predominate.

Example 5.6. *exo and endo ring closure in radical cyclization.*

exo adduct

exo adduct   *endo* adduct

Julia, M.; Descoins, C.; Baillarge, M.; Jacquet, B.; Uguen, D., Graeger, F. A. *Tetrahedron* **1975**, *31*, 1737–1745.

Example 5.7. *Intramolecular cyclization of a vinyl radical.*

The overall reaction is as follows:

5-5   **5-6**

Stork, G.; Baine, N. H. *J. Am. Chem. Soc.* **1982**, *104*, 2321–2323.

The vinyl radical, **5-7**, is formed by the process shown in Section 2. Cyclization of **5-7** can give an *exo* adduct or an *endo* adduct. The resulting radical abstracts hydrogen from *n*-Bu$_3$SnH to form products, **5-5** and **5-6**, as well as tri-*n*-butyltin radical, which continues the chain. As is predicted by the guidelines, the *exo* product (**5-5**) predominates, in this case by a factor of 2.

5-7

+ *n*-Bu$_3$Sn·

PROBLEM 5.5   **Consider the following reaction:**

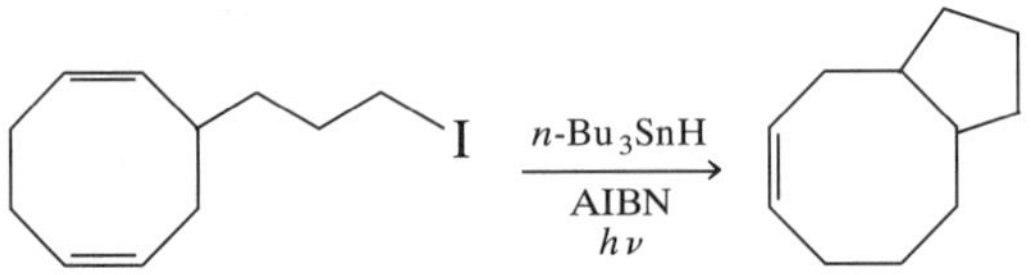

a. Write the initiation and propagation steps.
b. The following structure represents a minor product isolated from the reaction mixture. Show how it might have been formed.

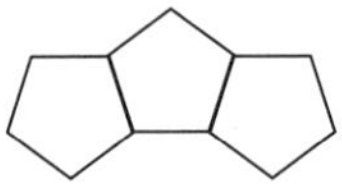

Winkler, J. D.; Sridar, V. *J. Am. Chem. Soc.* **1986**, *108*, 1708–1709.

PROBLEM 5.6   **Write the initiation and propagation steps for the following reaction:**

37%

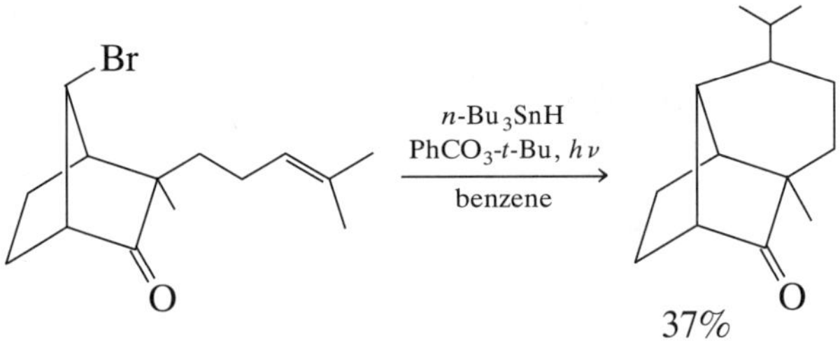

Hart, D. J. *Science* **1984**, *223*, 883–887.

PROBLEM 5.7   **Propose a mechanism for the following transformation.**

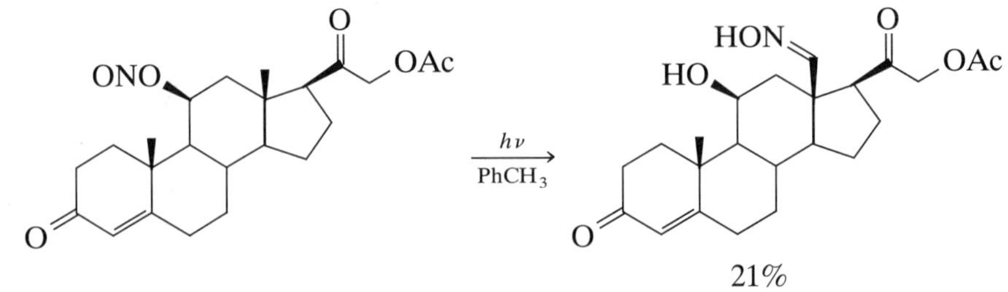

21%

Barton, D. H. R.; Beaton, G. M.; Geller, L. E.; Bechet, M. M. *J. Am. Chem. Soc.* **1960**, *82*, 2640.

7. FRAGMENTATION REACTIONS

Many radical processes involve the loss of small, stable molecules, such as carbon dioxide, nitrogen, or carbon monoxide. These kinds of reactions are called fragmentations.

Hint 5.6

A. Loss of CO_2

The radical initiator, diacetyl peroxide, homolyzes at the $O-O$ bond to form carboxy radicals, which then readily lose CO_2 to give methyl radicals. In fact, between 60°C and 100°C, acetyl peroxide can be a convenient source of methyl radicals.

(See Walling, C. *Free Radicals in Solution*; John Wiley: New York, 1957; p. 493.) Aryl-substituted carboxy radicals also lose CO_2, but they do so much less readily (e.g., the initial radical formed from benzoyl peroxide, Section 2). Fragmentation with loss of CO_2 also occurs in the Hunsdiecker reaction, in which a silver salt of a carboxylic acid reacts with bromine to produce an alkyl halide. The reaction results in shortening of the carbon chain by one carbon. The overall reaction is as follows:

$$RCO_2Ag + Br_2 \xrightarrow{\Delta} RX + CO_2 + AgBr$$

The mechanism is

$$RCO_2Ag + Br_2 \longrightarrow RCO_2Br + AgBr$$

Initiation:

$$RCO_2Br \longrightarrow RCO_2\cdot + Br\cdot$$

Propagation 1:

$$RCO_2\cdot \longrightarrow R\cdot + CO_2$$

Propagation 2:

$$R\cdot + RCO_2Br \longrightarrow RBr + RCO_2\cdot$$

B. Loss of a Ketone

The radical initially produced by homolytic decomposition of a dialkyl peroxide can undergo further scission. The rate of scission depends on the temperature and the stability of the resulting radical(s). For example, *t*-butoxy radicals decompose on heating to methyl radicals and acetone.

$$t\text{-BuO}\cdot \longrightarrow \cdot CH_3 + CH_3COCH_3$$

C. Loss of N$_2$

Azo compounds often decompose with loss of nitrogen. The decomposition of the initiator, AIBN, is an example (see Section 2).

D. Loss of CO

At elevated temperatures, the radical, **5-8**, generated by hydrogen abstraction from the aldehyde functional group fragments to give a new carbon radical and carbon monoxide. The fragmentation is temperature-dependent; the higher the temperature, the more favorable the loss of CO from the initial radical.

In the presence of a suitable compound, hydrogen abstraction may compete with loss of CO, as in the following example.

Example 5.8. ***Addition followed by fragmentation.***

The radical reaction of carbon tetrachloride with aliphatic double bonds involves addition of the trichloromethyl radical to the double bond, followed by chlorine atom abstraction from carbon tetrachloride by the intermediate radical to give the product. After the addition of the trichloromethyl radical to β-pinene, a fragmentation occurs prior to formation of the product.

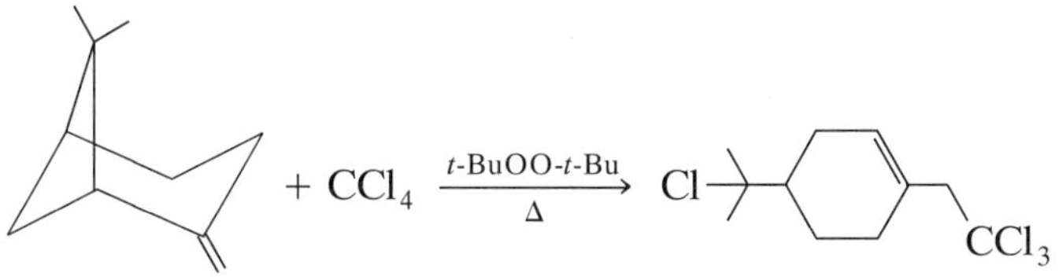

The mechanism for this reaction starts with the generation of the trichloromethyl radical:

$$t\text{-BuOO-}t\text{-Bu} \xrightarrow{\Delta} 2t\text{-BuO}\cdot$$

$$t\text{-BuO}\cdot \quad Cl{-}CCl_3 \longrightarrow t\text{-BuOCl} + \cdot CCl_3$$

The trichloromethyl radical now adds regiospecifically to the double bond to form a new carbon radical, **5-9**:

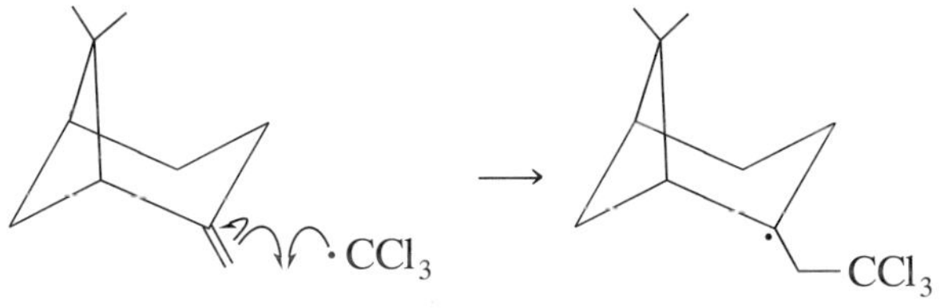

5-9

This radical can now fragment to give another radical, **5-10**, which then abstracts a chlorine atom from another molecule of CCl_4 to give the product:

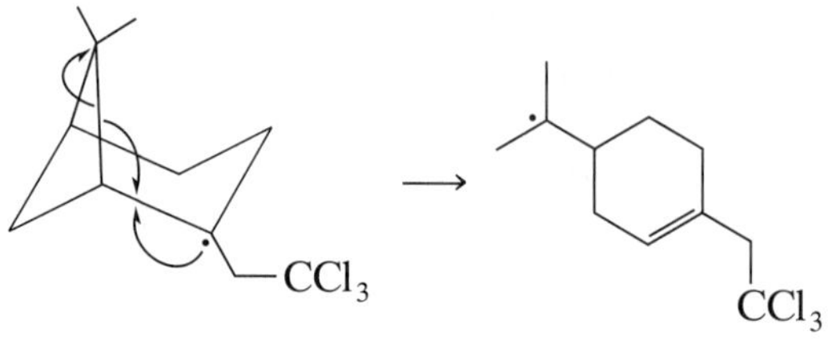

5-10

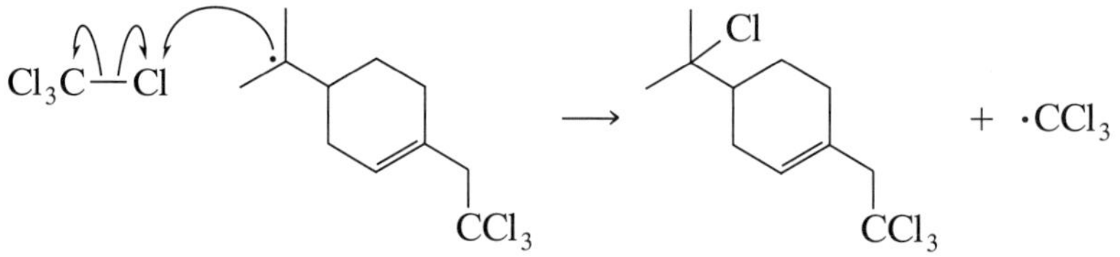

The radical addition of thiolacetic acid to β-pinene gives unrearranged product. This result is evidence for the discrete existence of radical **5-9**. That is, the rate of abstraction of a hydrogen atom from thiolacetic acid by **5-11** is faster than its rate of fragmentation. Note that the product is a mixture of stereoisomers, as indicated by the wavy bond lines.

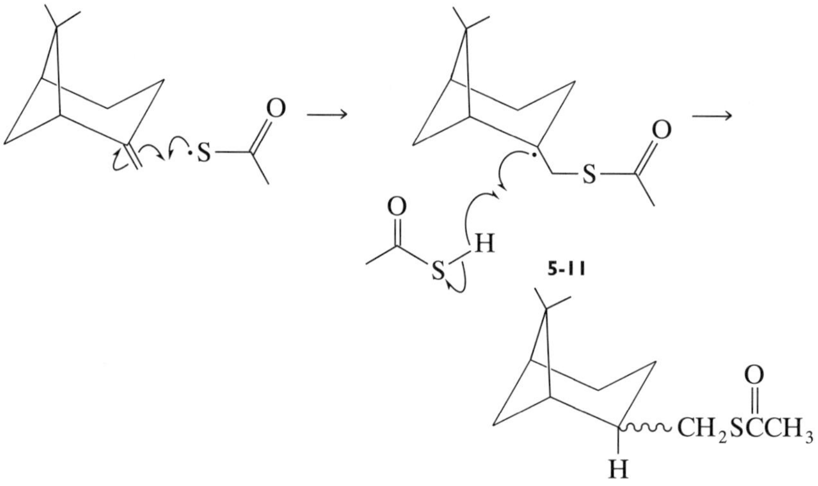

5-11

(See Claisse, J. A.; Davies, D. I.; Parfitt, L. T. *J. Chem. Soc. C* **1970**, 258–262.)

Write step-by-step mechanisms for the following reactions.   **PROBLEM 5.8**

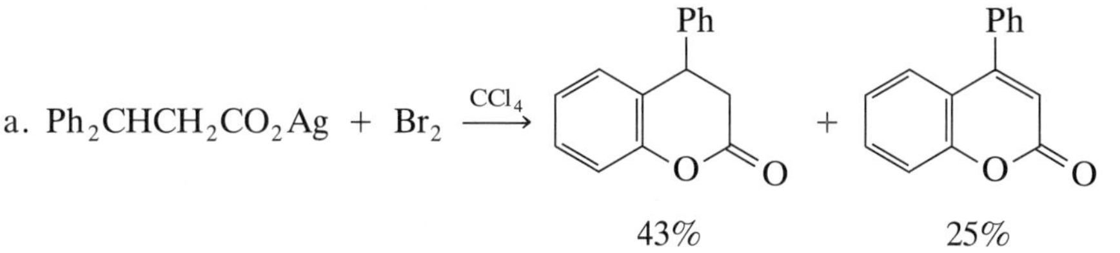

a. $Ph_2CHCH_2CO_2Ag + Br_2 \xrightarrow{CCl_4}$

43%   25%

5-12   5-13

Pandet, U. K.; Dirk, I. P. *Tetrahedron Lett.* **1963**, 891–895.

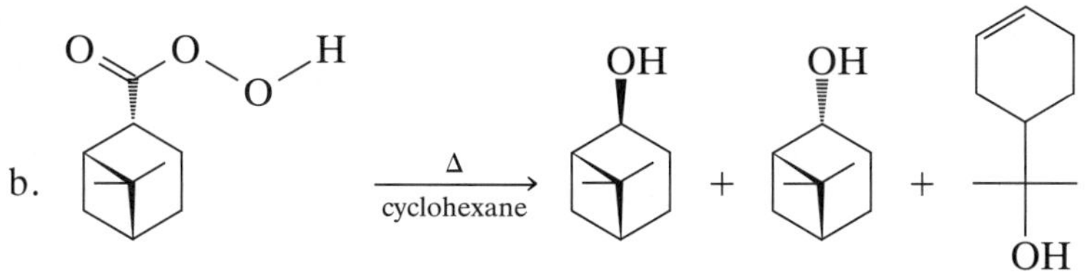

b.

5-14   5-15   5-16   5-17

In writing your mechanism, take into account the following experimental observations: (i) the other isomer (with the peracid group up) of the starting material reacts to give roughly the same ratio of **5-15** and **5-16**, and (ii) the ratio of bicyclic to ring-opened products increases with increasing peracid concentration.

Fossey, J.; Lefort, D.; Sorba, J. *J. Org. Chem.* **1986**, *51*, 3584–3587.

8. REARRANGEMENT OF RADICALS

Rearrangements of radicals are much less common than rearrangements of carbocations.

In radical rearrangements, the migrating groups are those that can accommodate electrons in a π system (vinyl, aryl, carbonyl) or atoms that can expand their valence shell, i.e., all halogens but fluorine. *Hydrogen and alkyl do not migrate to*   *Hint 5.7* radicals. However, an addition – elimination pathway could give the appearance of alkyl migration (see Example 5.11). For a discussion of the reasons why alkyl and hydrogen do not migrate, see Carey, F. A.; Sundberg, R. J. *Advanced Organic Chemistry*, 3rd ed.; Plenum Press: New York, 1990; p. 704.

Example 5.9.   *Aryl migration.*

In the following reaction, **5-19** is a product resulting from migration of an aryl group, and **5-18** is a nonrearranged product. Thus, there is competition between rearrangement and hydrogen abstraction by a radical intermediate.

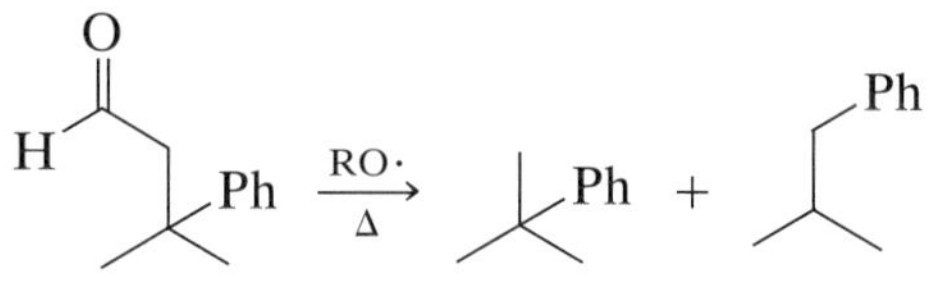

The mechanism for formation of **5-18** and **5-19** involves formation of a hydrocarbon radical from the starting aldehyde. The radical from the initiator abstracts the aldehyde proton to give a carbonyl radical. This loses carbon monoxide to give **5-20**.

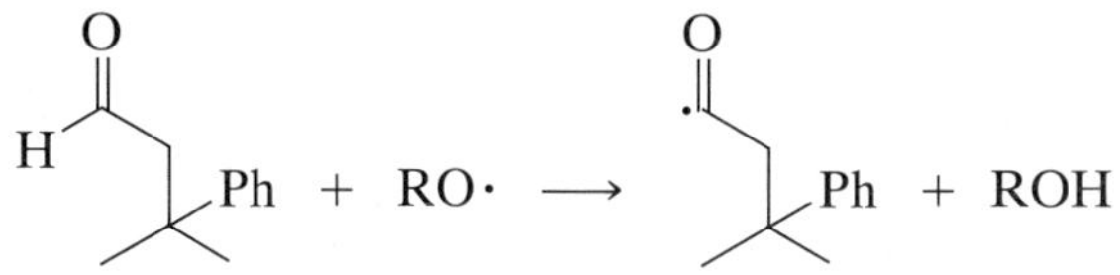

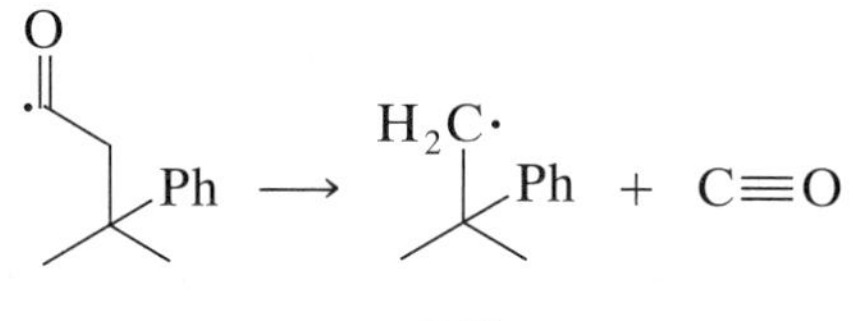

Radical **5-20** can either abstract a hydrogen atom from the starting aldehyde to give nonrearranged product, **5-18**, or rearrange via phenyl migration to **5-21**, which then abstracts a hydrogen atom to give **5-19**.

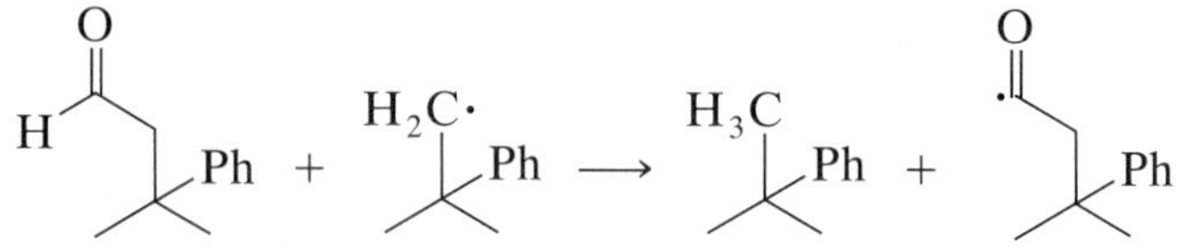

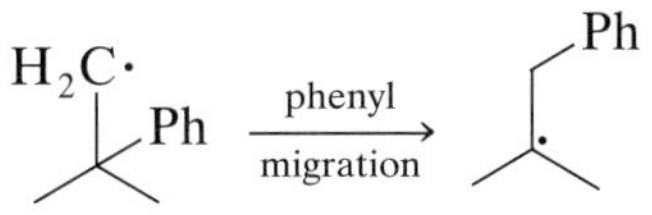

5-20 **5-21**

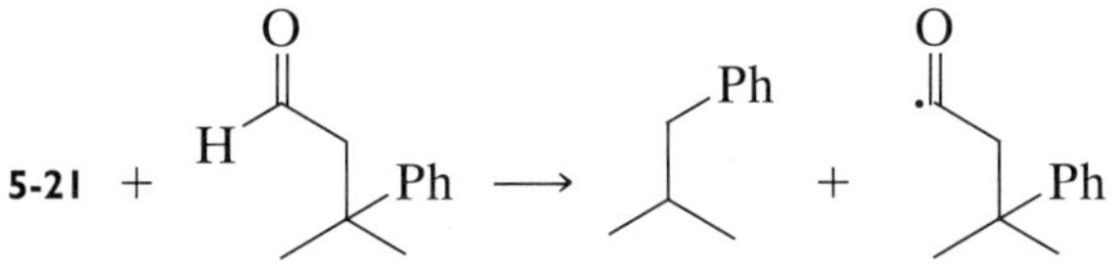

5-21 + **5-19**

Example 5.10. *Halogen migration.*

The radical addition of HBr to 3,3,3-trichloropropene, involves migration of a chlorine atom in an intermediate step.

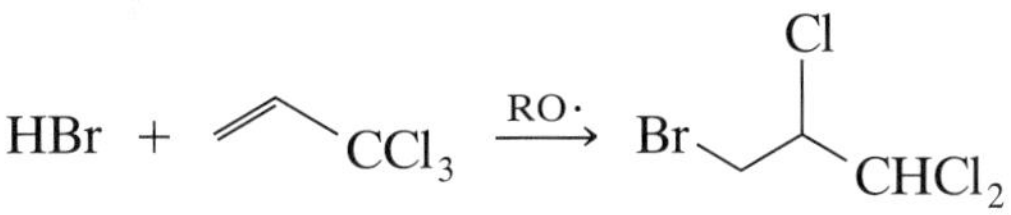

The mechanism for the reaction involves a regiospecific addition of bromine radical to the double bond.

$$RO\cdot \; + \; HBr \longrightarrow ROH + Br\cdot$$

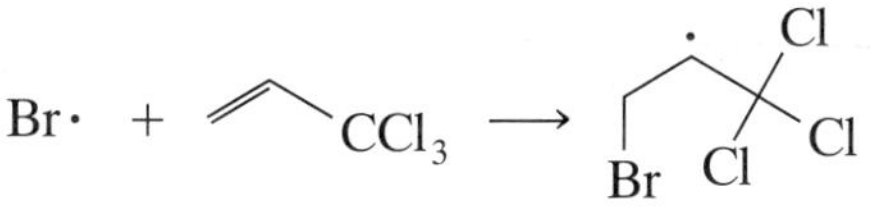

A chlorine atom then migrates to the adjacent radical, giving **5-22**. Finally, **5-22** abstracts a hydrogen from HBr to form the product plus another bromine radical to continue the chain.

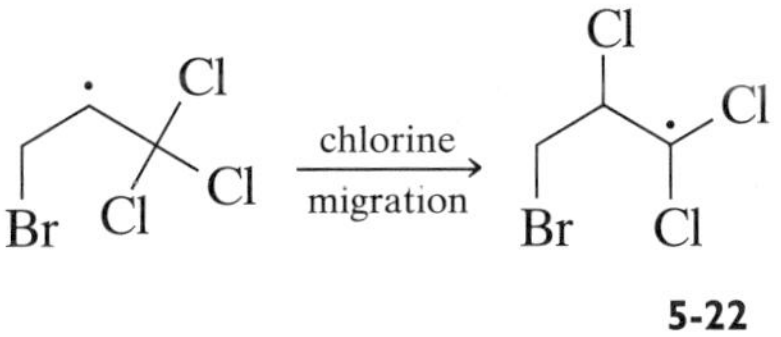

5-22

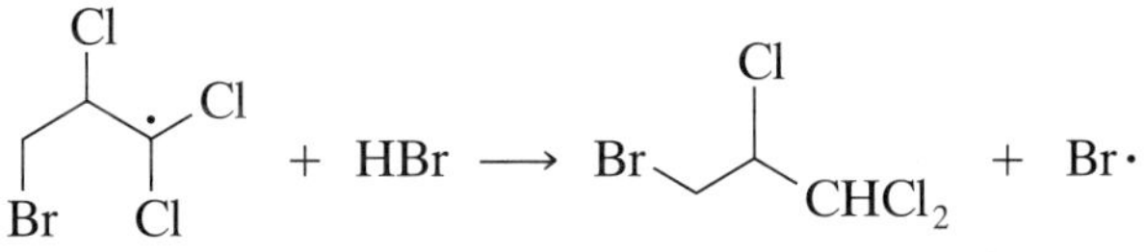

5-22

Example 5.11.   *An apparent acyl migration mediated by radical addition to a carbonyl group.*

The overall reaction is as follows:

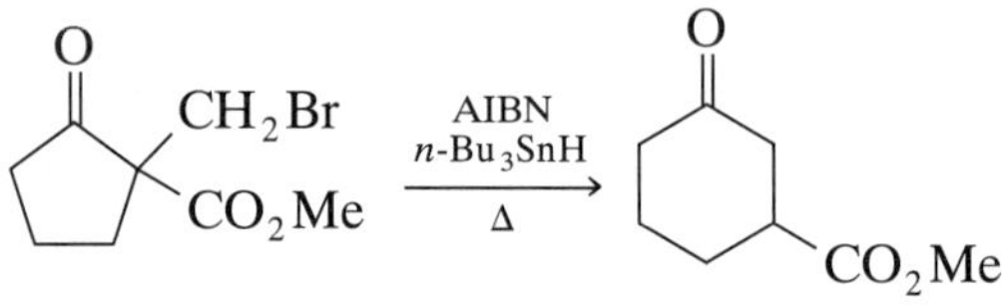

In this reaction, the tri-*n*-butyltin radical, formed by the usual initiation steps, removes the bromine atom from the starting material to give **5-23**. This radical then adds to the carbonyl carbon, forming a three-membered ring that then opens at a different bond to give **5-24**. On the basis of Hint 5.7, we would not expect direct formation of **5-24** by migration of carbon C-1 in **5-23** to the radical center.

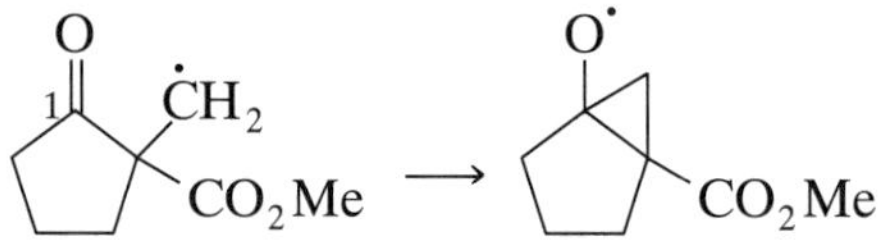

5-23

Radical **5-24** can abstract a hydrogen atom from tri-*n*-butyltin hydride to give the product and a new radical to propagate the chain:

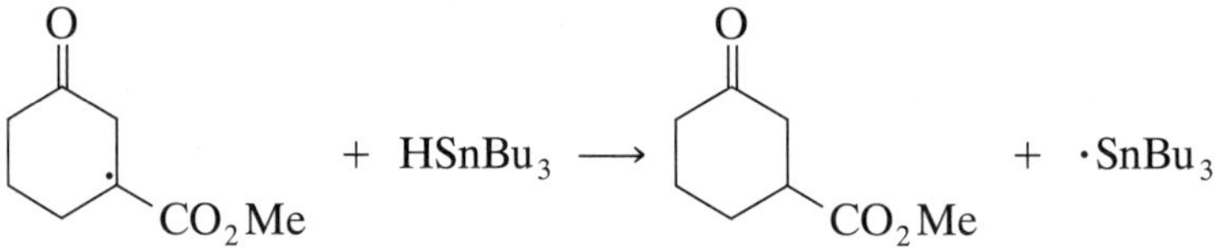

5-24

Dowd, P.; Choi, S.-C. *J. Am. Chem. Soc.* **1987**, *109*, 3493–3494.

PROBLEM 5.9   **Compare the two reactions shown and explain why one process gives rearrangement and the other does not. Write step-by-step mechanisms for both processes.**

PROBLEM 5.9
continued

Weinstock, J.; Lewis, S. N. *J. Am. Chem. Soc.* **1957**, *79*, 6243–6247.

Write a complete mechanism for the following process.   **PROBLEM 5.10**

Beckwith, A. L. J.; O'Shea, D. M.; Westwood, S. W. *J. Am. Chem. Soc.* **1988**, *110*, 2565–2575.

9. THE S$_{RN}$I REACTION

The S$_{RN}$1 reaction, a versatile synthetic tool, is initiated by generation of a radical anion. The designation S$_{RN}$1 indicates that the reaction is a nucleophilic substitution proceeding through a radical intermediate and that the rate-limiting step is unimolecular decay of the radical anion intermediate formed from the substrate (see a review by Bunnett, J. F. *Acc. Chem. Res.* **1978**, *11*, 413–420). S$_{RN}$1 reactions occur with both aliphatic and aromatic compounds.

For the general case, the propagation steps for the S$_{RN}$1 reaction can be represented as follows:

$$[RX]^{\cdot -} \longrightarrow R\cdot + X^-$$
$$R\cdot + Y^- \longrightarrow [RY]^{\cdot -}$$
$$[RY]^{\cdot -} + RX \longrightarrow RY + [RX]^{\cdot -}$$

Note that the radical anion consumed in the first propagation step is regenerated in the third propagation step.

S$_{RN}$1 reactions can be initiated by photochemical excitation, electrochemical reduction, and solvated electrons (alkali metal in ammonia). In some cases, spontaneous thermal initiation can also take place. The leaving group, X$^-$, is often a halide—frequently bromide or iodide, never fluoride. The nucleophile, Y$^-$, is commonly a nitroalkane anion (**5-25**) or another anion such as thiolate (RS$^-$), phenolate (PhO$^-$), or various enolates.

5-25

Because these are free radical reactions, they are inhibited by the addition of free radical inhibitors such as di-*t*-butyl nitroxide and 1,4-dinitrobenzene.

Addition of the propagation steps of the general mechanism for an S$_{RN}$1 reaction gives the following equation for the overall reaction:

$$RX + Y^- \longrightarrow RY + X^-$$

This equation also describes the overall reaction of either an S$_N$2 or a nucleophilic aromatic substitution process. In some cases, the only way to distinguish an S$_{RN}$1 reaction from these processes is that an S$_{RN}$1 is inhibited by radical inhibitors. Another distinguishing feature is that the order of the relative leaving group abilities of halides are opposite that found for nucleophilic aromatic substitution by the addition–elimination mechanism (see Chapter 3).

Example 5.12.   *Reaction of an enolate with an aromatic iodide.*

The overall reaction is as follows:

70%

The mechanism involves photochemical excitation of the enolate. (An excited state is indicated by an asterisk, *.) The excited enolate transfers an electron to the aromatic π system. Because a single electron has been added, there must be an unpaired electron present; thus, the new intermediate, **5-26**, is a radical. Because an electron has been added to a neutral system, there must be a negative charge. Thus, **5-26** is both an anion and a radical or what is usually called a radical anion or an anion radical.

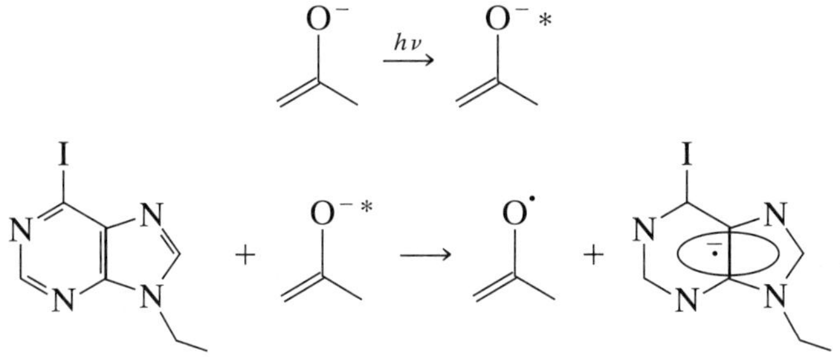

5-26

The radical anion now loses iodide ion to give a radical.

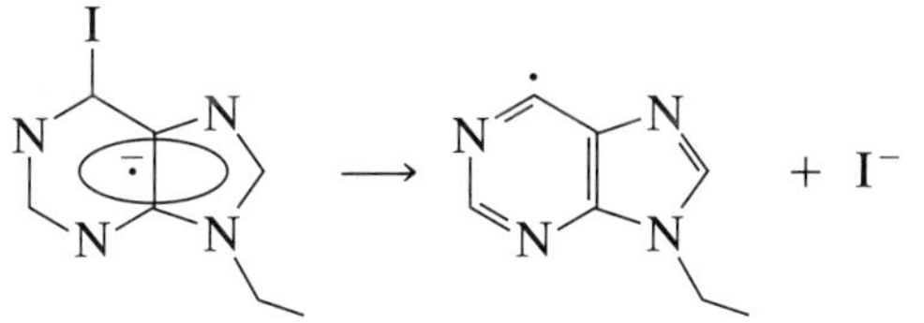

This radical couples with the enolate to give a new radical anion, **5-27**.

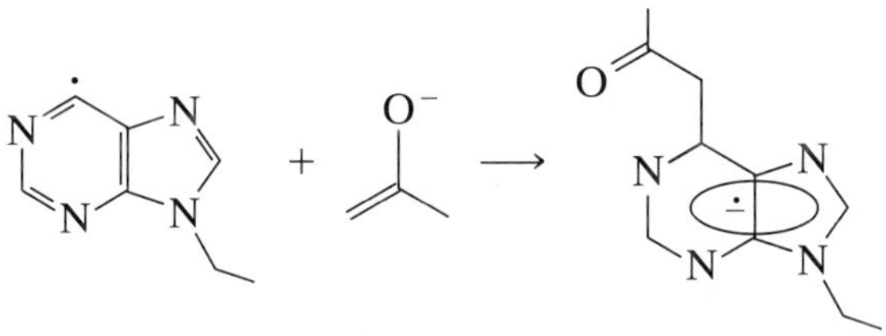

5-27

Intermediate **5-27** transfers an electron to starting material to give a molecule of product and a new molecule of **5-26** to propagate the chain.

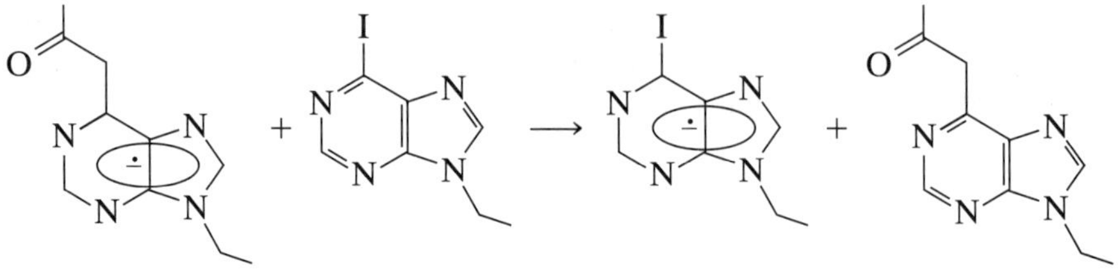

5-27   **5-26**

Nair, V.; Chamberlain, S. D. *J. Am. Chem. Soc.* **1985**, *107*, 2183–2185.

PROBLEM 5.11   **Write step-by-step mechanisms for the following reactions:**

a.

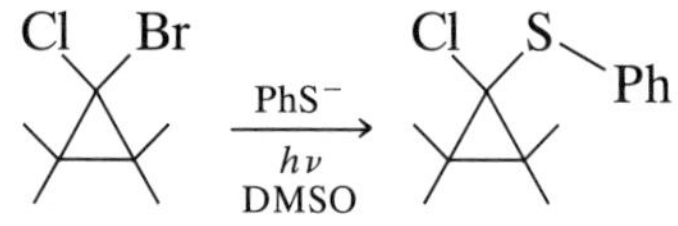

Meijs, G. F. *J. Org. Chem.* **1986**, *51*, 606–611.

b.

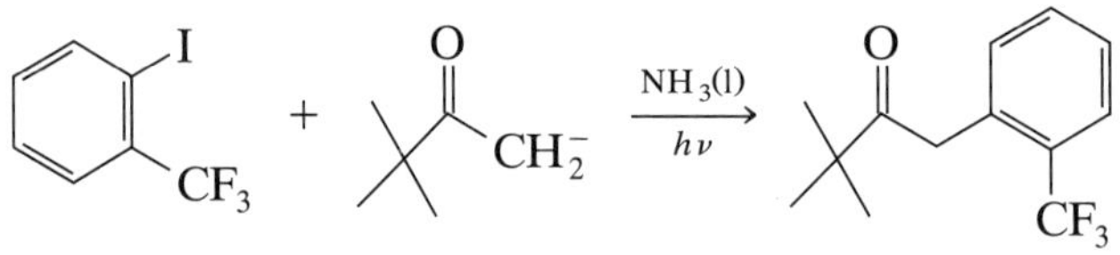

Bunnett, J. F.; Galli, C. *J. Chem. Soc., Perkin Trans. I* **1985**, 2515–2519.

---

10. THE BIRCH REDUCTION

The typical conditions for the Birch reduction are sodium in liquid ammonia that contains a small amount of ethanol. Workup generally is in acid.

Example 5.13. *Birch reduction of benzoic acid.*

90%

Under the basic reaction conditions, the carboxylate ion is formed from benzoic acid. Sodium then transfers a single electron to the aromatic system to produce a pentadienyl radical anion, **5-28**. Protonation of **5-28** gives the radical, **5-29**, which will be reduced to the corresponding anion, **5-30**, by another sodium atom. Finally, **5-30** is protonated, giving the product.

Molecular orbital (MO) calculations indicate that the intermediate radical anions have more anionic character at the positions *ortho* or *meta* to a π-donating substituent and *para* or *ipso* (*ipso* means at the substituent-containing carbon) to a π-accepting substituent. (See Birch, A. J.; Hinde, A. L.; Radom, L. *J. Am. Chem. Soc.* **1980**, *102*, 3370–3376). The pentadienyl anions formed after transfer of the second electron have more anionic character on the central carbon, and that is where the second protonation occurs.

The result of the various factors is that in the product 1,4-cyclohexadienes, π-withdrawing groups (e.g., $-CO_2H$) are found on the saturated carbons, whereas π-donating groups (e.g., $-OCH_3$) are found on the unsaturated carbons.

PROBLEM 5.12   **Write a mechanism for the following reaction that is consistent with the regiochemistry observed.**

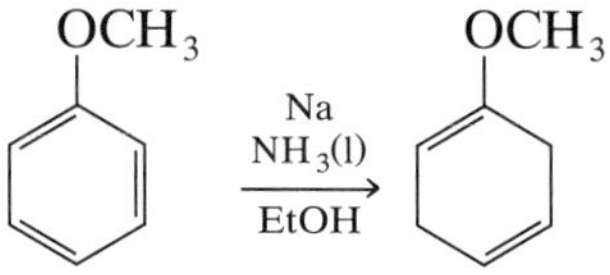

I I. A RADICAL MECHANISM FOR THE REARRANGEMENT OF SOME ANIONS

In the Wittig rearrangement, an anion derived from an ether rearranges to the salt of an alcohol.

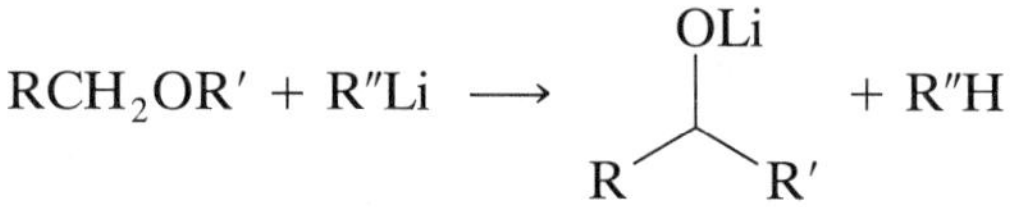

In some cases, the mechanism for the formation of at least some of the product(s) of reaction appears to be radical scission of the anion produced initially.

The radicals can now recombine to form the product:

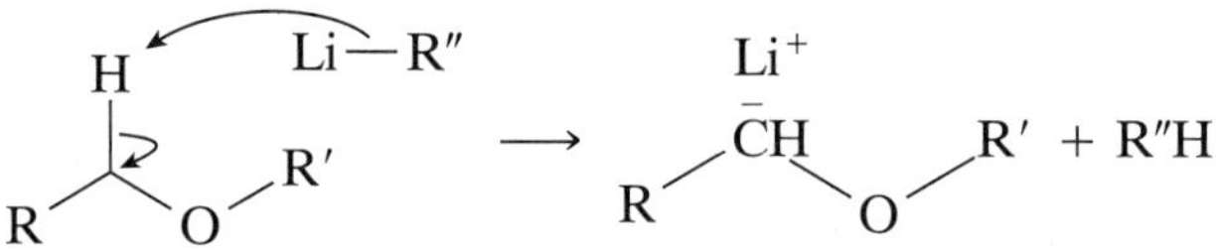

5-31

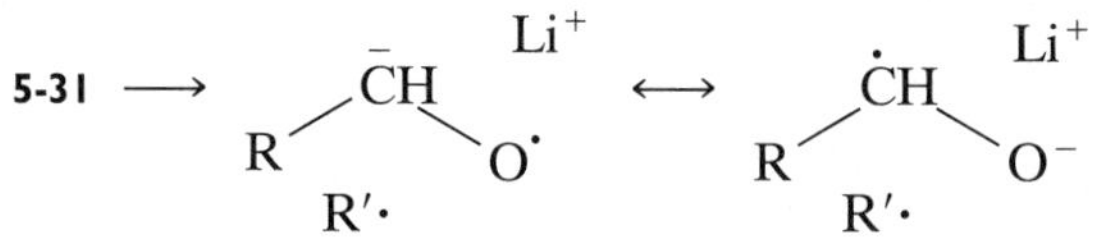

There is evidence to suggest that some of these reactions go by an ionic

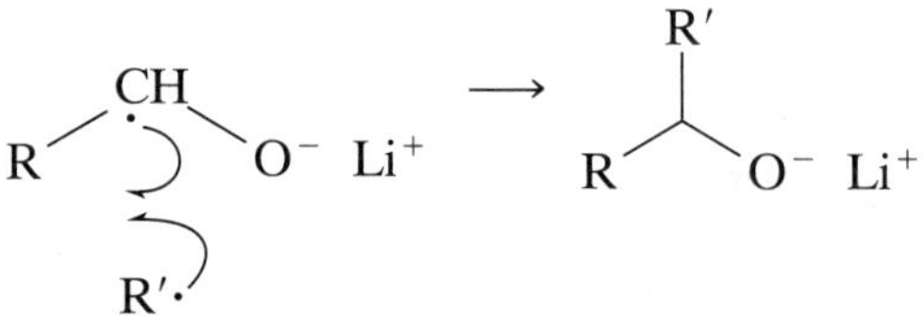

There is evidence to suggest that some of these reactions go by an ionic mechanism. The purpose in introducing a radical mechanism here is to indicate that when there are several possible mechanisms for a reaction, some may be radical in character.

Write reasonable mechanisms for the formation of each of the products in the following reaction.   **PROBLEM 5.13**

$$(PhCH_2)_2S \xrightarrow[\text{THF}]{\text{BuLi}} \xrightarrow{\text{MeI}} PhCH_2CH_2Ph \; + \; \underset{Ph \quad SMe}{\overset{MeS \quad Ph}{\diagup\!\!\diagdown}} \; + \; \underset{Ph \quad SMe}{\diagup\!\!\diagdown}$$

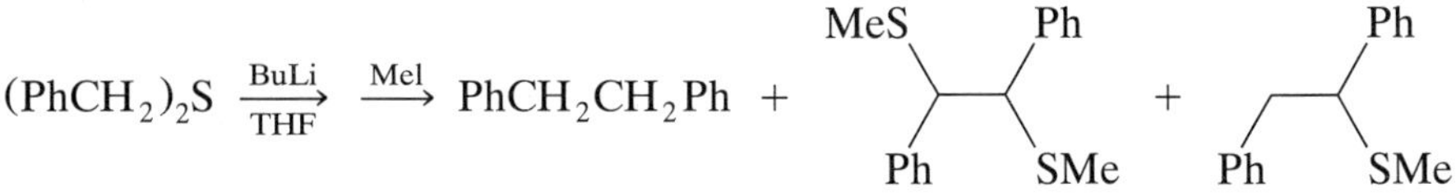

Biellmann, J. F.; Schmitt, J. L. *Tetrahedron Lett.* **1973**, 4615–4618.

Other reactions that may involve radical scission of anion intermediates include the rearrangement of anions adjacent to either trivalent or tetravalent nitrogen.

Example 5.14. *Unusual rearrangement of an anion α to a nitrogen.*

In this example, an anion formed initially undergoes radical scission, recombination, and then anionic rearrangement to the final product.

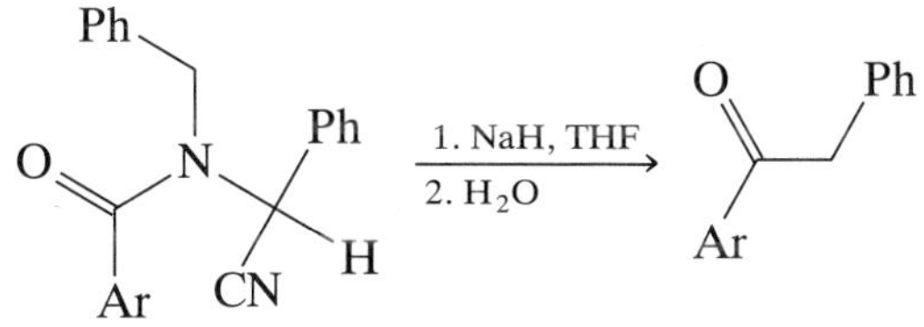

When Ar = *p*-tolyl, the yield is 57%.

Stamegna, A. P.; McEwen, W. E. *J. Org. Chem.* **1981**, *46*, 1653–1655.

In the paper cited, the following mechanism was proposed. The hydride removes the proton on the carbon α to the nitrile, leaving a carbanion. Then a benzyl radical cleaves, leaving a resonance-stabilized radical anion, **5-32**.

5-32-1   **5-32-2**   **5-32-3**

The benzyl radical then recombines at the carbon of the carbonyl group in **5-32**, and subsequent elimination of benzonitrile and cyanide ion gives the product.

PROBLEM 5.14   Consider the following data and then write reasonable mechanisms for the formation of 5-33 and 5-34.

5-33   **5-34**

(a) 10 mol% di-*t*-butyl nitroxide completely suppressed formation of **5-33**.

(b) The yield of **5-34** was higher in the dark or in the presence of nitroxide. **PROBLEM 5.14**
(c) The yield of **5-33** was lower in the dark than with sunlamp irradiation. *continued*

Russell, G. A.; Ros, F. *J. Am. Chem. Soc.* **1985**, *107*, 2506–2511.

Write step-by-step mechanisms for the following reactions. **PROBLEM 5.15**

a.

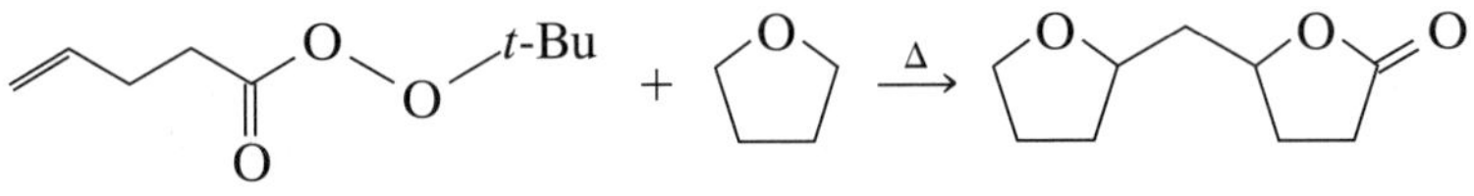

Kharrat, A.; Gardrat, C.; Maillard, B. *Can. J. Chem.* **1984**, *62*, 2385–2390.

b.

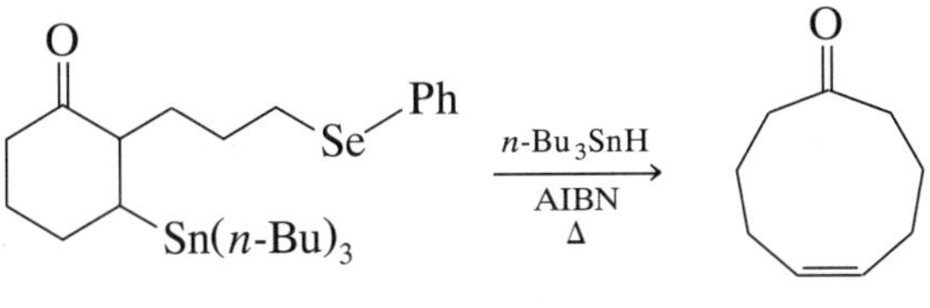

Only 10 mol% tin hydride is used.

Baldwin, J. E.; Adlington, R. M.; Robertson, J. *J. Chem. Soc., Chem. Commun.* **1988**, 1404–1406.

c.

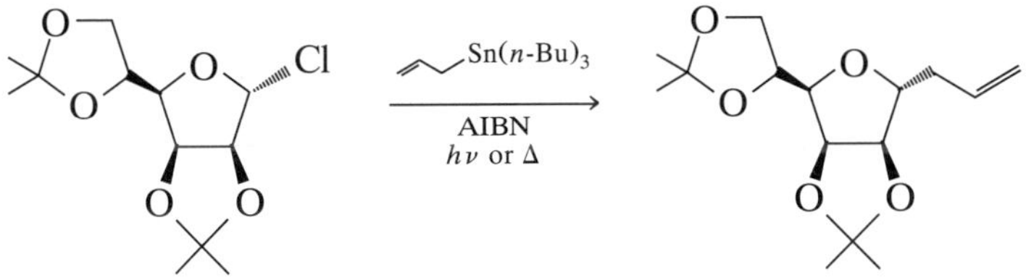

79%

Hart, D. J. *Science* **1984**, *223*, 883–887.

d.

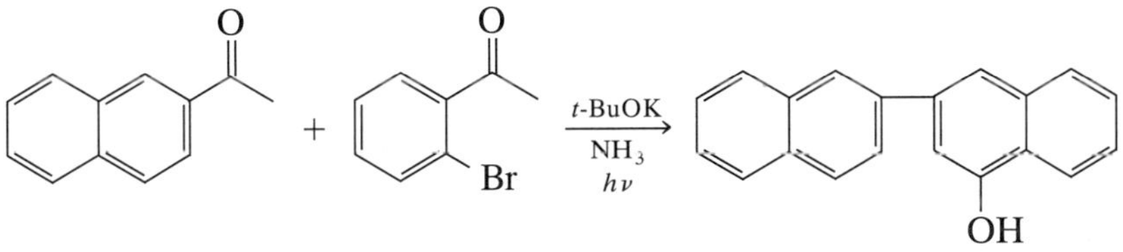

5-35

The yield of **5-35**, when the reaction mixture is irradiated, is 72%; in the dark, none of this product is formed.

Beugelmans, R.; Bois-Choussy, M.; Tang, Q. *J. Org. Chem.* **1987**, *52*, 3880–3883.

PROBLEM 5.15
continued

e. MeO ⟶ (I) ⟶ OMe + (C₄H₉ alkene with H) $\xrightarrow[\;h\nu\;]{\text{Bu}_3\text{SnSnBu}_3,\ \text{benzene}}$

MeO_2C / MeO_2C ... C₄H₉, I   **5-36** $\xrightarrow{\Delta}$ O, O, C₄H₉, MeO_2C **5-37**

Hexabutylditin was present in 10 mol%. The yield of **5-37** from 1-hexene and iodomalonate was 69%.

Curran, D. P.; Chen, M.-H.; Spletzer, E.; Seong, C. M.; Chang, C.-T. *J. Am. Chem. Soc.* **1989**, *111*, 8872–8878.

ANSWERS TO PROBLEMS

Problem 5.1

The radical formed is stabilized by resonance and by two alkyl groups. In addition, the small, stable molecule N_2 is formed in the process (see Hints 2.13 and 2.14).

$$\text{CH}_3-\overset{\cdot}{\underset{\text{CH}_3}{\text{C}}}-\text{C}{\equiv}\text{N}\colon \quad \longleftrightarrow \quad \text{CH}_3-\underset{\text{CH}_3}{\text{C}}{=}\text{C}{=}\overset{\cdot}{\text{N}}\colon$$

Problem 5.2

$$t\text{-BuO}\cdot \ + \ \text{(radical)} \ \longrightarrow \ \text{(product)}{-}\text{O-}t\text{-Bu}$$

$$t\text{-BuO}\cdot \ + \ \text{Cl}\cdot \ \longrightarrow \ t\text{-BuOCl}$$

The last coupling is the reverse of an initiation process. It removes *t*-butoxy radical from the chain propagation steps and, thus, is a termination process.

Problem 5.3

The three equations shown add up to the same overall reaction as equations 1 and 2 in Example 5.3. Thus, the total enthalpy change, −47 kcal/mol is the same as the total enthalpy change calculated in Example 5.3. Why then do these equations not represent a mechanism for the reaction? There are two reasons. First, the equations in the problem do not represent a chain process, so the thermochemistry of each step, *including the first one*, must be considered. Because this first reaction is highly endothermic, the overall process would be very slow. Second, the third step involves the coupling of two radicals. Whereas the enthalpy for this reaction is very favorable, both radicals will be present in very low concentrations. Thus, the probability that they will collide and react is very low. In contrast, both propagation steps in Example 5.3 have a radical colliding with a stable molecule whose concentration is much higher than that of any radical intermediate.

Problem 5.3
continued

Problem 5.4

a. Step 1

$$PhC{-}O{-}O{-}CPh \longrightarrow 2PhCO\cdot$$

Step 2

$$PhC{-}O\cdot \; + \; Br{-}CBr_3 \longrightarrow PhC{-}O{-}Br \; + \; \cdot CBr_3$$

Step 3

$$\cdot CBr_3 \; + \; CH_2{=}CHC_6H_{13} \longrightarrow Br_3CCH_2\dot{C}HC_6H_{13}$$

Step 4

$$Br_3CCH_2\dot{C}HC_6H_{13} \; + \; Br{-}CBr_3 \longrightarrow Br_3CCH_2\dot{C}HBrC_6H_{13} \; + \; \cdot CBr_3$$

5-38

b. The initiation steps are steps 1 and 2. The propagation steps are steps 3 and 4.

c. Addition of the tribromomethyl radical to the double bond could give either a primary or a secondary radical. Regiospecific addition leads to the more stable secondary radical, **5-38**.

a. *Initiation steps*

Problem 5.5

Thermal decomposition of AIBN and then hydrogen abstraction from tri-*n*-butyltin hydride.

Problem 5.5
continued

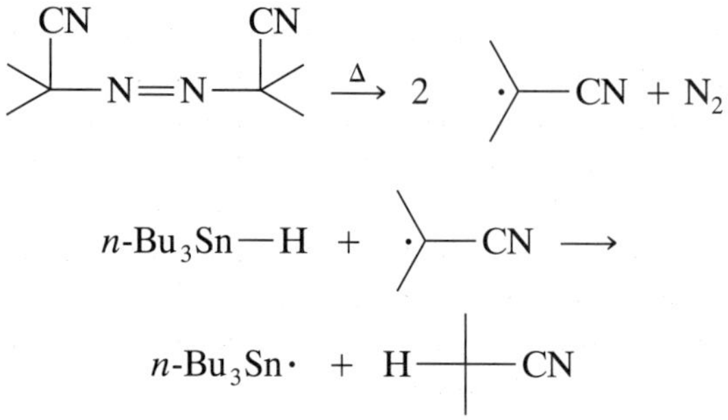

Propagation steps

(1) The tin radical abstracts an iodine atom.

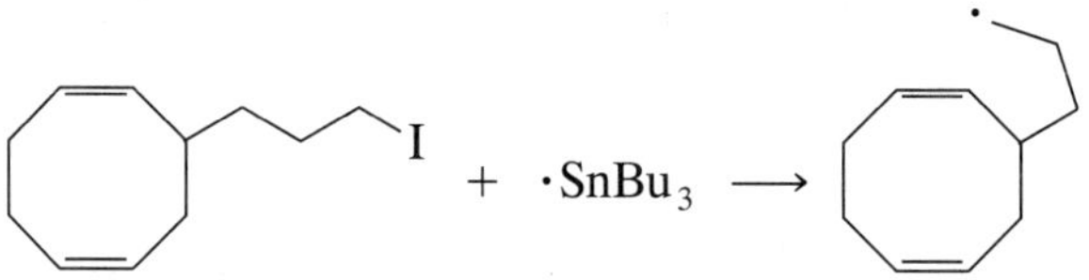

(2) The resulting radical cyclizes.

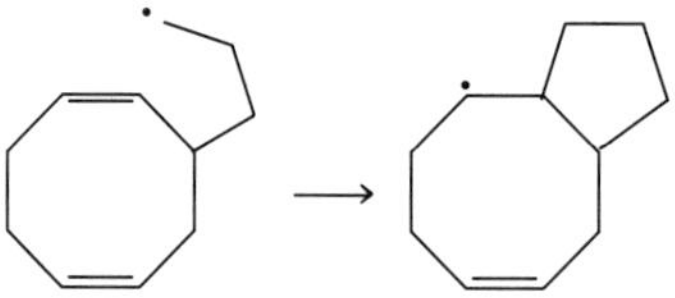

5-39

(3) The cyclized radical abstracts a hydrogen atom from tri-*n*-butyltin hydride to give the product and a new tri-*n*-butyltin radical to continue the chain.

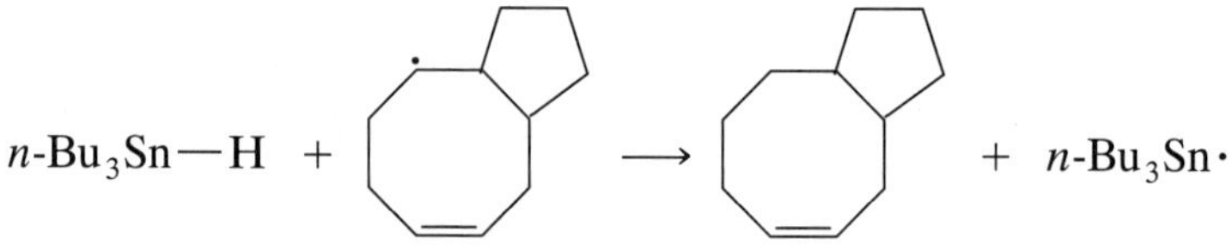

5-39

b. Radical **5-39** can undergo further cyclization across the ring. Such transannular reactions are common in medium-sized rings.

Problem 5.5
continued

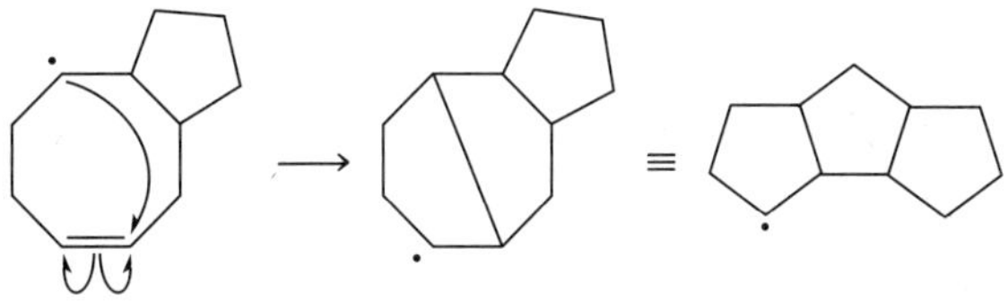

The resulting radical also can abstract a hydrogen from tri-*n*-butyltin hydride to produce the hydrocarbon and a tin radical to continue the chain.

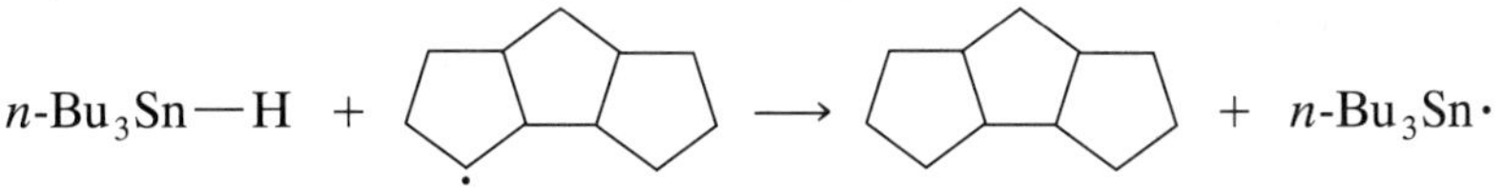

The relative ease with which transannular reactions occur in medium-sized rings becomes apparent when we look at the actual three-dimensional structure of these rings. In their most stable conformations, the hydrogen atoms attached to the ring are staggered, and this means that the carbon atoms across the ring are sufficiently close to form bonds when the appropriate functionality is present. Although medium-sized rings are often represented as regular polygons, alternative representations give a much more realistic picture of the molecular configuration.

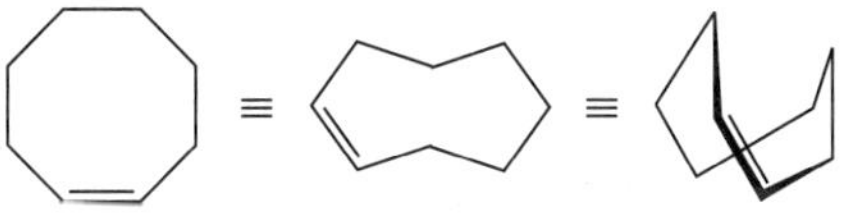

These kinds of pictures make it easier to rationalize transannular reactions.

Initiation steps

Problem 5.6

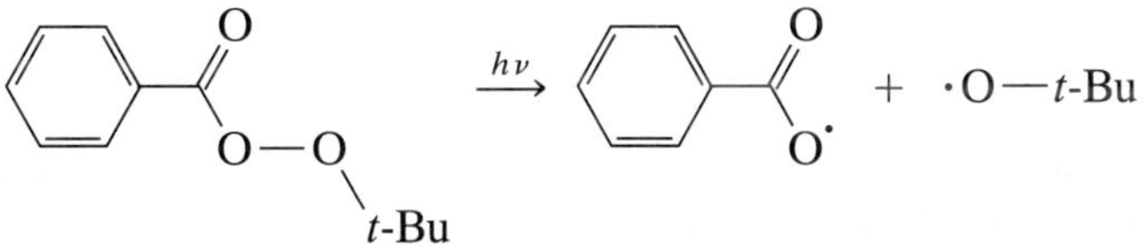

Problem 5.6
continued

Either of the resulting radicals can remove a hydrogen from tri-*n*-butyltin hydride.

$$t\text{-BuO}\cdot \; + \; \text{H}\!-\!\text{SnBu}_3 \; \longrightarrow \; t\text{-BuOH} \; + \; \cdot\text{SnBu}_3$$
$$\text{PhCO}_2\cdot \; + \; \text{H}\!-\!\text{SnBu}_3 \; \longrightarrow \; \text{PhCO}_2\text{H} \; + \; \cdot\text{SnBu}_3$$

Propagation steps

The tributyltin radical abstracts a bromine atom from the starting material.

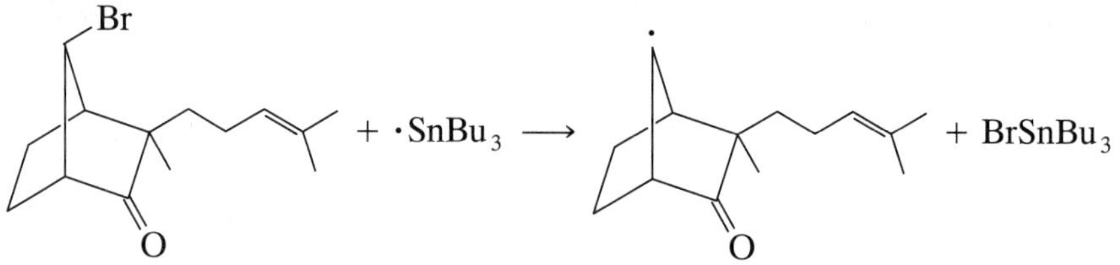

The radical undergoes intramolecular cyclization. This cyclization is regiospecific to form a new radical on the exocyclic carbon. This is expected on the basis of the Baldwin–Beckwith guidelines (Section 6.B).

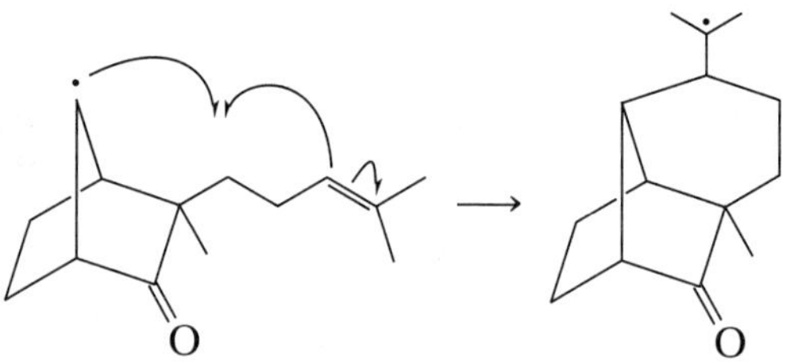

The cyclized radical abstracts a hydrogen from another molecule of tributyltin hydride to generate the product and another tributyltin radical.

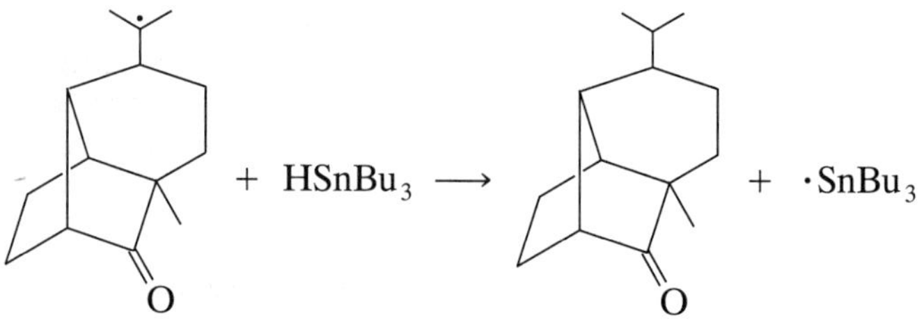

Problem 5.7

Note that this is not a chain process. The reaction, known as the Barton nitrite photolysis, was developed to functionalize steroids in positions that are difficult to activate by other routes. The first step is photolysis of the nitrite ester, cleaving the relatively weak N—O bond.

Problem 5.7
continued

Because of the conformational rigidity of the steroid ring system, the nitrite ester and the methyl group that becomes functionalized are in close proximity, and the oxyradical formed by photolysis of the nitrite ester is well-positioned to abstract a hydrogen from the methyl group via a six-membered cyclic transition state. (The formation of a five- or six-membered ring is a common feature of the transition state for many reactions because bond lengths, bond angles, and entropy all combine to stabilize these ring sizes.) Presumably the intramolecular hydrogen abstraction is so rapid that the nitroso radical is still close enough to recombine with the primary carbon radical. The resulting nitroso compound tautomerizes to the oxime on workup.

Problem 5.8

Bromine reacts with the silver salt to form silver bromide and the acyl hypobromite. The latter undergoes homolytic scission to give the carboxy radical, **5-40**.

$$Ph_2CHCH_2CO_2Ag + Br_2 \longrightarrow Ph_2CHCH_2CO_2Br + AgBr$$

$$Ph_2CHCH_2CO_2Br \xrightarrow{\Delta} Ph_2CHCH_2CO_2\cdot + Br\cdot$$

5-40

Hydrogen abstraction by the bromine radical gives the radical **5-41**, which cyclizes, regenerating a bromine radical.

$$Ph_2CHCH_2CO_2Br + Br\cdot \longrightarrow Ph_2\overset{\cdot}{C}HCH_2CO_2Br + HBr$$

5-41

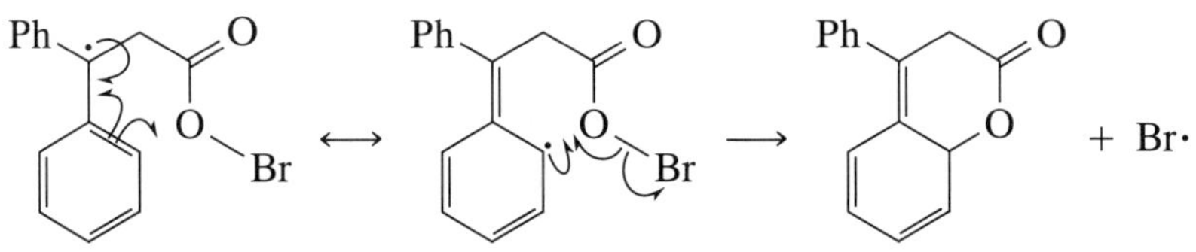

5-41-1   **5-41-2**   **5-42**

The intermediate **5-42** can isomerize to product **5-12** by a protonation-deprotonation mediated by the strong acid HBr formed in the reaction medium.

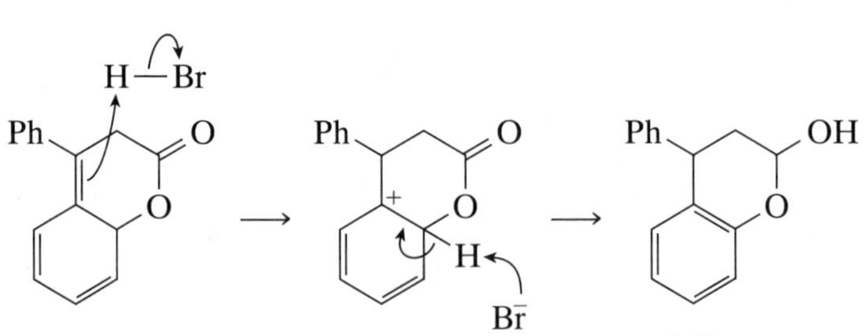

5-42   **5-12**

An alternative mechanism can be written in which **5-40** cyclizes to **5-43**, which is then oxidized by bromine to the corresponding carbocation. The product is then obtained by abstraction of a proton by bromide ion.

Problem 5.8
continued

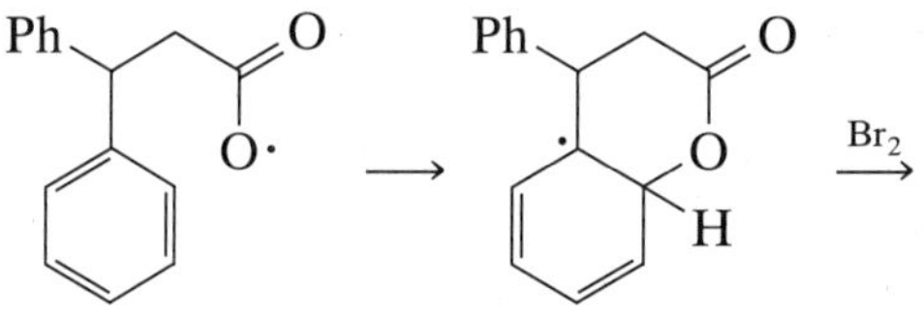

5-40   **5-43**

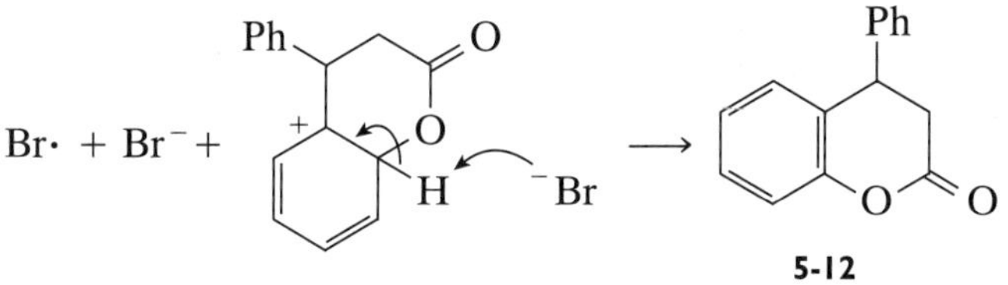

5-12

This second mechanism is less satisfactory because cyclization of the carboxy radical **5-40** is unlikely. Alkyl carboxy radicals are highly unstable and they decarboxylate so rapidly that fragmentation would be expected to occur faster than cyclization. The previous mechanism has at least two other advantages. First, it proceeds through the radical **5-41**, which is stabilized by resonance with two phenyl substituents, so that hydrogen abstraction to form **5-41** should compete effectively with homolytic scission of the acyl bromide to form **5-40**. Second, it will occur rapidly because it is a radical chain process; the bromine radical formed in the cyclization of **5-41** regenerates **5-41** from the acyl hypobromite.

Product **5-13** can be produced via hydrogen atom abstraction

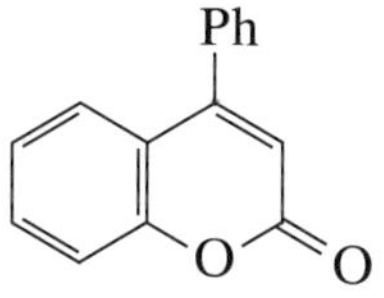

5-13

from **5-12** to form radical **5-44**, followed by mechanistic steps analogous to those for the transformation of **5-43** to **5-12**.

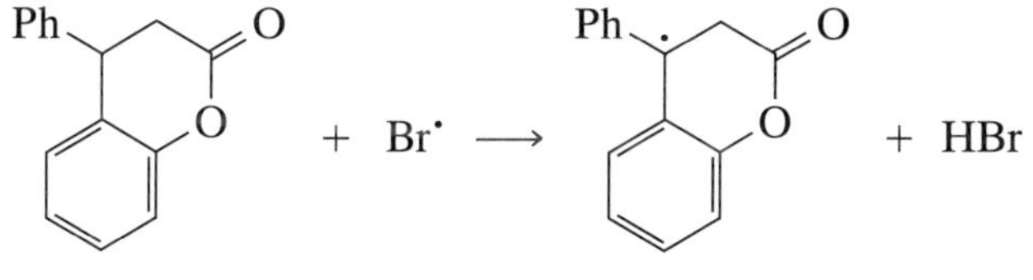

5-12   **5-44**

b. Note that the bicyclic structure of the starting material is the same as the bicyclic structure in Example 5.8. It has been written in a different (more old-fashioned) way. Once this is recognized, a mechanism can be written for the reaction, which is analogous to Example 5.8.

   The first step is homolysis of the weakest bond in **5-14**, the O—O bond. This is followed by loss of CO_2.

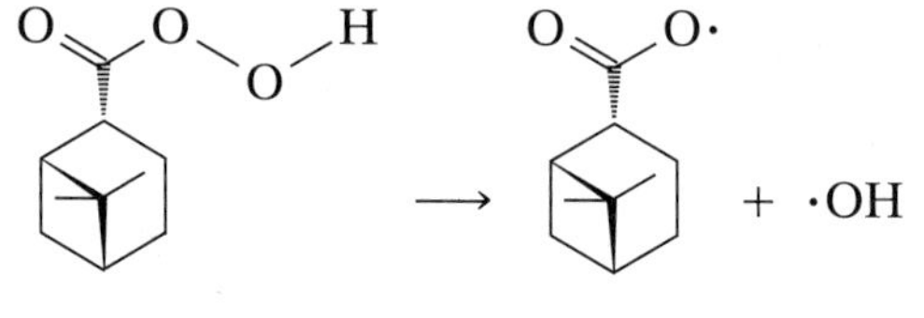

5-14

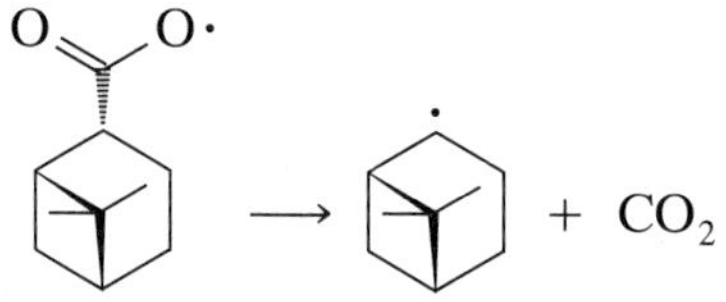

5-45

Because the ratio of bicyclic to ring-opened products increases with increasing peracid concentration, the bicyclic products (**5-15** and **5-16**) must be formed by reaction of the intermediate **5-45** with **5-14**. This bimolecular process competes with the unimolecular fragmentation of **5-45** that leads to the ring-opened product **5-17**.

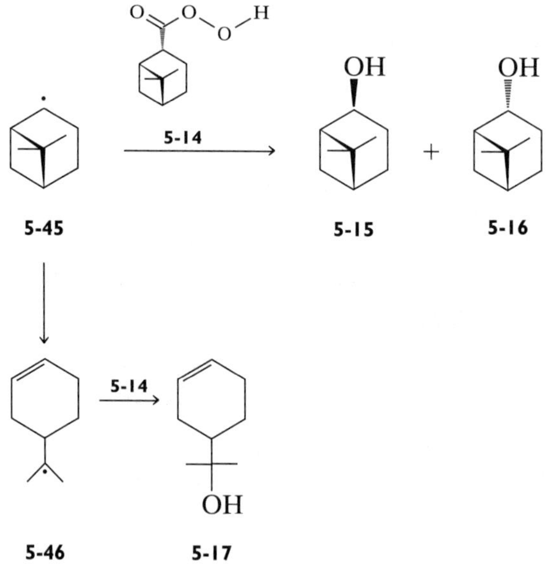

The intermediate radicals **5-45** and **5-46**, symbolized by ·R′ in the following equation, can abstract OH from **5-14** to produce the product alcohols.

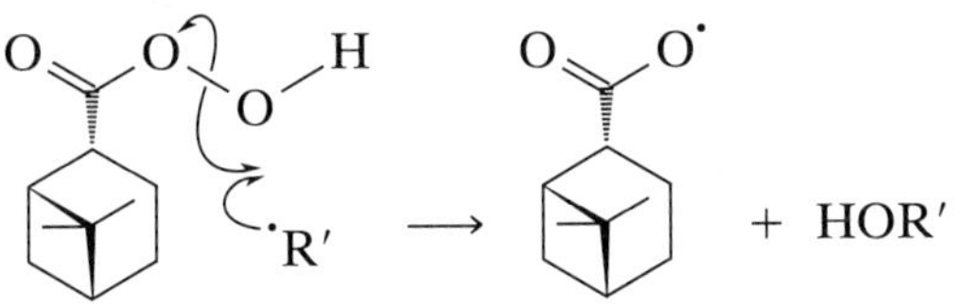

Problem 5.8
continued

Because roughly the same amounts of **5-15** and **5-16** are produced from either isomer of starting material, **5-14**, the mechanism of the reaction cannot be concerted:

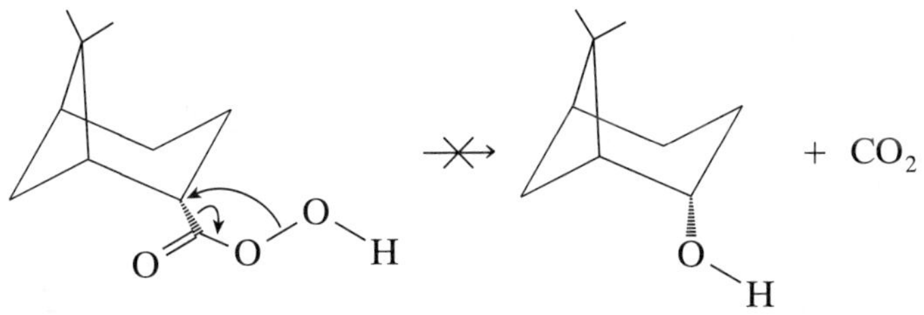

n-Butylmercaptan is a better chain-transfer agent than the aldehyde, because hydrogen abstraction from the mercaptan is much easier. Thus, the radical, formed by addition of *n*-butylthio radical to the aliphatic double bond, is captured by mercaptan before it can rearrange. This is not the case for reaction with aldehyde.

Problem 5.9

$$t\text{-BuOOH} \longrightarrow t\text{-BuO·} + ·OH$$

$$t\text{-BuO·} + n\text{-BuSH} \longrightarrow t\text{-BuOH} + n\text{-BuS·}$$

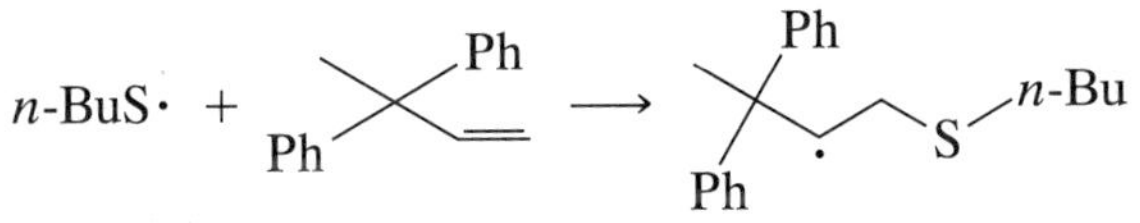

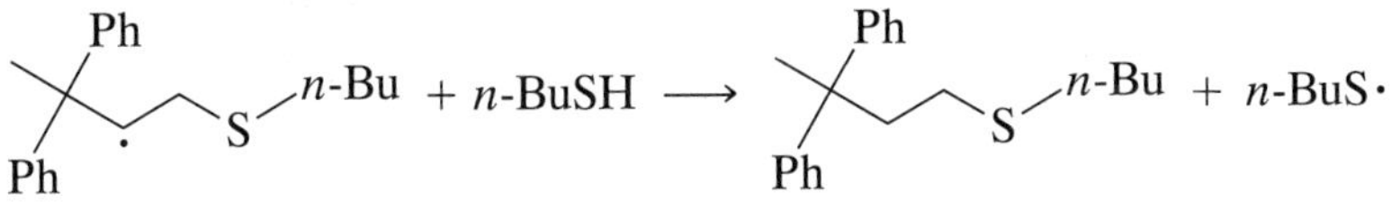

Problem 5.9
continued

The reaction with aldehyde is as follows:

$$t\text{-BuO} \cdot + n\text{-PrCHO} \longrightarrow t\text{-BuOH} + n\text{-Pr}\overset{\cdot}{\text{C}}{=}\text{O}$$

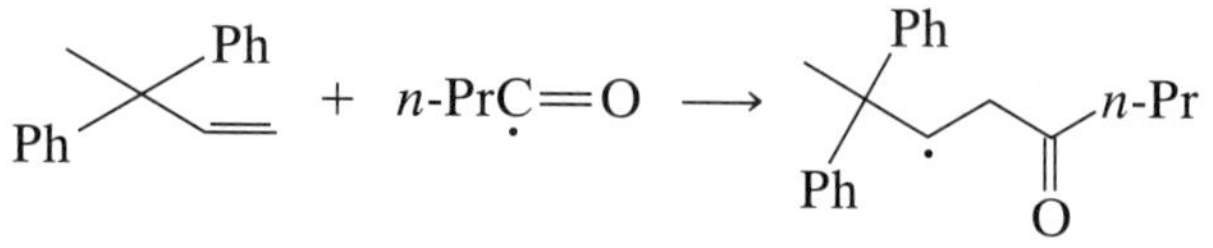

5-47

In this case, rearrangement of **5-47** competes effectively with its abstraction of hydrogen from starting material.

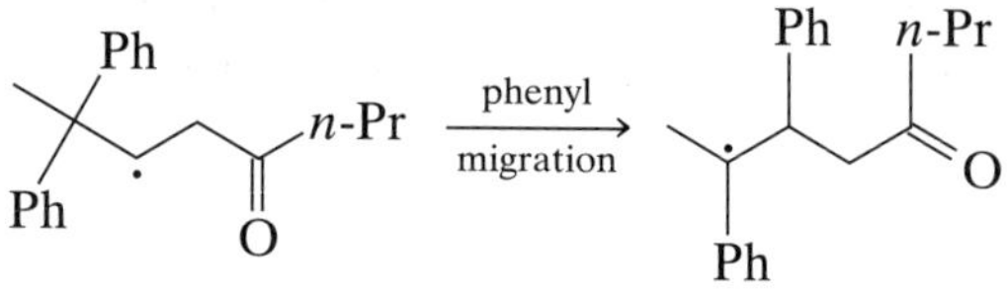

5-47

The rearranged radical abstracts a hydrogen from aldehyde to continue the chain.

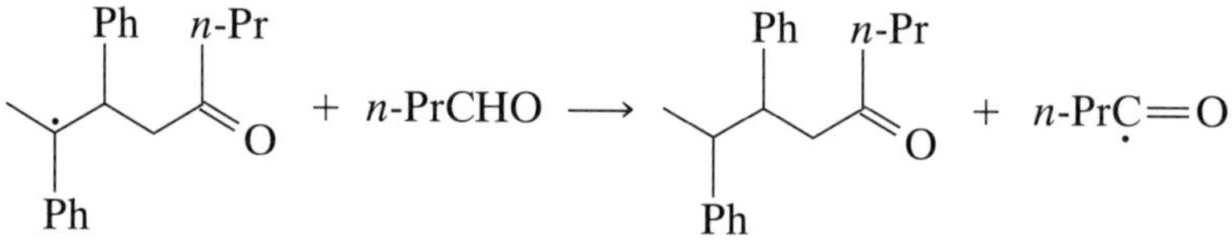

Problem 5.10

Numbering of the critical carbon atoms in the starting material and product gives a good indication of the course of the reaction.

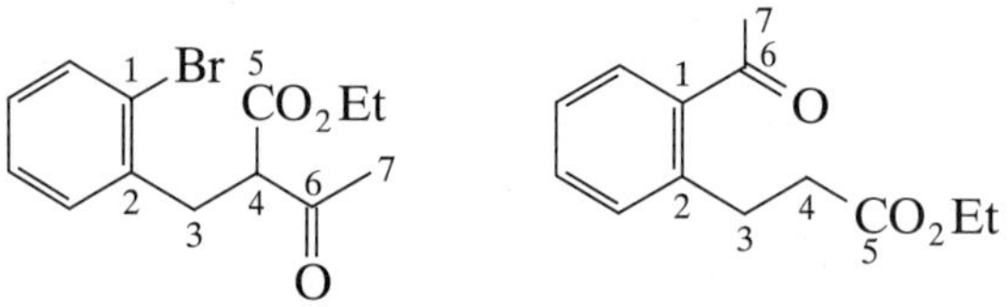

The acetyl group has "migrated" from C-4 in the starting material to C-1 in the product. Thus, the following sequence of events can be anticipated: (i) removal of the bromine atom from C-1, (ii) addition of the resulting radical to C-6, (iii) cleavage of the C-6 to C-4 bond, and (iv) reaction with tri-*n*-butyltin hydride to give the product and a radical to continue the chain.

Problem 5.10
continued

The initiation steps, to form the tri-*n*-butyltin radical, are the same as those in Problem 5.5. This radical then abstracts bromine from the starting material.

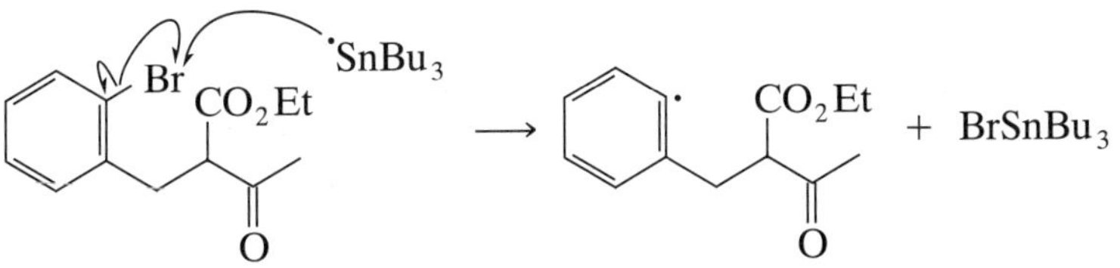

The aryl radical adds to the carbonyl group, a bond cleaves, and the resulting radical abstracts hydrogen from tri-*n*-butyltin hydride to form the product and a new chain-carrying radical.

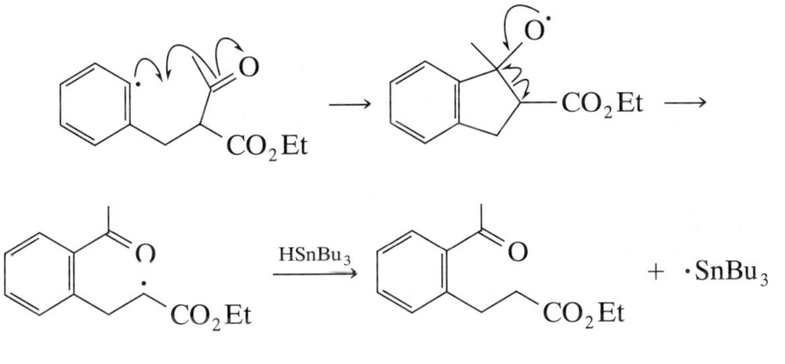

Problem 5.11

The fact that reactions in both (a) and (b) require light suggests that a radical mechanism is involved for each of them. This rules out a simple S_N2 substitution in part a or a nucleophilic aromatic substitution by an addition–elimination reaction in part b. When substitution occurs under basic conditions in the presence of light, the most likely mechanism is $S_{RN}1$.

Problem 5.11
continued

a. The thiolate anion is excited by the light.

$$PhS^- \xrightarrow{h\nu} PhS^{*-}$$

The excited thiolate anion transfers an electron to cyclopropane.

The better leaving group, bromide, leaves.

Thiolate reacts, as a nucleophile, with the electrophilic radical to produce the radical anion of the product.

Electron transfer to another molecule of starting cyclopropane gives the product and another cyclopropane radical anion to continue the chain.

The following is not a proper product-forming step because it does not also produce a radical to continue the chain. It actually is a termination step, involving the coupling of two radicals. Formation of a major product by the coupling of two radicals is unlikely because the concentration of radical intermediates usually is quite low.

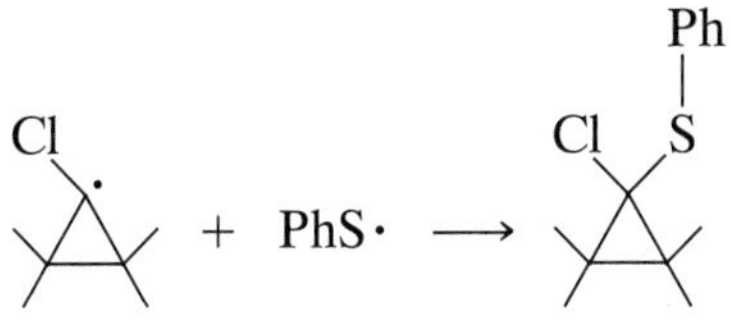

Problem 5.11
continued

b. Each step in the mechanism for this reaction has a direct counterpart in the mechanism for part a.

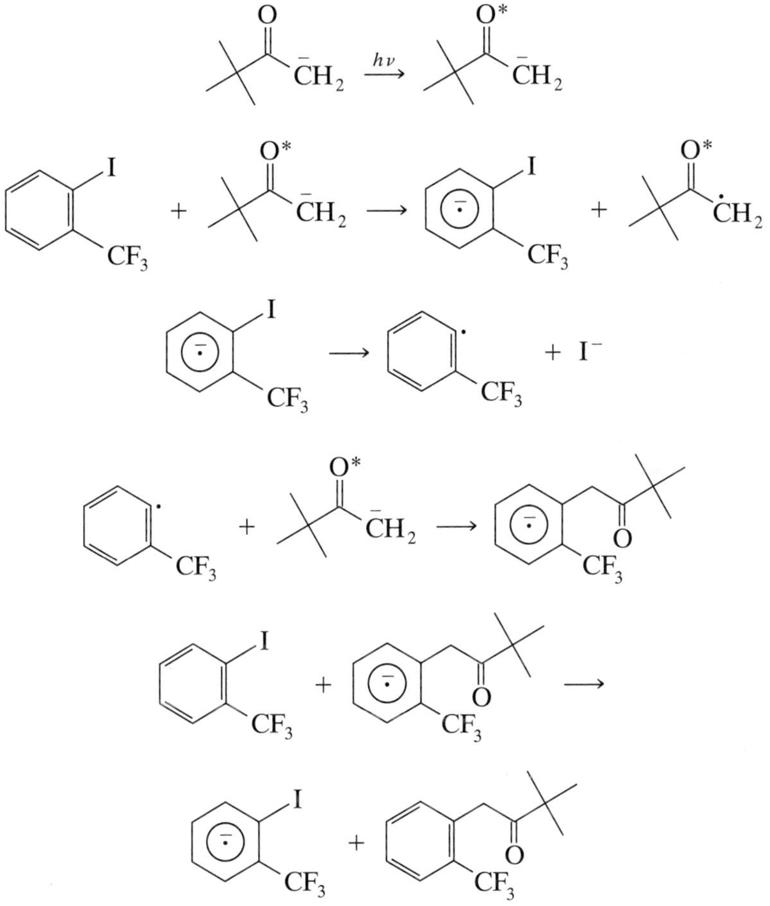

Sodium transfers a single electron to the aromatic system to produce a pentadienyl radical anion, **5-48**. This anion is protonated at the position either *ortho* or *meta* to the π-donating methoxy group to give an interme-

Problem 5.12

Problem 5.12
continued

diate radical. When a second electron is added to this radical, the pentadi-enyl anion is again protonated at its central carbon, *ortho* or *meta* to the methoxy substituent.

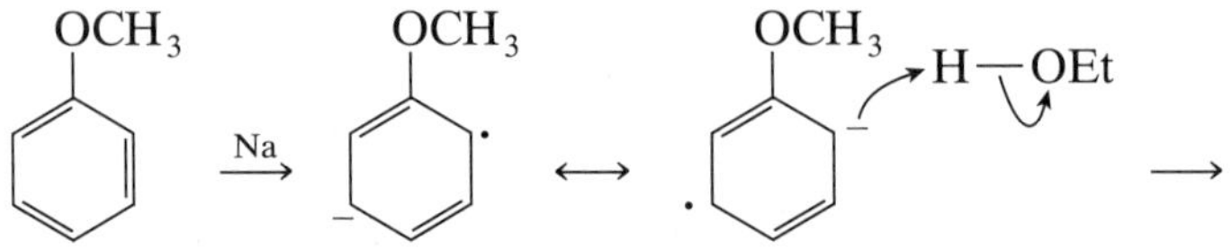

5-48-1   **5-48-2**

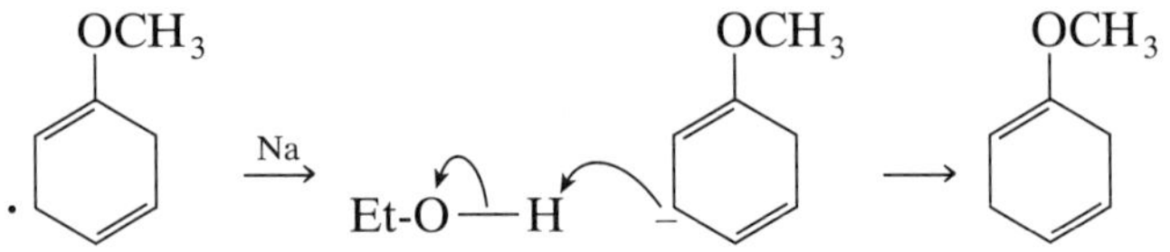

Problem 5.13

Comparison of the starting material with products shows that methyl iodide is used to methylate sulfur. The reaction takes place in base, so that the most likely mechanism for the methylation is an S_N2 reaction with thiolate ion:

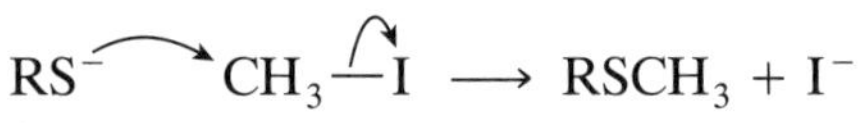

$$RS^- \quad CH_3-I \longrightarrow RSCH_3 + I^-$$

Despite the fact that radical coupling is rarely a significant source of reaction products, the structural symmetry of the first two products sug-gests that they are formed by this route. Butyllithium can remove a proton from the carbon α to the sulfur. The resulting anion could then fragment into a radical and a radical anion by scission of the S—C bond.

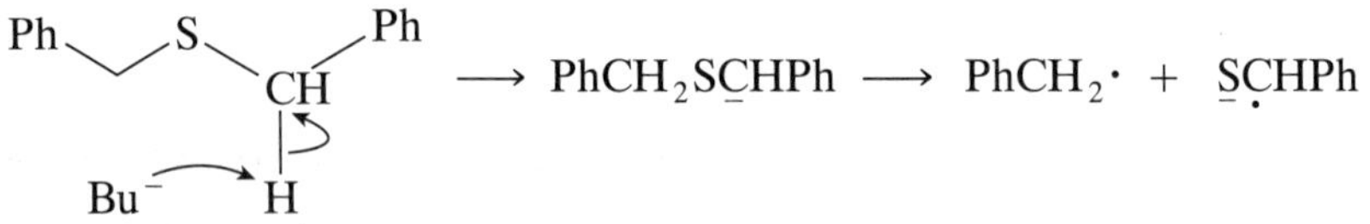

Problem 5.13
continued

Two benzyl radicals can couple to give the first product.

$$PhCH_2\cdot \ + \ \cdot CH_2Ph \ \longrightarrow \ PhCH_2CH_2Ph$$

Two thiolate radical anions can couple to give the dithiolate precursor, **5-49**, of the second product. This is analogous to the well-documented coupling of ketyls (radical anions derived from one-electron reduction of carbonyl compounds), which results in the formation of pinacols (1,2-diols).

Finally, the two radicals formed initially can couple, at carbon, to give the thiolate precursor, **5-50**, of the third product.

This is one of those relatively few reactions that proceed through radical coupling. The reaction occurs in part because the radicals involved are sufficiently stable to remain in existence until they can couple with another radical.

Problem 5.14

The evidence suggests that **5-33** and **5-34** are formed by different types of mechanisms: structure **5-33** by a radical process and structure **5-34** by a non-radical process. Formation of **5-33** is suppressed completely by addition of a free radical inhibitor, but is stimulated by light. These observations strongly support a radical pathway to **5-33**. Other factors also make ionic S_N pathways unlikely. An S_N2 reaction is ruled out because **5-33** is formed by substitution at a tertiary carbon. An S_N1 ionization also is unlikely because formation of a carbocation next to the partially positive

carbonyl would be required. Therefore, we are left with the strong possibility that **5-33** is formed by an $S_{RN}1$ mechanism. On the other hand, **5-34**
forms in better yield in the dark or in the presence of a radical inhibitor,
data which support a nonradical reaction. Note the regiochemistry of
addition is different in the formation of **5-33** and **5-34**.

An $S_{RN}1$ mechanism for the formation of **5-33** can be written as follows:

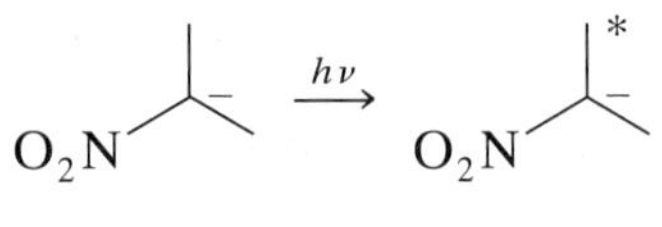

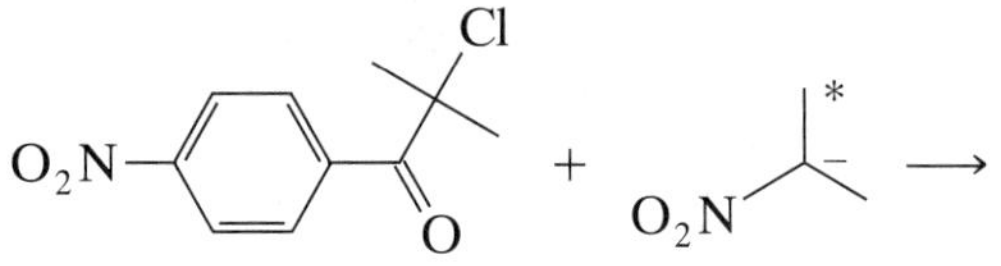

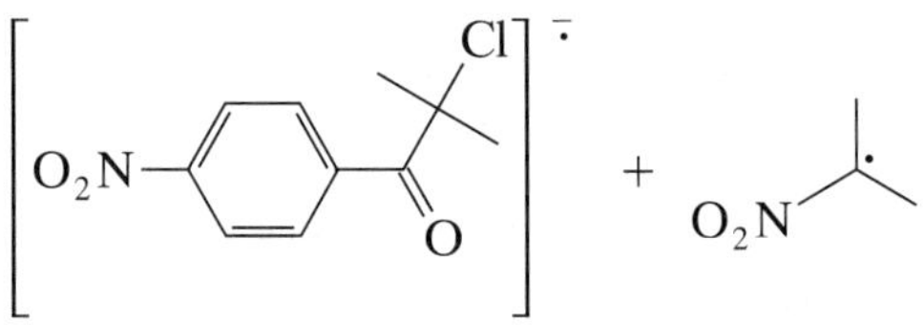

5-51

$$\textbf{5-51} \longrightarrow O_2N\text{—}\bigcirc\text{—}C(\cdot) + Cl^-$$

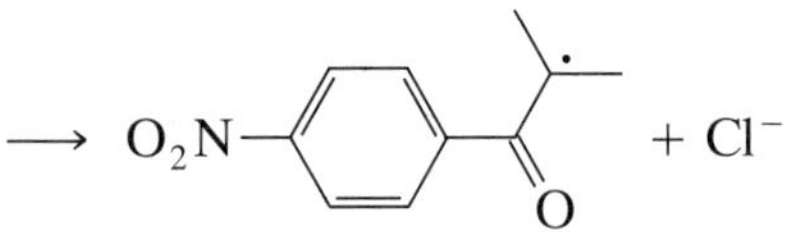

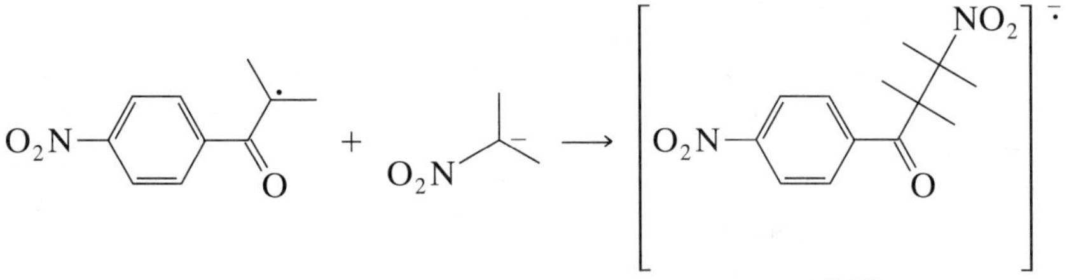

5-52

$$\textbf{5-52} + O_2N\text{—}\bigcirc\text{—}C\text{—}CCl \longrightarrow$$

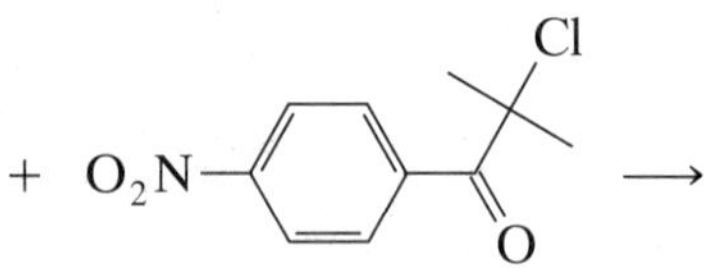

Problem 5.14
continued

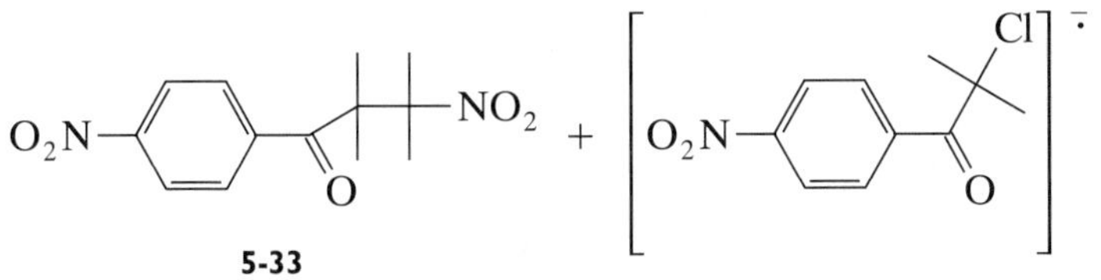

5-33

It is tempting to write a radical coupling to form product **5-33**, because it involves fewer steps. However, this mechanism would be energetically inefficient because it does not continue the chain process. Thus, the following *is not* the major product-forming step:

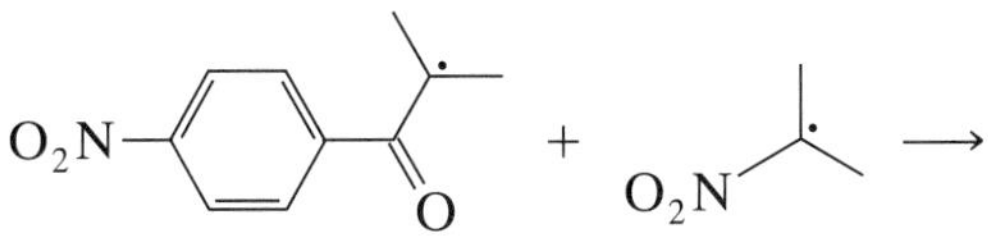

Noting the formation of the epoxide group and numbering the atoms in the starting materials and in **5-34** help to indicate how **5-34** must be formed.

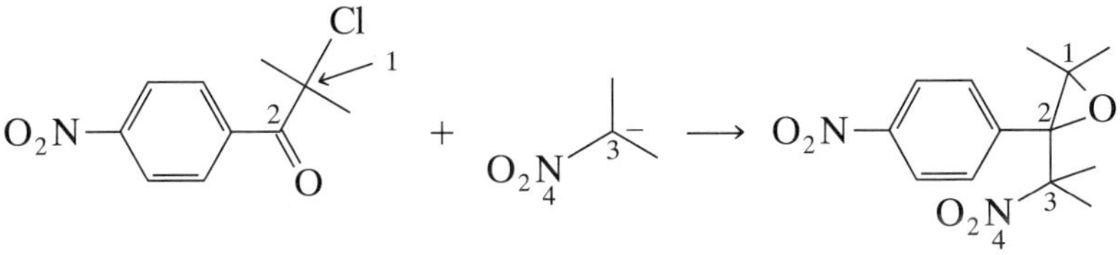

5-34

Because C-3 becomes attached to C-2, a nucleophilic reaction of the nitranion at the electrophilic carbonyl carbon must take place. Sterically this carbon is the most accessible. This step is followed by an intramolecular nucleophilic reaction of the alcoholate with the adjacent carbon (numbered 1 in the previous equation).

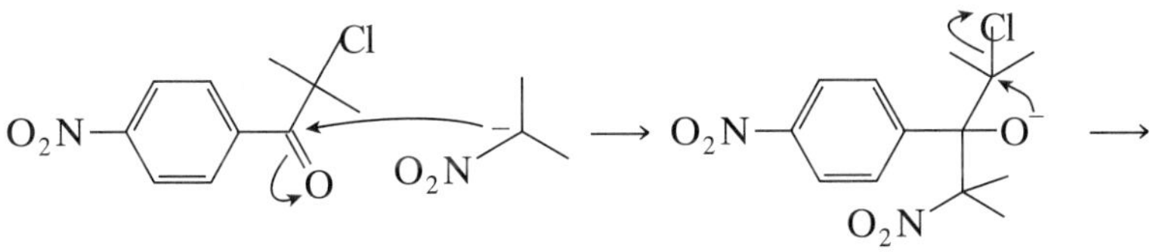

Problem 5.15

a. The perester could serve as a free radical initiator. At first glance, it appears that two tetrahydrofuran molecules form the two rings of the product, but the product contains one additional carbon atom. Therefore, the perester must be the source of the atoms of one ring, as well as a source of radicals. Because the perester contains a carbonyl group, the simplest explanation is that the lactone ring is derived from the perester. These considerations suggest the following chain mechanism:

Initiation

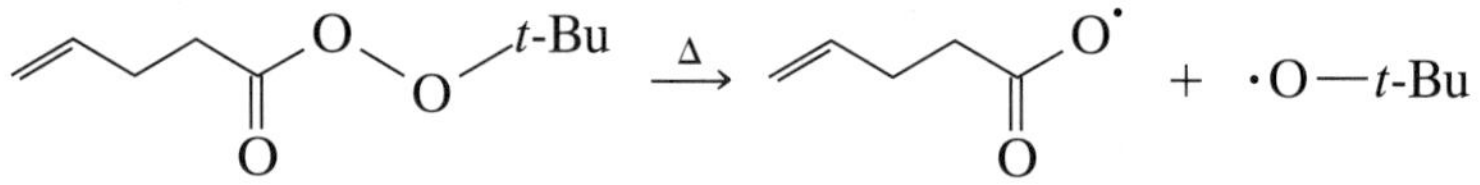

Propagation

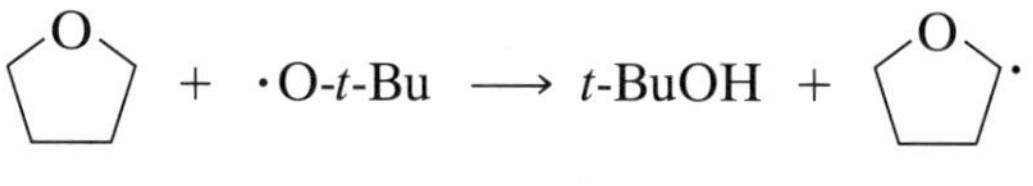

5-53

The tetrahydrofuranyl radical, **5-53**, adds to the carbon–carbon double bond of the perester.

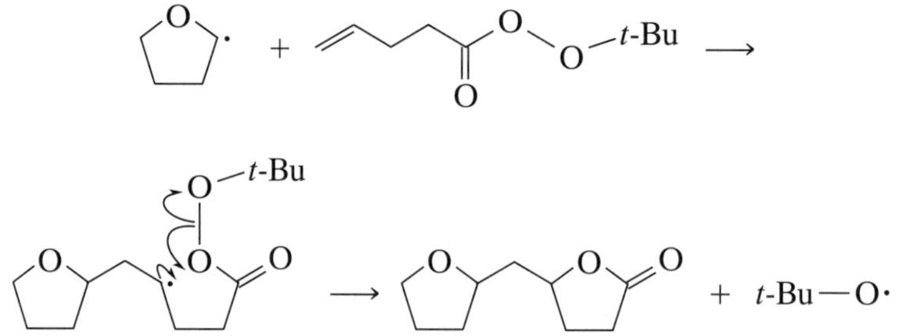

The following mechanism is not as satisfactory.

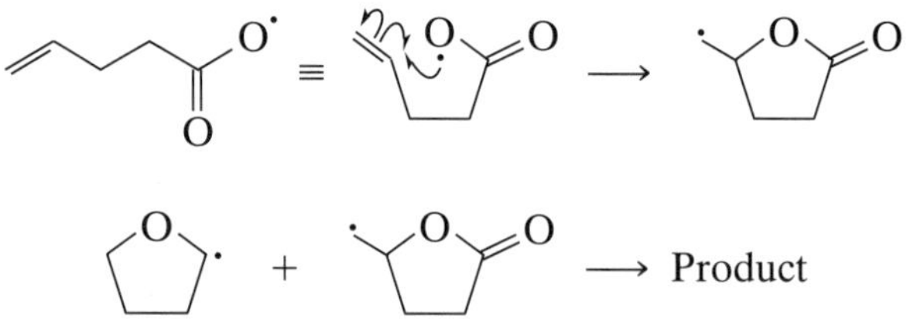

Rapid decarboxylation of the alkyl carboxyl radical to the corresponding alkyl radical would be expected to compete with intramolecular cyclization. Moreover, the reaction scheme shown would not be efficient. Not only would the reaction be slow because two radicals must collide in order to form the product, but the step fails to regenerate a radical to continue the reaction. (Note: Aryl carboxyl radicals lose carbon dioxide more slowly than alkyl carboxyl radicals.)

Problem 5.15
continued

b. From the reaction conditions, we can make the following speculations and/or conclusions: (i) the tri-*n*-butyltin radical will be the chain carrier. (ii) Tri-*n*-butyltin will be lost from the starting material. (iii) A bond in the six-membered ring breaks at some point in the mechanism.

In Section 2, we learned that tri-*n*-butyltin radicals react with selenium compounds to produce radical intermediates. Therefore, this might be a logical first step, after the initiation reaction between AIBN and tri-*n*-butyltin hydride has produced tri-*n*-butyltin radicals.

5-54

The carbonyl group of radical **5-54** can undergo an intramolecular cyclization reaction with the alkyl radical. The shared bond between the five- and six-membered rings then breaks to form the product and a new tri-*n*-butyltin radical, which continues the chain.

Product + ·Sn(*n*-Bu)$_3$

The major difference between this reaction and that of Example 5.11 is that only 10 mol% tin hydride is present. This is enough to initiate the reaction, but not enough to reduce intermediate radicals significantly.

Problem 5.15
continued

c. Clues to the reaction mechanism are (i) chlorine is missing from the product, (ii) AIBN is present as an initiator, and (iii) the tributyltin group is present. Thus, it appears that tributyltin radicals abstract chlorine from the starting material. In the initiation steps, AIBN forms 2-cyano-2-propyl radicals in the usual way, and these radicals react with allyltri-*n*-butyltin to produce tri-*n*-butyltin radicals by an addition–elimination mechanism.

Initiation

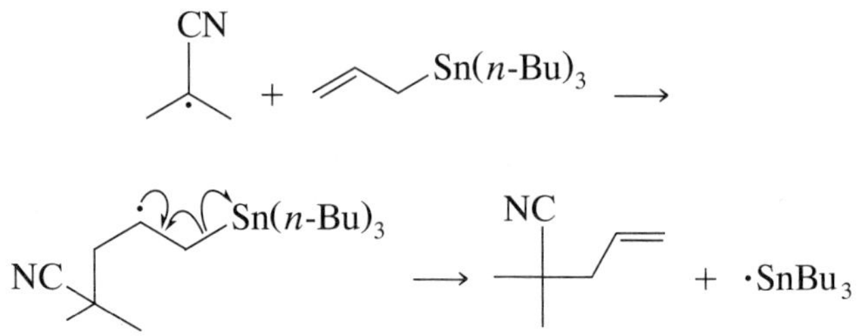

Propagation

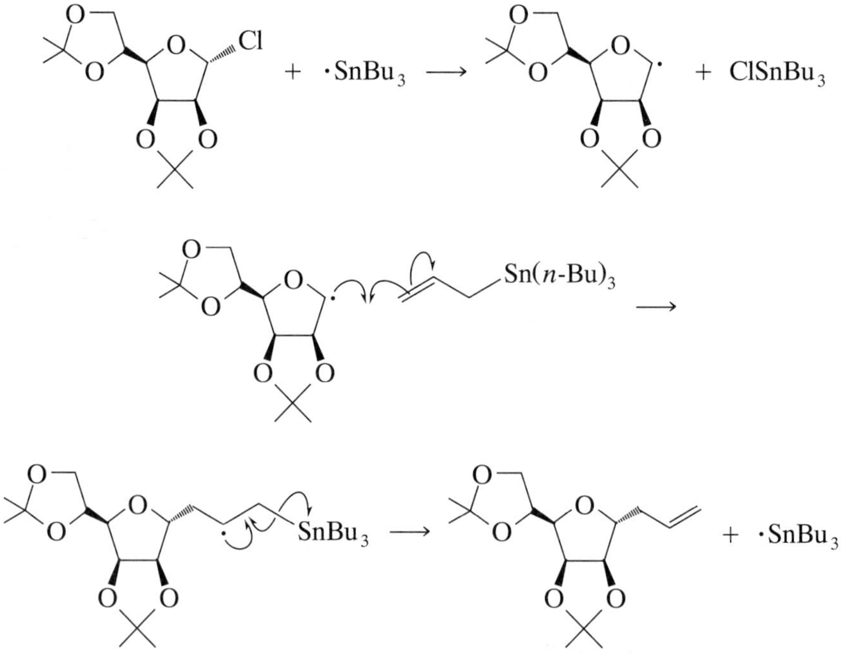

d. Reaction in the presence of light and strong base suggests the involvement of a radical anion intermediate and the $S_{RN}1$ mechanism for at least part of the reaction pathway. In the initiation steps, the anion of the naphthyl ketone, excited by light, donates an electron to the bromo ketone to form a radical anion. The product of an $S_{RN}1$ reaction would be **5-55**.

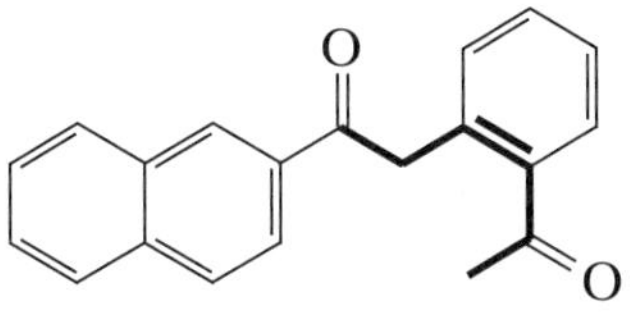

Problem 5.15
continued

5-55

If the six highlighted carbon atoms in **5-55** form a six-membered ring, the correct carbon skeleton for the final product is obtained. In fact, an intramolecular aldol condensation (discussed in Chapter 3), followed by elimination of water, gives the final product.

$S_{RN}1$ Reaction

(1) Initiation Steps

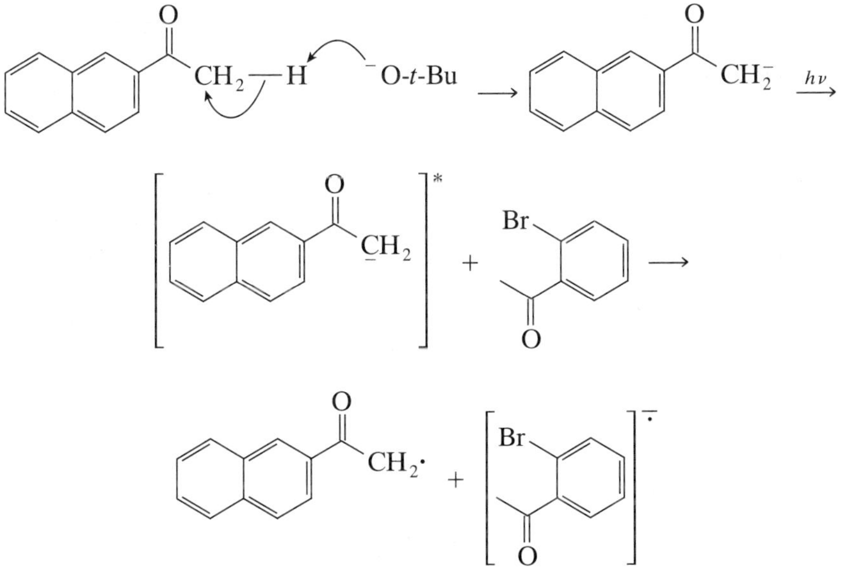

Problem 5.15
continued

(2) Propagation Steps

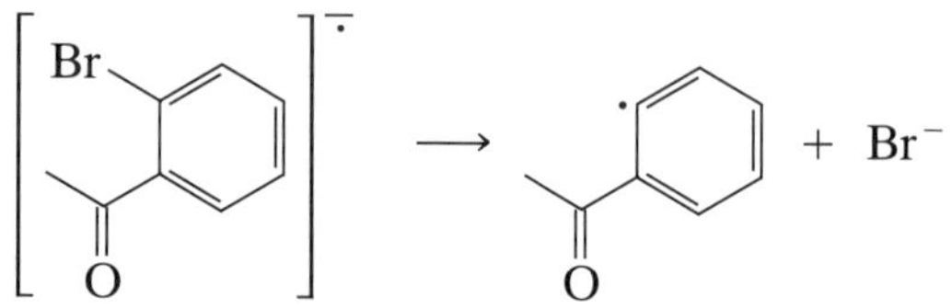

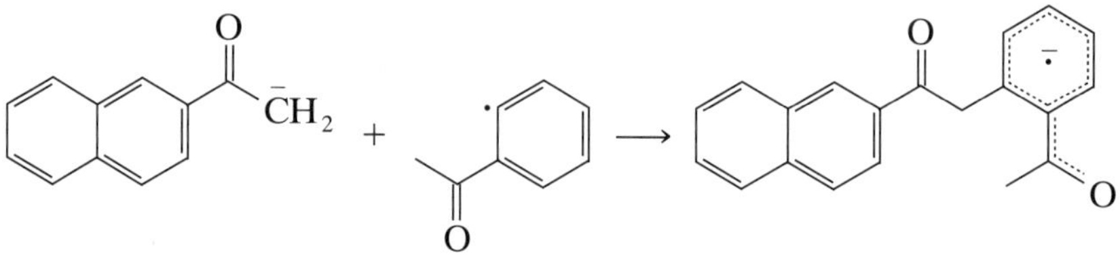

5-56

In the last step, a new radical anion of the starting bromo compound is formed, which continues the chain.

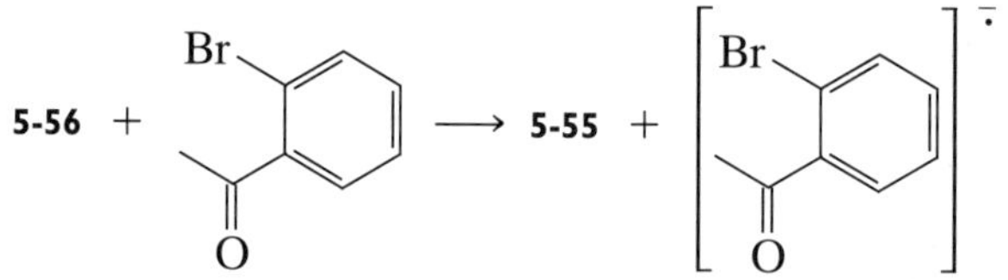

Aldol Condensation

The aldol condensation of **5-55** involves several steps. First, a proton is removed from the methyl group by the *t*-butoxide ion. (If the proton were removed from the methylene group α to the other carbonyl, condensation would be at the carbonyl attached to the methyl. This would produce a five-membered ring.) The resulting enolate adds to the other carbonyl group.

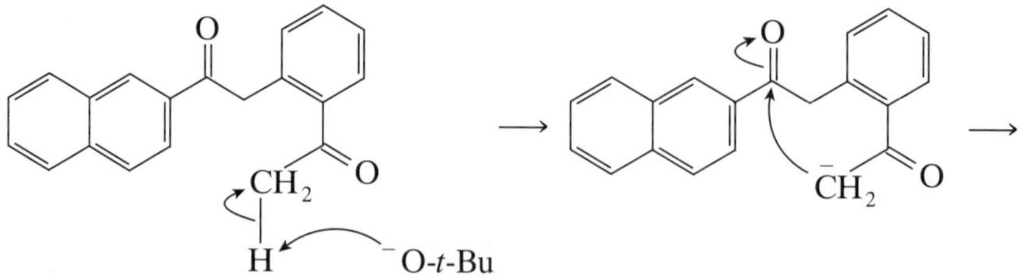

Protonation of the resulting alkoxide ion gives alcohol **5-57**.

Problem 5.15
continued

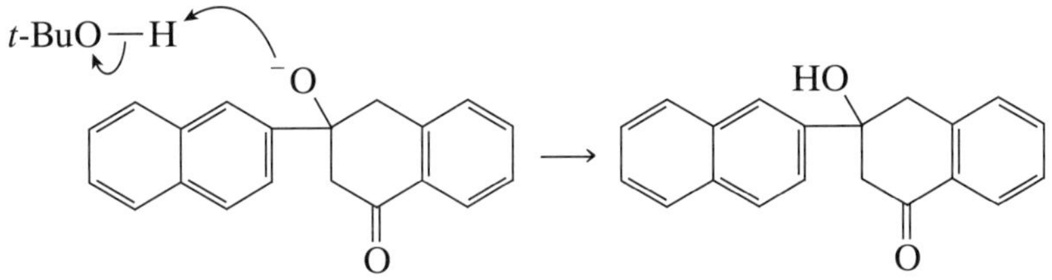

5-57

The alcohol can undergo base-promoted elimination to give two possible products, **5-58** or **5-59**. The new double bond in each of these products is stabilized by conjugation with the carbonyl group and at least one of the aromatic rings.

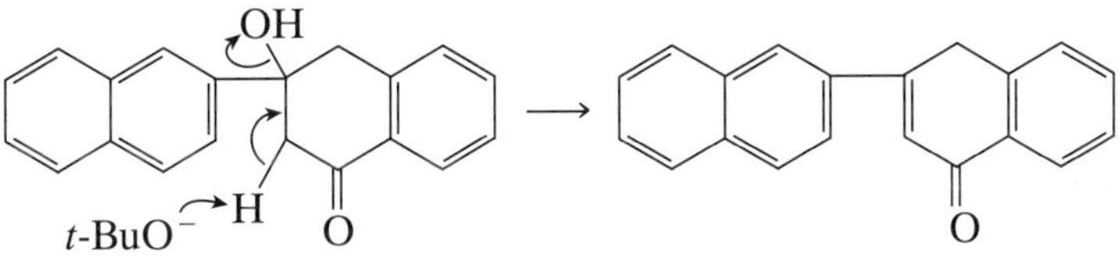

5-58

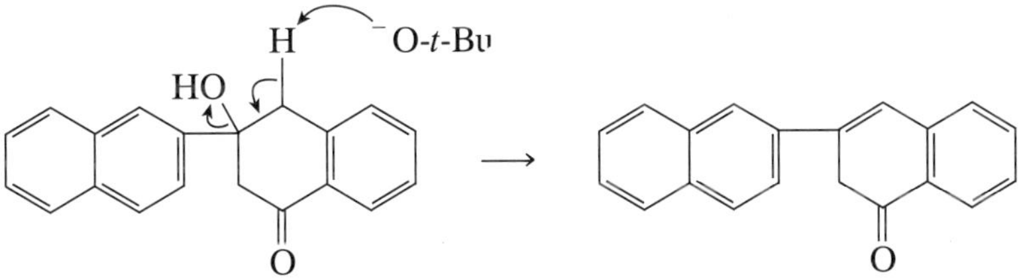

5-57   **5-59**

Either **5-58** or **5-59** can readily tautomerize to give the final product. Both mechanisms are very similar, so only one is shown.

Problem 5.15
continued

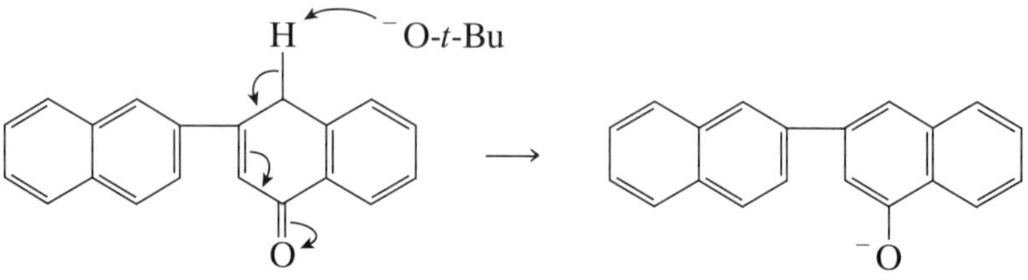

5-58   **5-60**

Upon acidic workup, the naphthoxide ion **5-60** is protonated to give **5-35**.

Another plausible mechanism for the final stages of the reaction is tautomerization of **5-57** to **5-61**, followed by elimination directly to the phenol. (Structure **5-62** would not be produced because the aromaticity of the right-hand ring is interrupted.) However, the mechanism previously discussed is preferable because alcohol **5-61** would not be as stable as **5-58** or **5-59**.

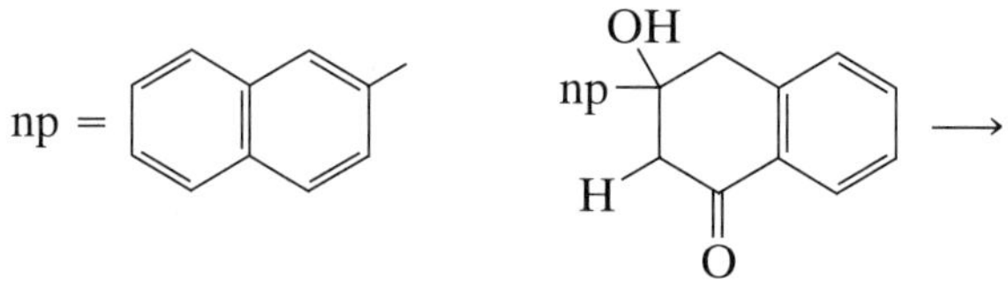

np =

5-57

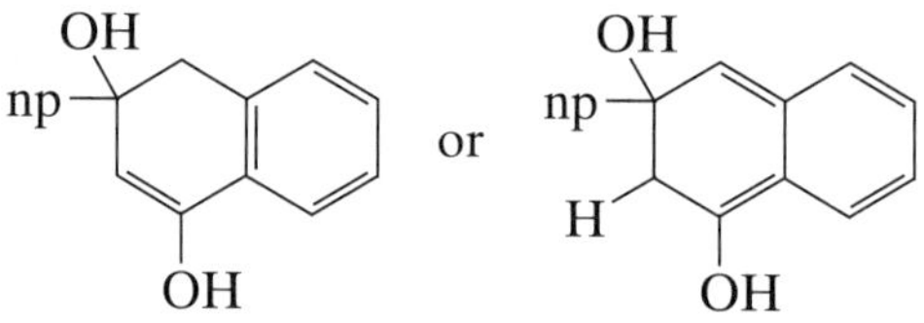

5-61   or   **5-62**

e. The overall reaction yield is 69%, with only 10 mol% hexabutylditin present, which indicates that this must be a chain process. Because iodide is present, probable initiation steps would be photochemical decomposition of the ditin compound to tri-*n*-butyltin radicals, followed by abstraction of iodine from the starting material.

Problem 5.15
continued

$$(n\text{-Bu})_3\text{SnSn}(n\text{-Bu})_3 \xrightarrow{h\nu} 2\,n\text{-Bu}_3\text{Sn}\cdot$$

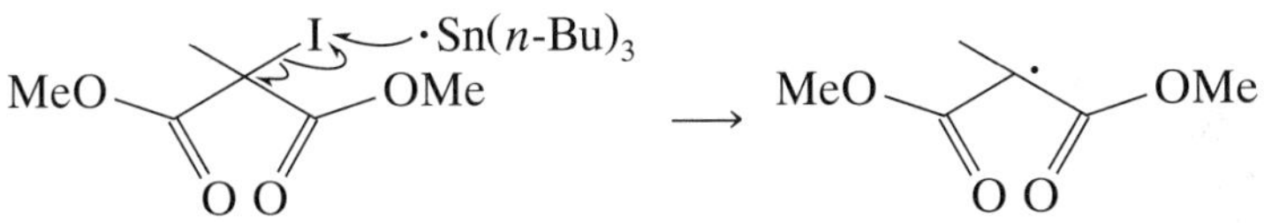

5-63

Radical **5-63** can add regiospecifically to the 1-alkene to produce an intermediate radical, which can then abstract iodine from the starting ester to produce **5-36** and a new radical to continue the chain.

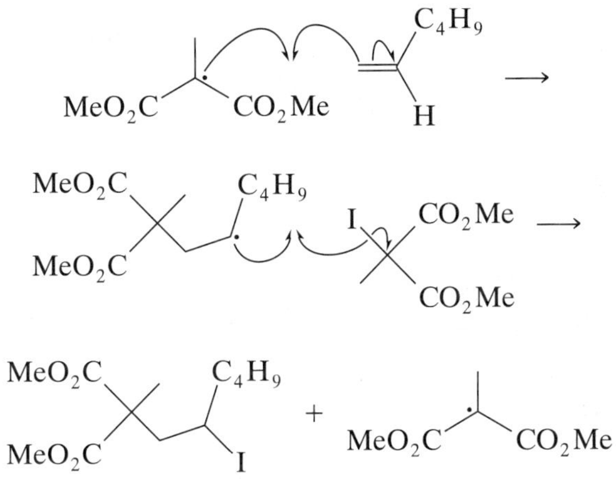

5-36

Because both homolytic and heterolytic cleavages readily occurs with C—I bonds, plausible reaction mechanisms can be written with either radical or ionic intermediates. However, in the nonpolar solvent benzene, the radical mechanism is more likely because the ionic mechanism requires the carbocation and anion to be separated, and these species cannot be stabilized by interaction with the nonpolar solvent.

6

Pericyclic Reactions

I. INTRODUCTION

Many reactions involve a cyclic transition state. Of these, some involve radical or ionic intermediates and proceed by stepwise mechanisms. *Pericyclic* reactions are *concerted*, and in the transition state the redistribution of electrons occurs in a single continuous process. In this chapter, we will consider several different types of pericyclic reactions, including electrocyclic transformations, cycloadditions, sigmatropic rearrangements, and the ene reaction.

A. Types of Pericyclic Reactions

• **Electrocyclic transformations** involve intramolecular formation of a ring by bond formation at the ends of a conjugated π system. The product has one more σ bond and one less π bond than the starting material. The reverse reaction, ring opening of a cyclic polyene, is also an electrocyclic process.

• **Cycloadditions** involve bonding between the termini of two π systems to produce a new ring. The product has two more σ bonds and two less π bonds than the reactants. Common examples are the Diels–Alder reaction and many 1,3-dipolar cycloadditions.

Diels–Alder reaction

• **Sigmatropic reactions.** In these, an allylic σ bond at one end of a π system appears to migrate to the other end of the π system. The π bonds change position, but the total number of σ and π bonds is the same, as in the Cope and Claison rearrangements.

Cope rearrangement

• **Ene reactions.** These combine aspects of cycloadditions and sigmatropic reactions. They may be inter- or intramolecular.

B. Theories of Pericyclic Reactions

Pericyclic reactions can be initiated either thermally or photochemically, but in either case they show great stereospecificity. The conditions under which pericyclic reactions occur and the stereochemistry of the products formed are dependent on the symmetry characteristics of the *molecular orbitals* involved. On this basis, pericyclic processes are classified as either *symmetry-allowed* or *symmetry-forbidden*. There are several theoretical approaches for deriving the *selection rules* governing concerted pericyclic reactions, and each of these theories represents the work of many contributors. All of these approaches assume that a cyclic transition state is formed by π orbital overlap.

• **Conservation of Orbital Symmetry.** This approach relies on a detailed analysis of the symmetry properties of the molecular orbitals of starting materials and products. Orbital correlation diagrams link the orbital characteristics of starting materials and products.
• **Frontier Orbital Theory.** This view focuses on the symmetry characteristics of the highest occupied and lowest unoccupied orbitals, particularly the symmetry at the termini of the systems.
• **Moebius–Huckel Theory.** The molecular orbital array of the transition state is analyzed in terms of aromaticity, which is determined by the number of π electrons.

Each of these theoretical approaches leads to the same predictions regarding reaction conditions and stereochemistry. For a wide range of reactions, *the selection rules can be used empirically, based on a simple method of electron counting, without regard to their theoretical basis.* The selection rules for pericyclic reactions relate three features:

1. The number of π electrons involved.
2. The method of activation.
3. The stereochemistry.

If any two of these features are specified, the third feature is determined by the selection rules.

Note: Reactions involving a cyclic transition state are not always concerted, and the selection rules and their stereochemical consequences apply only to *concerted* pericyclic processes. Indeed, *failure to conform to the selection rules is usually taken as proof that a reaction does not proceed by a concerted mechanism.* On the other hand, failure to react may simply mean that the reaction is symmetry-allowed, but does not occur because the thermodynamics is unfavorable. This point will be discussed further in the next section.

The following sections present an empirical approach to applying the selection rules. The chapter continues with a basic introduction to the analysis of symmetry properties of orbitals and the application of orbital correlation diagrams to the relatively simply cyclobutene-butadiene interconversion; it concludes with some examples of the frontier orbital approach to pericyclic reactions.

2. ELECTROCYCLIC REACTIONS

Electrocyclic reactions are intramolecular ring openings or ring closures. The interconversion of substituted cyclobutenes and butadienes under different conditions illustrates which modes of reaction are symmetry-allowed and the stereochemical consequences of the selection rules. (In the reactions that follow, because there are no polar substituents, the directions of electron flow are arbitrary.) By the principle of microscopic reversibility, the bonding changes involved in going from starting material to product are the exact reverse of the changes involved in going from product to starting material. Thus, for electrocyclic reactions, any analysis we make for cyclization also applies for the ring-opening reaction and vice versa.

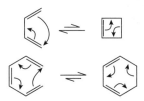

Whereas, in theory, symmetry-allowed electrocyclic reactions may proceed in either direction, in practice one side of the equation is usually favored over the other. Because of the large strain energy of the four-membered ring, the equilibrium for the thermal opening of cyclobutene rings is usually favored over the reverse reaction (ring closure). On the other hand, equilibrium usually favors the six-membered ring rather than the ring-opened compound.

A. Selection Rules for Electrocyclic Reactions

Like other pericyclic reactions, electrocyclic reactions may be initiated either thermally or photochemically. The selection rules enable us to correlate the stereochemical relationship of the starting materials and products with the method of activation required for the reaction and the number of π electrons in the reacting system.

Hint 6.1   To apply the selection rules for electrocyclic reactions, count the number of π electrons in the open-chain polyene.

Example 6.1.   *Ring opening of cis-3,4-dimethylcyclobutene.*

The cyclobutene–butadiene interconversion involves four π electrons and is designated a π^4 process. Note that by the principle of microscopic reversibility, the number of π electrons involved in the transformation is the same for ring opening as for ring closing. Once we know the number of π electrons involved in an electrocyclic reaction and the method of activation, the stereochemistry of the process is fixed according to the rules outlined in Table 6.1.

B. Stereochemistry of Electrocyclic Reactions (Conrotatory and Disrotatory Processes)

Example 6.2.   *The thermal ring opening of cis-3-chloro-4-methyl-cyclobutene, a conrotatory process.*

By applying Hint 6.1, we see that the number of π electrons involved in the transformation is four. The reaction takes place thermally, so that the selection rules as outlined in Table 6.1 indicate that a conrotatory process should occur. The following discussion will show what this means.

TABLE 6.1   Selection Rules for Electrocyclic Reactions

Number of electrons	Mode of activation	Allowed stereochemistry[a]
$4n$	Thermal	Conrotatory
	Photochemical	Disrotatory
$4n + 2$	Thermal	Disrotatory
	Photochemical	Conrotatory

[a]The terms *conrotatory* and *disrotatory* are explained in Section 6.B.

When the cyclobutene ring is transformed into a butadiene, the C-3—C-4 bond breaks. As it does so, there is rotation about the C-1—C-4 bond and the C-2—C-3 bond, so that the substituents on the breaking C-3—C-4 σ bond rotate into the plane of the conjugated diene system of the product. In the starting cyclobutene, the methyl and chloro groups lie above the plane of the four ring carbons; in the product, these groups lie in the same plane as the four carbon atoms of the butadiene. If all possible rotations about the C-3—C-4 σ bond were allowed, four different products would be formed, two by conrotatory processes and two by disrotatory processes. We will take a closer look at this process.

Conrotatory Process

In a conrotatory process, the substituents on C-3 and C-4 rotate *in the same direction*.

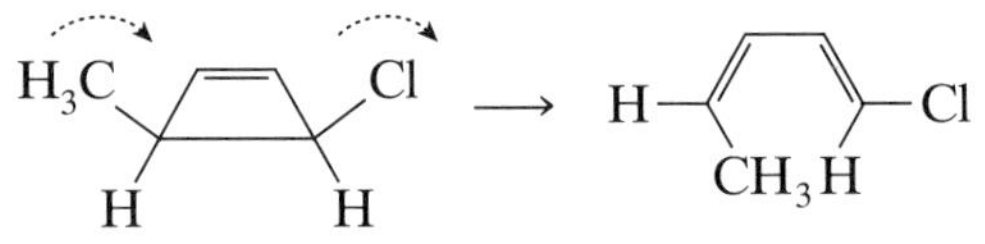

6-2

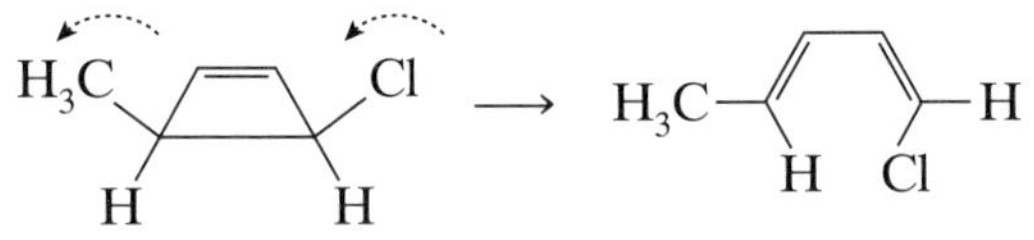

6-3

Clockwise rotation of both substituents gives product **6-2**, whereas *counterclockwise* rotation gives product **6-3**. (The arrows in the figure show the direction of rotation, not the movement of electrons.)

The selection rules predict that the thermal ring opening should be conrotatory for *cis*-3-chloro-4-methylcyclobutene, but they do not distinguish between products **6-2** and **6-3**. Both products are allowed by the selection rules, but experimentally only product **6-2** is formed. An explanation for this preference of one allowed process over the other is beyond the scope of this book; however, possible explanations can be found in the following references: Dolbier, W. R., Jr.; Koroniak, H.; Burton, D. J.; Bailey, A. R.; Shaw, G. S.; Hansen, S. W. *J. Am. Chem. Soc.* **1984**, *106*, 1871−1872; Rondan, N. G.; Houk, K. N. *J. Am. Chem. Soc.* **1985**, *107*, 2099−2111; Krimse, W.; Rondan, N. G.; Houk, K. H. *J. Am. Chem. Soc.* **1984**, *106*, 7989−7991.

Disrotatory Process

In a disrotatory process, the substituents on C-3 and C-4 rotate *in opposite directions*.

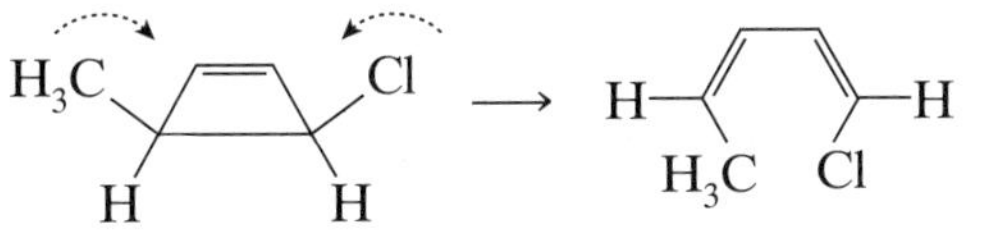

6-4

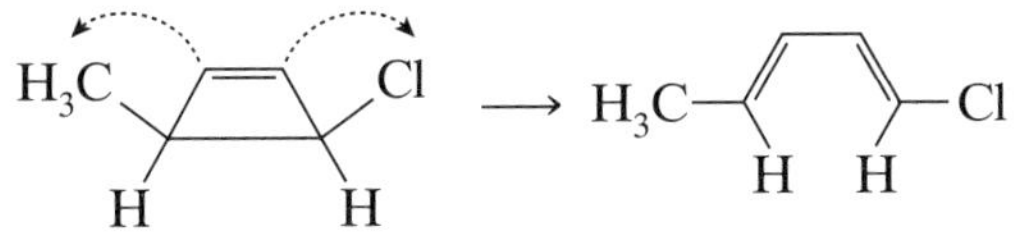

6-5

Rotation of the substituents *toward one another* would give the product **6-4**, whereas rotation of the substituents *away from one another* would give the product **6-5**. For the thermal reaction, neither of these rotations is allowed by the selection rules.

To summarize the results in the preceding example: Each of the four rotational modes of ring opening leads to a particular stereoisomer of the product. For the concerted ring opening of a π^4 system, only the two conrotatory modes are allowed. Of these, only one is observed experimentally.

By applying Hint 6.1 and the rules in Table 6.1, decide whether the following *thermal* reactions are symmetry-allowed or symmetry-forbidden.   **PROBLEM 6.1**

a.

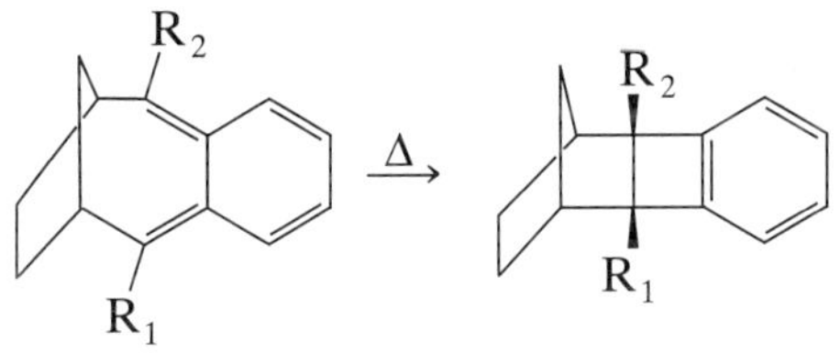

Spellmeyer, D. C.; Houk, K. N.; Rondan, N. G.; Miller, R. D.; Franz, L.; Fickes, G. N. *J. Am. Chem. Soc.* **1989**, *111*, 5356–5367.

b.

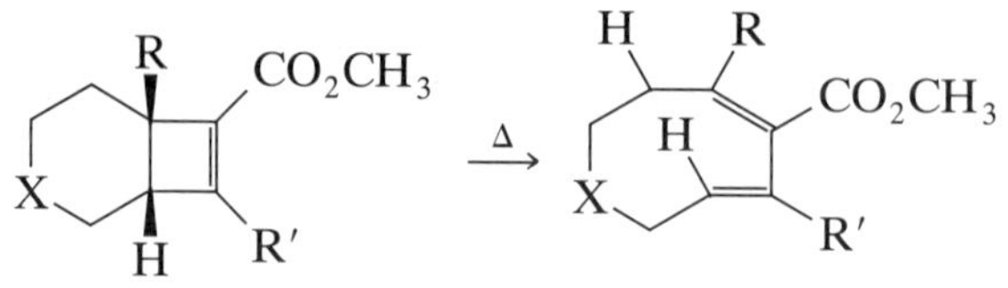

$$X = (CH_2)_3$$

Vos, G. J. M.; Reinhoudt, D. N.; Benders, P. H.; Harkema, S.; Van Hummel, G. J. *J. Chem. Soc., Chem. Commun.* **1985**, 661–662.

Hint 6.2

When applying the selection rules, only those π electrons taking part in the cyclization are counted.

Example 6.3.   *The thermal cyclization of cyclooctatetraene to bicyclo[4.2.0]-octatriene.*

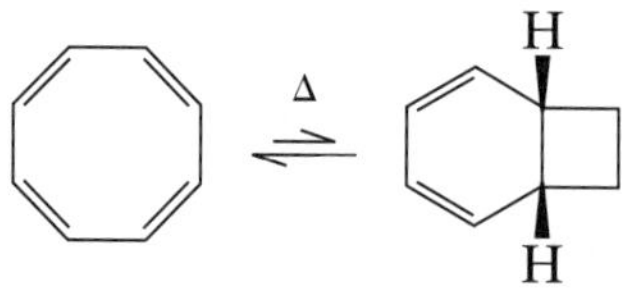

Because the cyclic transition state of the electrocyclic reaction involves joining the ends of the π system, this reaction may be regarded as a π^4 ring closure to a cyclobutene or a π^6 ring closure to a cyclohexadiene. Under thermal conditions, the selection rules require a conrotatory ring closure for the π^4 system, leading to a *trans* ring junction, which is unlikely for such a fusion because of ring strain. On the other hand, the π^6 system should close by a disrotatory process to give a *cis* ring junction. The product observed experimentally is consistent with a disrotatory closure of the π^6 system. Cyclooctatetraene is tub-shaped rather than planar, and the disrotatory process can be visualized as shown in the following drawing:

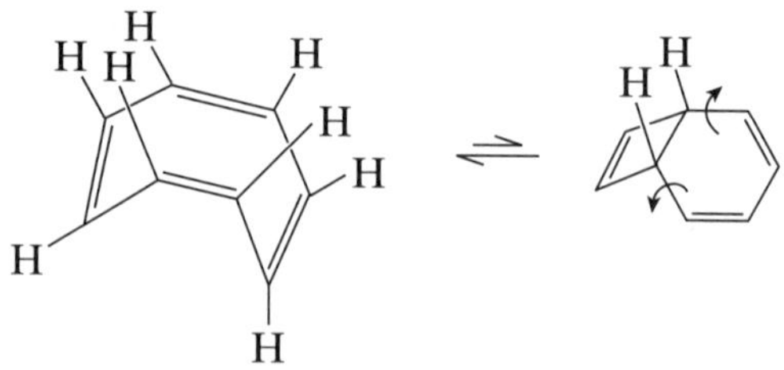

In practice, cyclooctatetraene is the more stable isomer, and the presence of the bicyclic compound has been demonstrated only through trapping experiments.

What is the relative stereochemistry of the ambiguous groups in the following concerted processes?   **PROBLEM 6.2**

a.

b.

Huisgen, R.; Dahmen, A.; Huber, H. *J. Am. Chem. Soc.* **1967**, *89*, 7130–7131.

c.

Sauter, H.; Gallenkamp, B.; Prinzbach, H. *Chem. Ber.* **1977**, *110*, 1382–1402.

PROBLEM 6.3   **Is the following reaction symmetry-allowed or symmetry-forbidden? Explain.**

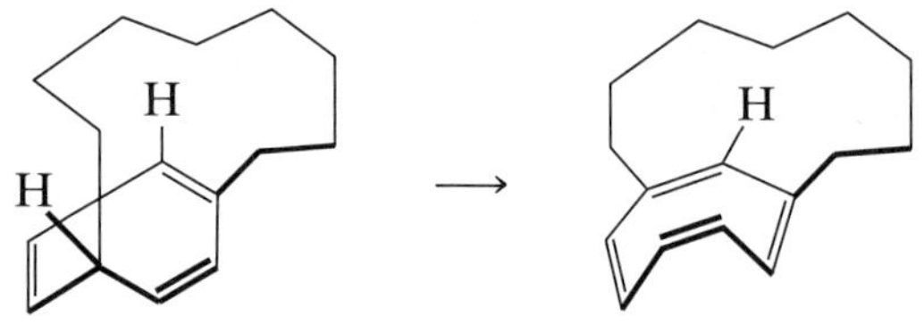

Paquette, L. A.; Want, T.-Z. *J. Am. Chem. Soc.* **1988**, *110*, 3663–3665.

PROBLEM 6.4   **Consider the following concerted process:**

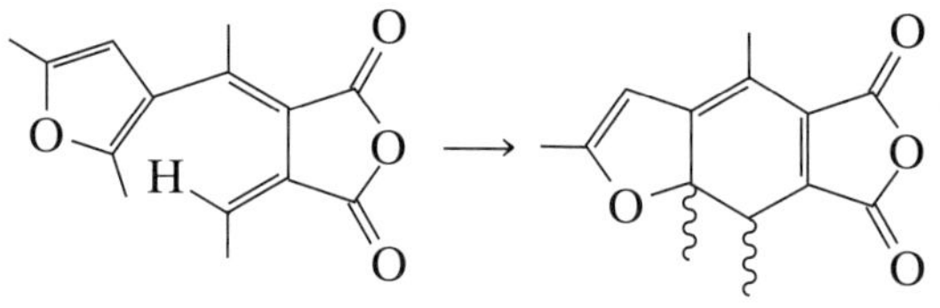

Darcy, P. J.; Heller, H. G.; Strydon, P. J.; Whitall, J. *J. Chem. Soc., Perkin Trans. I,* **1981**, 202–205.

a. What is the relative stereochemistry of the product formed when the reaction is initiated by heat?

b. What is the relative stereochemistry of the product formed when the reaction is initiated by light?

c. What is the relative stereochemistry of the product of the photochemical ring opening of the product formed in part a?

C. Electrocyclic Reactions of Charged Species (Cyclopropyl Cations)

The selection rules can be applied to charged species as well as to neutral molecules. The only requirement is that the reaction be a concerted process involving electrons in overlapping p orbitals. For example, the conversion of a cyclopropyl cation to the allyl cation can be considered as a π^2-electrocyclic process. For this process, the selection rules predict a disrotatory process.

Example 6.4. *The solvolysis of cyclopropyl tosylates.*

The relative rates of concerted ring opening for substituted cyclopropyl tosylates can be explained on the basis of the selection rules and the principle of maximum orbital overlap. The relative rates of solvolysis for two different dimethylcyclopropyl tosylates are shown in the accompanying diagram:

Rel. rate in
HoAc, 150°C

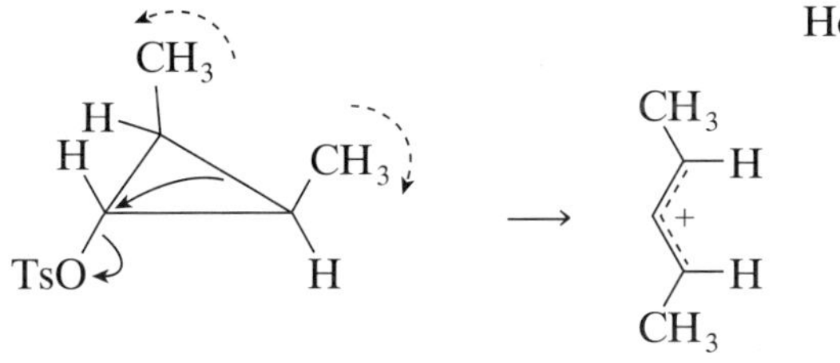

4500

A

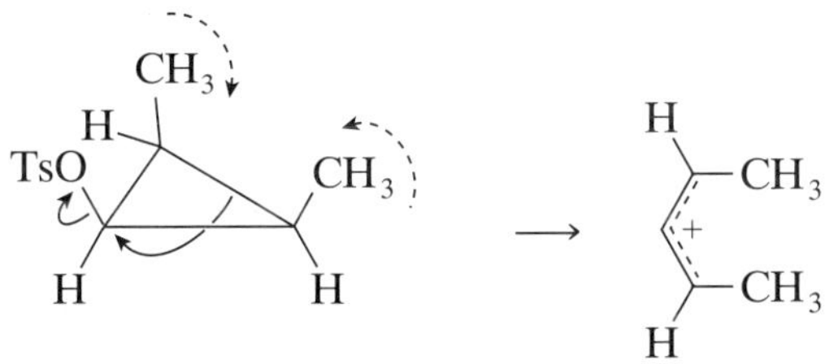

1

B

von R. Schleyer, P.; Van Dine, G. W.; Schollkopf, U.; Paust, J. *J. Am. Chem. Soc.* **1966**, *88*, 2868–2869.

The large difference in relative rates suggests that the rate-limiting step does not involve unimolecular ionization by loss of the tosyl group (an $S_N 1$

process). If it did, we would expect that the all-*cis* isomer would react faster due to relief of steric interaction between the tosyl and methyl groups. Instead, the results can be explained by assuming that loss of the tosyl group occurs with concerted ring opening to the allyl cation and that loss of the tosyl group is assisted by backside reaction of the electrons in the breaking C—C bond of the cyclopropyl group.

As shown in the diagram, opening of a cyclopropyl cation by a disrotatory process can occur in two ways (compare the discussion of the disrotatory and conrotatory openings of the cyclobutene system). The methyl groups can rotate outward (away from one another) or inward (toward one another). Disrotatory opening in which the methyl groups rotate away from each other gives the allyl cation A, whereas if the methyl groups rotate toward each other, the result is the allyl cation B. The opening in which the methyl groups rotate outward to give allyl cation A is expected to be more favorable due to lack of steric interaction. (Note that because of the delocalized π system in the allyl cation, there is restricted rotation around the C—C bonds and A and B are not interconvertible.) When the tosyl group is *trans* to the methyl groups, the electrons move so that they can assist the loss of the tosyl group by a backside reaction reminiscent of the neighboring group effect in the S_N2 reaction. For the isomer in which the tosyl group is *cis* to the methyl groups, outward rotation of the methyl groups would result in an increase in electron density in the vicinity of the tosyl group and a consequent increase in repulsion between centers of increased electron density.

Example 6.5. *Geometric constraint to disrotatory ring opening of a bicyclo[3.1.0]hexane system.*

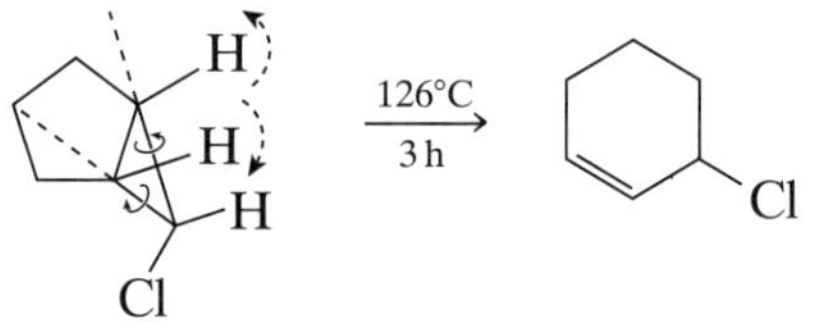

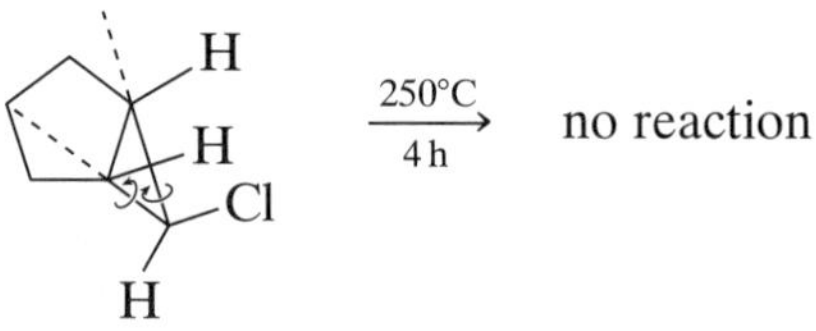

Baird, M. S.; Lindsay, D. G.; Reese, C. B. *J. Chem. Soc. C* **1969**, 1173–1178.

In small, fused cyclopropyl systems, only the disrotatory mode that moves the bridgehead hydrogens outward is geometrically feasible. This means that the electrons of the breaking cyclopropyl C—C bond can assist in the loss of the chloride group only when it is *trans* to the bridgehead hydrogens.

Explain why the reactivity is so different in the following two reactions: **PROBLEM 6.5**

$\xrightarrow[4\ h]{110°C}$ no reaction

Jefford, C. W.; Hill, D. T. *Tetrahedron Lett.* **1969**, 1957–1960.

---

3. CYCLOADDITIONS

A cycloaddition is the reaction of two (occasionally more) separate π systems, in which the termini are joined to produce a ring. Cycloadditions may be intermolecular or intramolecular. One way of describing a cycloaddition is to record separately the number of electrons in each component involved in the reaction.

A. Terminology of Cycloadditions

Number of Electrons

Cycloadditions can be described on the basis of the number of electrons of each of the components. Additional symbols are used to designate the type of orbital and the type of process involved.

Example 6.6. *The Diels–Alder reaction of (2E, 3E)-2,4-hexadiene and dimethyl maleate.*

This is a six-electron process because there are four π electrons from the diene and two π electrons from the dienophile. It is also referred to as a [4 + 2] cycloaddition. Note that the carbonyl groups in the dimethyl maleate starting material are conjugated with the π bond undergoing reaction. However, because the π electrons of these carbonyl groups are not forming new bonds in the course of the reaction, they are not counted.

Stereochemistry (Suprafacial and Antarafacial Processes)

A cycloaddition reaction can be classified not only by the number of electrons in the individual components, but also by the stereochemistry of the reaction with regard to the plane of the π system of each reactant. For each component of the reaction, there are two possibilities: the reaction can take place on only one side of the plane or across opposite faces of the plane. If the reaction takes place across only one face, the process is called *suprafacial*; if across both faces, *antarafacial*. The four possibilities are shown in the following diagram:

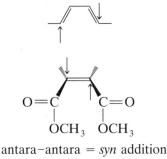

supra–supra = *syn* addition   antara–antara = *syn* addition

supra–antara = *anti* addition   antara–supra = *anti* addition

Essentially, a *suprafacial–suprafacial* or an *antarafacial–antarafacial* cycloaddition is equivalent to a concerted *syn* addition. A *suprafacial–antarafacial* or an *antarafacial–suprafacial* process is equivalent to a concerted *anti* addition. The Diels–Alder reaction is suprafacial for both components, so that the stereochemical relationships among the substituents are maintained in the product. In Example 6.6, suprafacial addition to the dienophile component means that the two carbomethoxy groups that are *cis* in the starting material also are *cis* in the product. Suprafacial reaction at the diene component leads to a *cis* orientation of the two methyl groups in the product.

When classifying cyclizations, the subscripts "s" and "a" are used to designate suprafacial and antarafacial, respectively. Thus, a more complete designation of the Diels–Alder reaction is $[\pi_s^4 + \pi_s^2]$. This type of designation can be applied to other concerted processes, such as electrocyclic reactions, sigmatropic rearrangements, and the ene reaction; however, we will use these designations only for cycloadditions.

Example 6.7. *A cycloaddition reaction with three π components.*

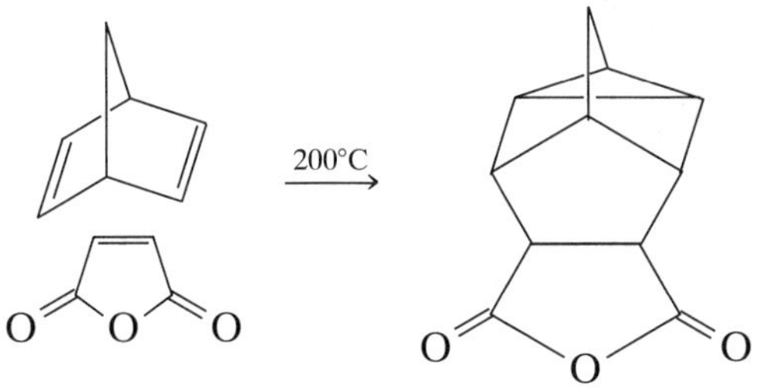

78%

Cookson, R. C.; Dance, J.; Hudec, J. *J. Chem. Soc.* **1964**, 5417–5422.

Because the two π bonds in the bicycloheptadiene are not conjugated, each is designated separately in the description of the reaction. Only the carbons, at each end of the C=C bond of maleic anhydride, are forming new bonds; thus, only the two π electrons of this bond are counted and this is a $[\pi_s^2 + \pi_s^2 + \pi_s^2]$ cycloaddition.

Number of Atoms

Some confusion has been occasioned by the introduction of another kind of terminology based on counting the number of atoms of the component systems taking part in the cycloaddition. In this system, the total number of atoms in each component is counted. This number includes both termini and all of the atoms in between. The description of a cycloaddition based on the number of atoms does not give the same designation as that based on the number of electrons.

Example 6.8. *Counting the atoms involved in a cycloaddition.*

Consider the following reaction:

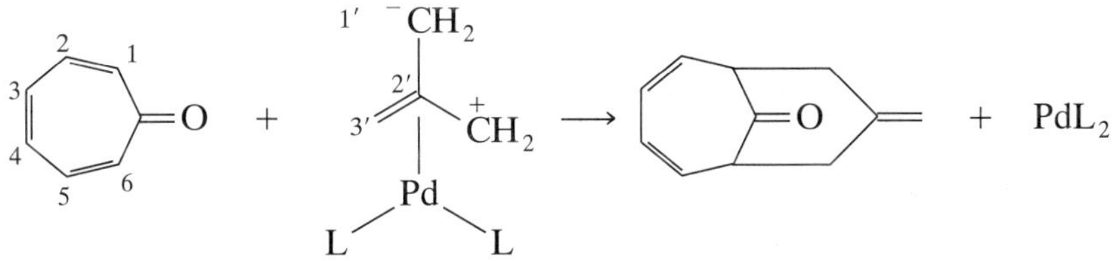

Trost, B. M.; Seoane, P. R. *J. Am. Chem. Soc.* **1987**, *109*, 615–617.

The authors call this a [6 + 3] cycloaddition, because the reacting portion of tropone (the ketone) contains six atoms and the alkene component contains three atoms. The mechanism could be shown with arrows as follows:

In tropone, 6π electrons are reacting, and in the alkene, 4π electrons are reacting. Therefore, if this is a concerted reaction, it also could be called a [6 + 4] cycloaddition. If the distinction is made that [6 + 3] refers to the cycloadduct and [6 + 4] refers to the cycloaddition, the two methods of nomenclature are compatible. However, this distinction is not always made in the literature, and unless one thinks about the details of the particular reaction, the descriptions may be confusing.

B. Selection Rules for Cycloadditions

Hint 6.3   To apply the selection rules for cycloadditions, add the number of π electrons from each component undergoing reaction and then apply the rules outlined in Table 6.2.

TABLE 6.2   Stereochemical Rules for Cycloaddition Reactions

Number of electrons	Mode of activation	Allowed stereochemistry
$4n$	Thermal	Supra–antara
		Antara–supra
	Photochemical	Supra–supra
		Antara–antara
$4n + 2$	Thermal	Supra–supra
		Antara–antara
	Photochemical	Supra–antara
		Antara–supra

Many reactions encountered are of the supra–supra variety. For these systems, a good rule of thumb is that $4n$ systems are activated photochemically and $4n + 2$ systems are activated thermally.

Designate the following cycloadditions according to the number of   **PROBLEM 6.6**
electrons contributed by each component. Is the stereochemistry shown
in accordance with that predicted by the selection rules?

a.

de Meijere, A.; Erden, I.; Weber, W.; Kaufmann, D. *J. Org. Chem.* **1988**, *53*, 152–161.

The reaction of the unstable intermediate is called a cycloreversion of
retro–Diels–Alder reaction. Ordinarily it is designated by considering the
reverse cycloaddition of the products of the reaction.

b. 2

Machiguchi, T.; Hasegawa, T.; Itoh, S.; Mizuno, H. *J. Am. Chem. Soc.* **1989**, *111*, 1920–1921.

c.

Takeshita, H.; Sugiyama, S.; Hatsui, T. *J. Chem. Soc., Perkin Trans. II* **1986**, 1491–1493.

PROBLEM 6.6
continued

d.

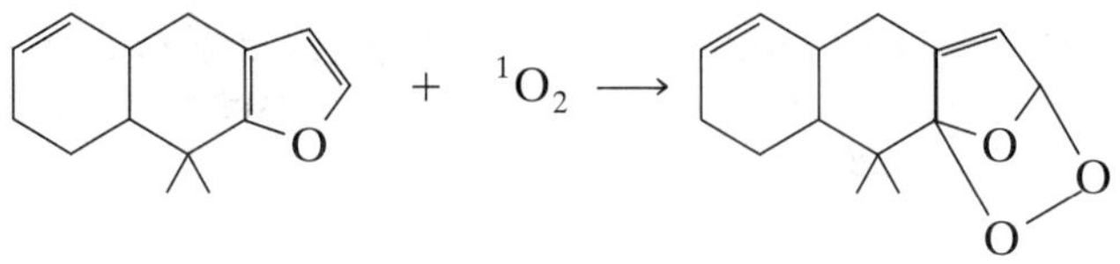

$^1O_2$ = singlet oxygen

Carte, B.; Kernan, M. R.; Barrabee, E. B.; Faulkner, D. J.; Matsumoto, G. K.; Clardy, J. *J. Org. Chem.* **1986**, *51*, 3528–3532.

PROBLEM 6.7 Give the complete formulation (π_s^2, etc.) for each of the following cycloadditions.

a.

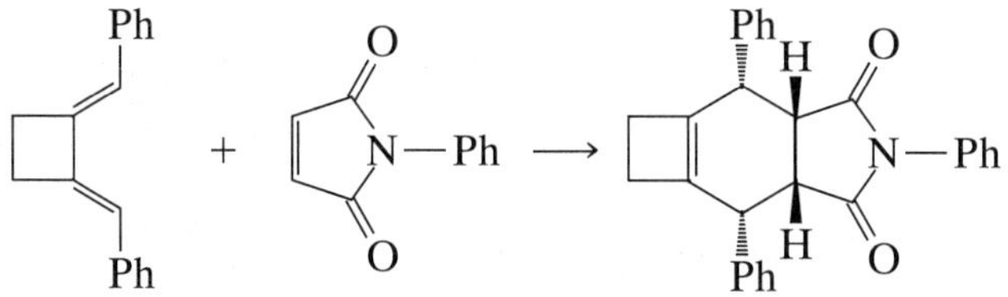

Minami, T.; Harui, N.; Taniguchi, Y. *J. Org. Chem.* **1986**, *51*, 3572–3576.

b.

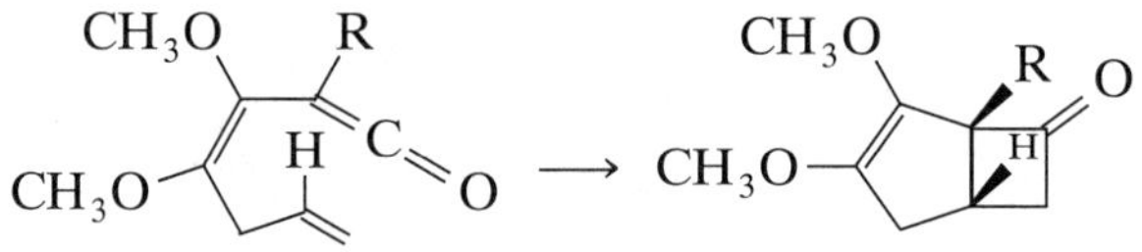

Xu, S. L.; Moore, H. W. *J. Org. Chem.* **1989**, *54*, 6018–6021.

PROBLEM 6.8 If it were concerted, how would the following reaction be described by both systems of nomenclature?

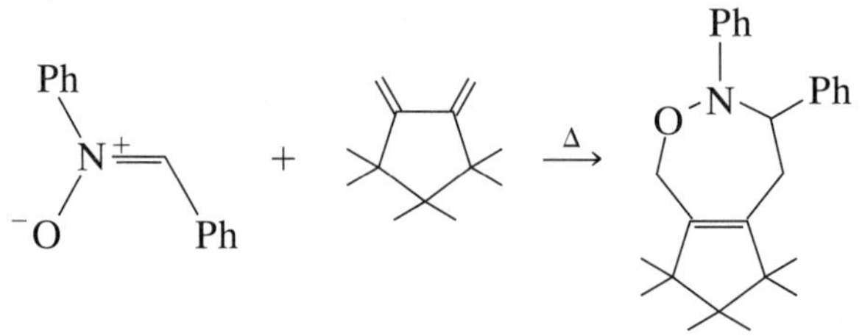

Baran, J.; Mayr, H. *J. Am. Chem. Soc.* **1987**, *109*, 6519–6521.

PROBLEM 6.9   **What is the product from an intramolecular [4 + 2] cycloaddition of the following molecule?**

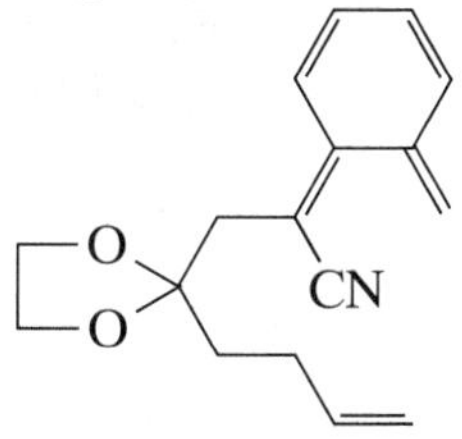

Kametani, T.; Suzuki, Y.; Honda, T. *J. Chem. Soc., Perkin Trans. I* **1986**, 1373–1377.

C. Secondary Interactions

In the Diels–Alder reaction of cyclopentadiene with dimethyl maleate, both *exo* and *endo* products are theoretically possible. Only the *endo* product **6-6** is found.

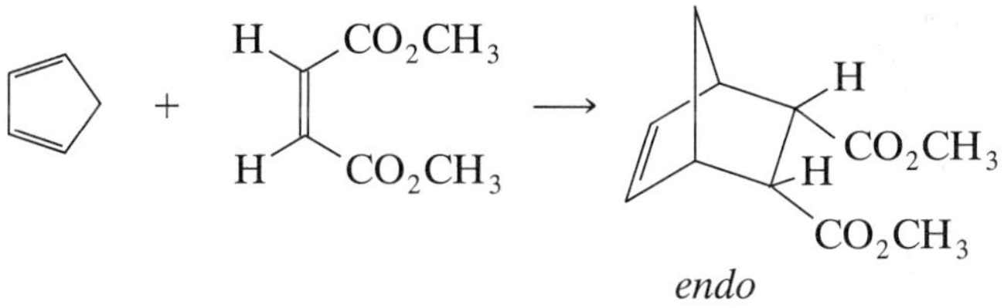

endo

6-6

→ **6-6**

exo

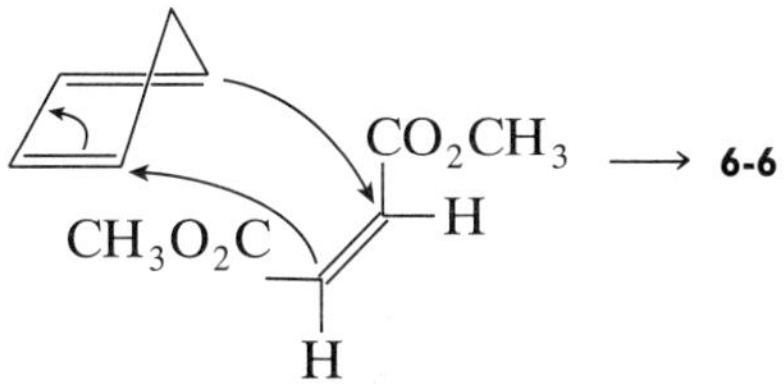

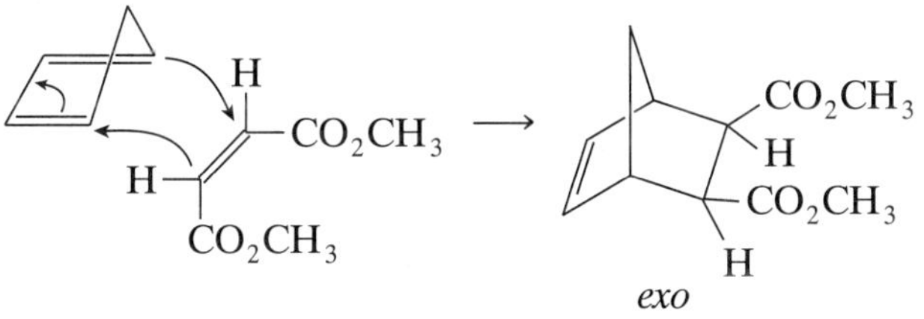

6-7

Because the carbonyl groups are part of the π system of the dienophile, there is an opportunity for secondary interactions between orbitals that are not involved in the bonding changes taking part in the cycloaddition. Molecular orbital calculations have indicated that stabilizing interactions can take place when the reactants are oriented in such a way as to produce the *endo* isomer.

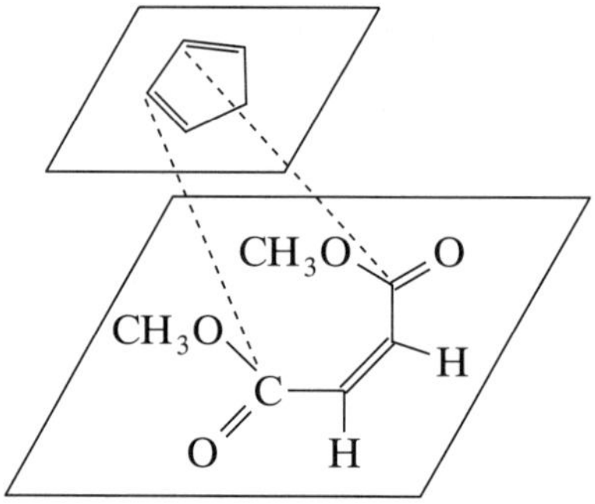

These secondary attractive interactions are represented by the dashed lines between the carbonyl groups and the diene system. Visual inspection of the orientation required for *endo* addition shows that there is greater overlap of the molecular orbitals of the two components than for the orientation that leads to *exo* addition.

Molecular orbital interactions are only one of the factors that can influence the *exo:endo* ratio. Solvent interactions and the structure of the starting materials also are important, and these factors can result in predominantly *exo* addition products.

D. Cycloadditions of Charged Species

Allyl Cations

Example 6.9. *The allyl cation as dienophile.*

A mechanism with several steps can be envisioned for the following reaction:

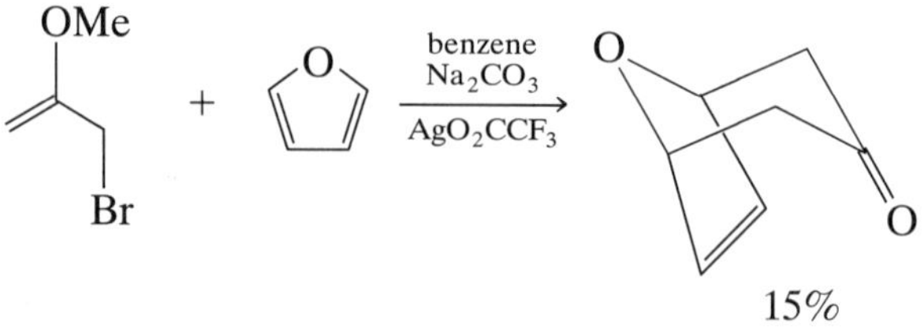

The first step, ionization of the bromide to the 2-methoxyallyl cation, **6-8**, is assisted by the silver ion:

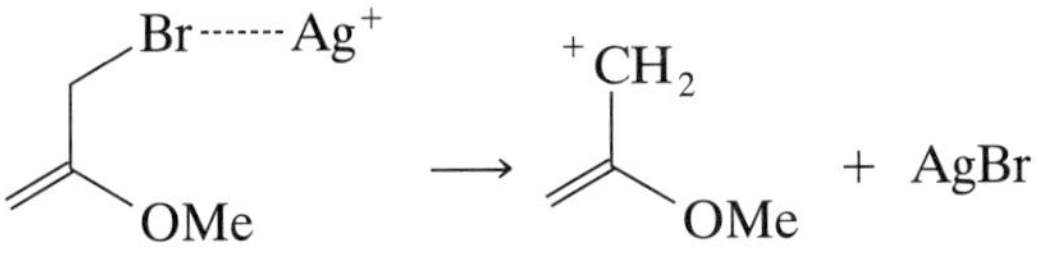

6-8

Cycloaddition occurs between cation **6-8**, which contains 2π electrons, and furan, which has 4π electrons, to give the cyclized cation, **6-9**:

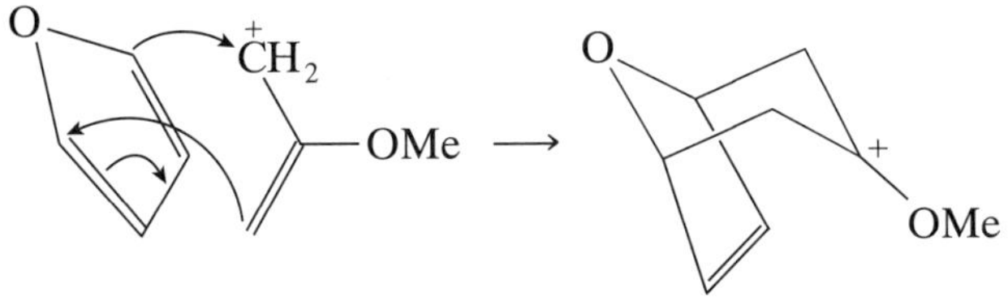

6-8   **6-9**

Under the workup conditions (dilute nitric acid) **6-9** hydrolyzes to the product ketone.

Hill, A. E.; Greenwood, G.; Hoffmann, H. M. R. *J. Am. Chem. Soc.* **1973**, *95*, 1338–1340.

1,3-Dipoles

From a synthetic standpoint, the 1,3-dipolar cycloaddition is a very important reaction. In this [4 + 2] cycloaddition, the four-electron component is dipolar in nature, and the two-electron component usually is referred to as the dipolarophile. When the thermal reaction is concerted, these reactions are suprafacial in both components, i.e., they are $[\pi_s^4 + \pi_s^2]$ cycloadditions.

Example 6.10. *1,3-Dipolar cycloaddition of a nitrile oxide and an alkyne.*

In the following reaction, the 4π-electron system of the nitrile oxide reacts with the 2π system of one of the acetylenic bonds.

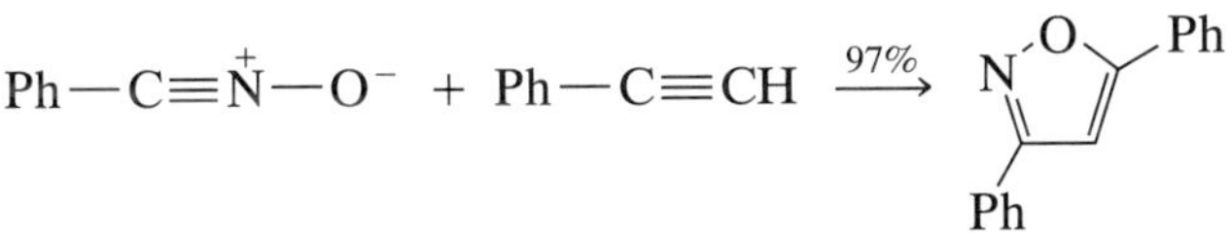

Huisgen, R. *Angew. Chem., Int. Ed. Engl.* **1963**, *2*, 565–598.

According to the newer terminology, they react to form what can be called a [3 + 2] adduct. If the mechanism is written with arrows, the usual rules apply. That is, the flow of electrons is away from negative charge and toward positive charge.

The regiochemistry of 1,3-dipolar additions can be explained on the basis of frontier orbital theory. In the examples and problems, the major products of the reactions are given. For further information, see Houk, K. N. *Acc. Chem. Res.* **1975**, 361–369.

Example 6.11.   *The intramolecular 1,3-dipolar cycloaddition of an azide.*

Tsai, C.-Y.; Sha, C.-K. *Tetrahedron Lett.* **1987**, *28*, 1419–1420.

This is a concerted reaction in which four of the π electrons of the azide group undergo cycloaddition with the 2π electrons of the carbon–carbon double bond of the α-β-unsaturated ester:

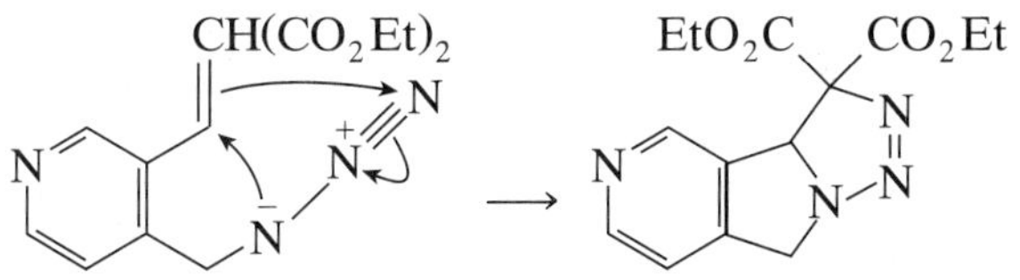

These kinds of intramolecular cycloadditions are very powerful synthetic tools for the synthesis of fused ring systems.

Write the possible products of the 1-3,-dipolar cycloaddition of 6-10   PROBLEM 6.10 with acrylonitrile. Many 1,3-dipoles, like the nitrile ylide 6-10, are unstable and therefore are formed *in situ* when needed, as shown in the first step of the following reaction sequence.

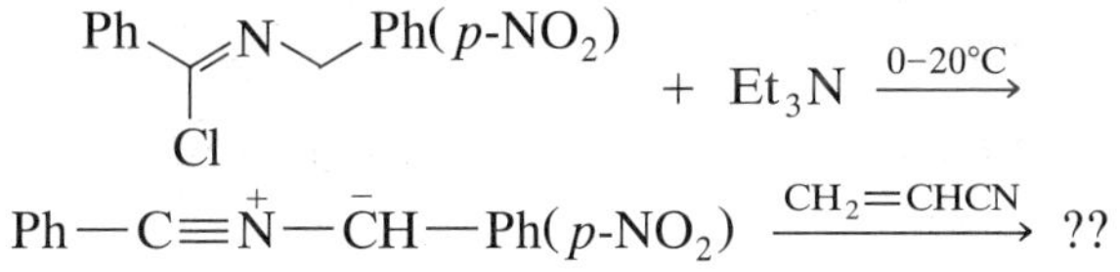

6-10

Huisgen, R. *Angew. Chem., Int. Ed. Engl.* **1963**, 2, 565–598.

Show how the following reaction might occur.   **PROBLEM 6.11**

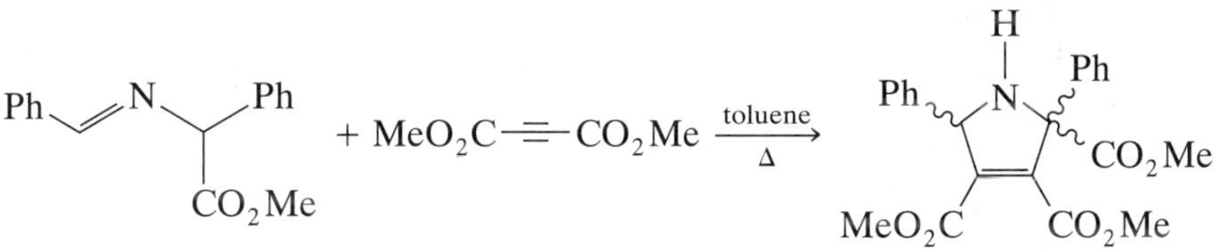

Grigg, R.; Kemp, J.; Sheldrick, G.; Trotter, J. *J. Chem. Soc., Chem. Commun.* **1978**, 109–111.

---

4. SIGMATROPIC REARRANGEMENTS

A. Terminology

In a sigmatropic reaction, movement of a σ bond takes place, producing rearrangement (tropic is from the Greek word "tropos," to turn). Common types of sigmatropic reactions are the familiar 1,2-hydride or alkyl shifts in carbocations and the Cope rearrangement.

Sigmatropic reactions can be designated by the same scheme we have used for π electron reactions, i.e., $[\pi_s^4 + \sigma_s^2]$, etc. However, sigmatropic reactions are more often designated in a different way.

Hint 6.4

To determine the order of a sigmatropic reaction, first label both atoms of the original (breaking) σ bond as 1. Then count the atoms along the chains on both sides until you reach the atoms that form the new σ bond. The numbers assigned to these atoms are given in brackets, separated by a comma, e.g., [1,5] or [3,3]. This nomenclature is illustrated in Examples 6.12 through 6.14.

Example 6.12.   *A [1,2] sigmatropic shift in a carbocation.*

The σ bond that is broken in the starting material and the σ bond formed in the product are highlighted. Hydrogen (1′) has moved from carbon 1 to carbon 2, so that the reaction is designated as a [1,2] sigmatropic shift. The 1 of this designation does not refer to the number on carbon, but to the number on hydrogen. This indicates that the same atom (hydrogen) is at one end of the σ bond in both starting material and product. The 2 of the designation, [1,2], is the number of the carbon where the new σ bond is formed, relative to the number 1 for the carbon where the old σ bond was broken.

This reaction can also be designated as $[\pi_s^0 + \sigma_s^2]$. This is more appropriate than labeling the reaction as $[\pi_s^2 + \sigma_s^0]$, because this is considered to be a hydride shift, not a proton shift.

Example 6.13. *A [3,3] sigmatropic shift.*

Consider the classical Cope rearrangement:

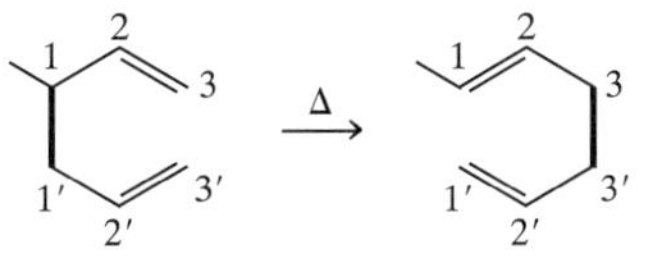

The atoms at the ends of the σ bond being broken are numbered 1 and 1'. Atoms are then numbered along each chain until the atoms at the ends of the new σ bond are reached. The atoms of the new σ bond are numbered 3 and 3'. Formally, the σ bond, originally at the 1,1' position, has moved to the 3,3' position. Thus, this is called a [3,3] sigmatropic shift.

Example 6.14. *A [1,5] sigmatropic shift.*

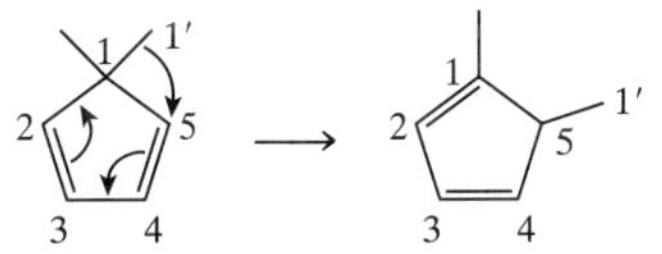

All of the carbons in the ring must be counted as part of the process. That is, this reaction is *not* a simple [1,2] shift, because the π electrons in the ring also must rearrange. Thus, this is a [1,5] shift.

Designate the type of sigmatropic shift that occurs in each of the following reactions (e.g., [1,3]). Also give the appropriate reaction designation. **PROBLEM 6.12**

a.

b.

Curran, D. P.; Jacobs, P. B.; Elliott, R. L.; Kim, B. H. *J. Am. Chem. Soc.* **1987**, *109*, 5280–5282.

c.

50%

Vedejs, E. *Acc. Chem. Res.* **1984**, *17*, 358–364.

d.

Wu, P.-L.; Chu, M.; Fowler, F. W. *J. Org. Chem.* **1988**, *53*, 963–972.

B. Selection Rules for Sigmatropic Rearrangements

For sigmatropic reactions, as for electrocyclic reactions and cycloadditions, the course of reaction can be predicted by counting the number of electrons involved and applying the selection rules. A comprehensive rationalization of all the stereochemical aspects of these reactions requires application of the frontier orbital or orbital symmetry approaches, and, at this point, we will content ourselves with pointing out the salient features of the more common reactions of this class.

TABLE 6.3 Selection Rules for Sigmatropic Hydrogen Shifts

Order	Number of electrons	Mode of activation	Allowed stereochemistry
[1,3]	$4n$	Thermal	Antarafacial
		Photochemical	Suprafacial
[1,5]	$4n + 2$	Thermal	Suprafacial
		Photochemical	Antarafacial
[1,7]	$4n$	Thermal	Antarafacial
		Photochemical	Suprafacial

For a sigmatropic reaction, the number of electrons involved is the number of π electrons plus the pair of electrons in the migrating σ bond.

Hint 6.5

Hydrogen Shifts

Thermal [1,3] hydrogen shifts are unknown. The migrating hydrogen must maintain overlap with both ends of the π system, but the geometry required by antarafacial migration makes this impossible. A few photochemical [1,3] hydrogen shifts are known. Because the hydrogen moves suprafacially with photochemical activation, the reaction is geometrically feasible, as well as being symmetry-allowed.

The following compound has been prepared and is stable at dry ice temperatures ($-78°C$), even though its isomer, toluene, an aromatic compound, is much more stable. Why is it possible to isolate the nonaromatic triene?

PROBLEM 6.13

Bailey, W. J.; Baylouny, R. A. *J. Org. Chem.* **1962**, *27*, 3476.

Example 6.15. *The stereochemical consequences of a concerted, thermally allowed [1,5] sigmatropic shift.*

An elegant demonstration of the stereochemistry of a thermally allowed [1,5] sigmatropic shift was reported by Roth and coworkers in 1970. They studied the stereochemistry of the reaction of the optically active starting material,

6-11. There are four possible products: two arise from suprafacial reaction and two more from antarafacial reaction.

Suprafacial rearrangement of hydrogen across the *top* face of the π system gives **6-12**.

Rotation about the C-5—C-6 bond axis gives a new conformation of **6-11**. The suprafacial movement of hydrogen across the *bottom* face of the π system in this conformation produces **6-13**, a geometric isomer of **6-12**.

Antarafacial movement of hydrogen also gives two possible products, **6-14** and **6-15**. In the reaction to give **6-14**, hydrogen moves from the top face to the bottom face; in the reaction to give **6-15**, hydrogen moves from the bottom face to the top face.

In agreement with the theoretical prediction of suprafacial reaction, compounds **6-12** and **6-13** were produced rather than **6-14** and **6-15**. Because of the favorable geometry for suprafacial migration, there are many examples of thermal [1,5] sigmatropic shifts, which occur with ease.

These elegant experiments are reported by Roth, W. R.; Konig, J.; Stein, K. *Chem. Ber.* **1970**, *103*, 426–439.

The industrial synthesis of vitamin D₂ involves two pericyclic processes. **PROBLEM 6.14**
Identify these processes in the following transformations and comment
on the stereochemistry observed.

Ergosterol   Precalciferal

Vitamin D₂
(calciferal)

Schlatmann, J. L. M. A.; Pot J.; and Havinga, E. *Recl. Trav. Chim. Pays-Bas* **1964**, *83*, 1173.

Alkyl Shifts

When an alkyl group migrates, there is an additional stereochemical aspect to consider, namely, the carbon atom can migrate with inversion or retention of configuration. Commonly encountered processes include suprafacial [1,3] shifts with inversion and suprafacial [1,5] shifts with retention. (Note that this contrasts with the case for hydrogen migration, where the [1,3] shift is antarafacial and the [1,5] shift is suprafacial.) Inversion at carbon is an antarafacial process because the bond formed is on the opposite side of the carbon atom to the bond broken; retention at carbon is a suprafacial process.

TABLE 6.4   Selection Rules for Sigmatropic Alkyl Shifts

Number of electrons	Mode of activation	Allowed stereochemistry
$4n$	Thermal	Suprafacial with inversion
	Photochemical	Antarafacial with retention
$4n + 2$	Thermal	Suprafacial with retention
	Photochemical	Antarafacial with inversion

Example 6.16.   *Sigmatropic shifts in the* exo-6-methylbicyclo[3.1.0]hexenyl cation.

This is an interesting example of the stereochemical consequences of orbital symmetry. In this cation, the *exo*-6-methyl group remains *exo* as the migration of carbon-6 proceeds around the ring.

(Note that the numbers shown in the equation are not those used to name the compound. Thus, the 1′ carbon is numbered 6 for nomenclature purposes, and the methyl on this carbon is referred to as the *exo*-6-methyl.)

Inspection of the structures involved might suggest that the electrons of the C-1—C-1′ bond can "slide over" to form the C-4—C-1′ bond, as represented by the arrows in the structure. If this is the reaction pathway, the configuration at C-1′ is maintained because both bond breakage and bond formation occur at the same location relative to C-1′, and we would expect to see the methyl group change from an *exo* to an *endo* orientation because of the rotation required about the C-5—C-1′ bond. Consideration of the selection rules leads to a different prediction.

By applying Hint 6.4, we can classify the reaction as a [1,4] sigmatropic shift. Counting the electrons in the bonds undergoing rearrangement shows that there are four electrons involved; consequently, a thermal process should proceed suprafacially with inversion at the migrating carbon. This means that in the bicyclo[3.1.0]hexenyl cation, bond breaking and bond formation occur on opposite sides of C-1′, so that the methyl group remains in the *exo* position. Because these transformations frequently are difficult to visualize, molecular models are very useful for studying these kinds of reactions.

For the actual structures studied, see Hart, H.; Rodgers, T. R.; Griffiths, J. *J. Am. Chem. Soc.* **1969**, *91*, 754–756.

Show how the following concerted processes can be explained on the basis of one or more sigmatropic shifts. What are the designations for the shift(s) involved?   **PROBLEM 6.15**

a.

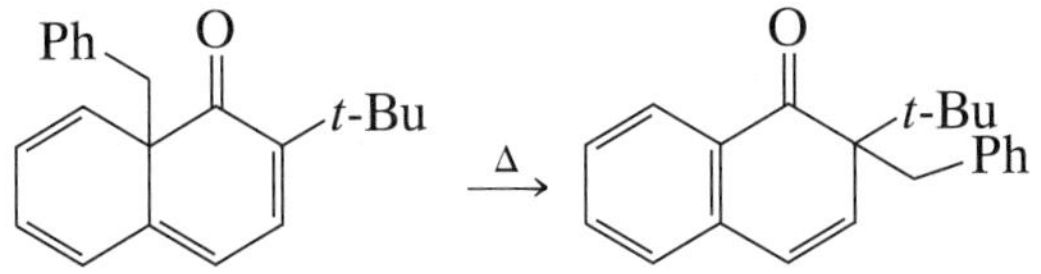

Miller, B.; Baghdadchi, J. *J. Org. Chem.* **1987**, *52*, 3390–3394.

b.

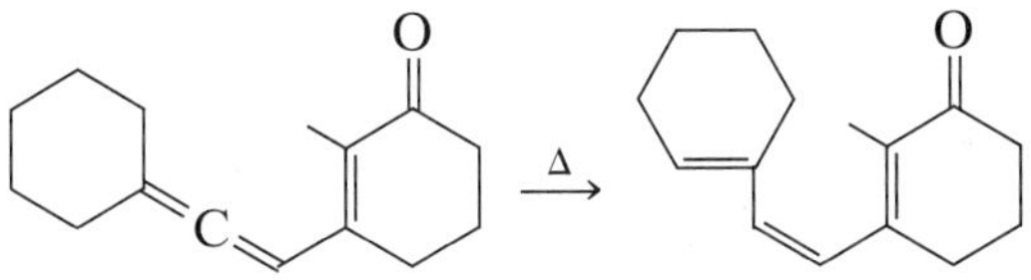

Barrack, S. A.; Okamura, W. H. *J. Org. Chem.* **1986**, *51*, 3201–3206.

---

5. THE ENE REACTION

The ene reaction appears to combine the characteristics of cycloaddition and sigmatropic reactions.

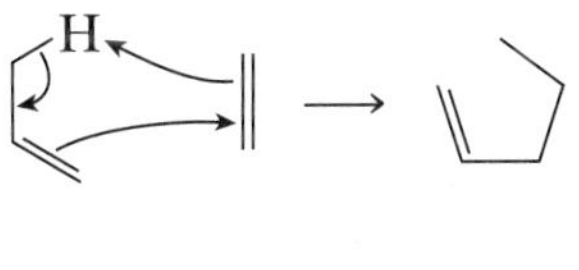

This looks similar to the Diels–Alder reaction, in which a C—H bond replaces the double bond of the diene component. In some cases, the ene reaction actually competes with the Diels–Alder reaction. Because of the similarities, the allyl component often is called the enophile, and the other component is called the ene.

The ene reaction also can take place intramolecularly and, thus, lead to new rings.

Example 6.17.   *An intramolecular ene reaction.*

The unconjugated diene, **6-16**, reacts to form a cyclic molecule.

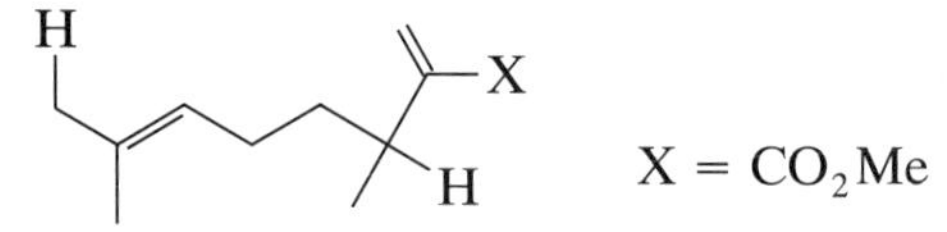

6-16

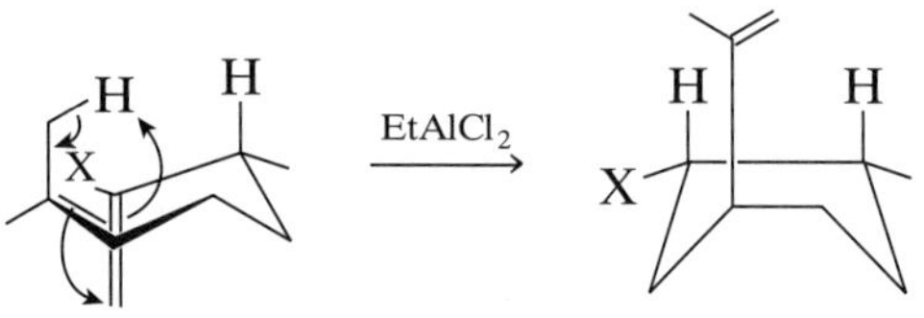

$$X = CO_2Me$$

6-16   82%

Because the proton is transferred to the top of the double bond, the carbomethoxy group, X, is forced down. Molecular models are extremely useful for visualizing the conformation of **6-16** needed for cyclization to take place. *Cautionary note:* there is no evidence that this is a concerted reaction.

Snider, B. B.; Phillips, G. B. *J. Org. Chem.* **1984**, *49*, 183–185.

PROBLEM 6.16   Draw the product for the ene reaction of the following components:

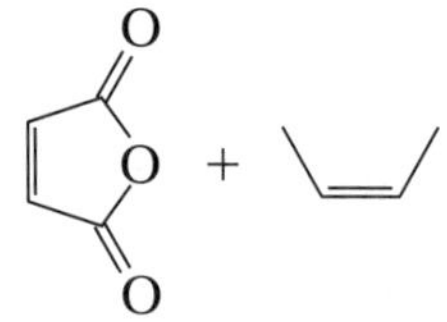

Propose a mechanism for the following cyclization. **PROBLEM 6.17**

350°C
100%

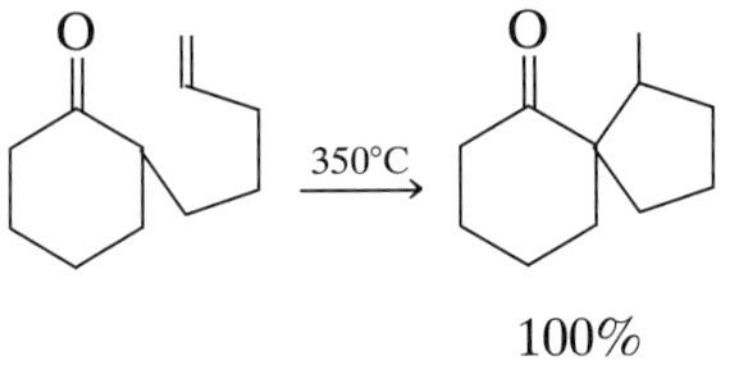

Conia, J.; Robson, M. J. *Angew. Chem., Int. Ed. Engl.* **1975**, *14*, 473.

Write appropriate mechanisms for each of the following reactions. **PROBLEM 6.18**

a.

Δ
89% of product
11% of product

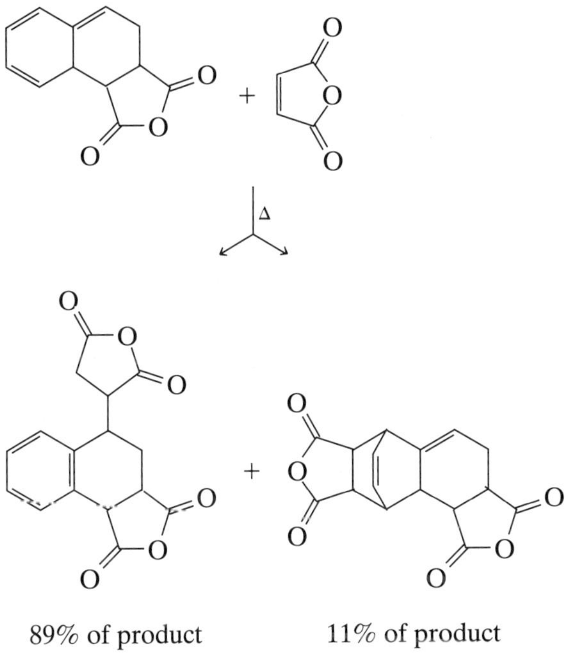

Alder, K.; Schmitz-Johnson, R. *Ann.* **1955**, *595*, 1–37; Hoffmann, H. M. R. *Angew. Chem., Int. Ed. Engl.* **1969**, *8*, 556–577.

PROBLEM 6.18
continued

b.

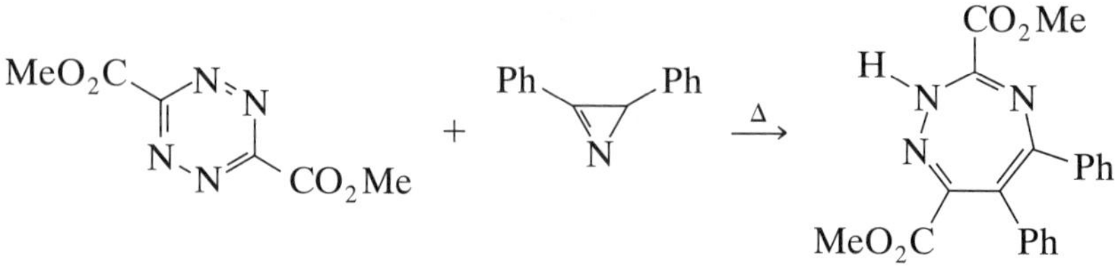

Johnson, G. C.; Levin, R. H. *Tetrahedron Lett.* **1974**, 2303–2307.

c.

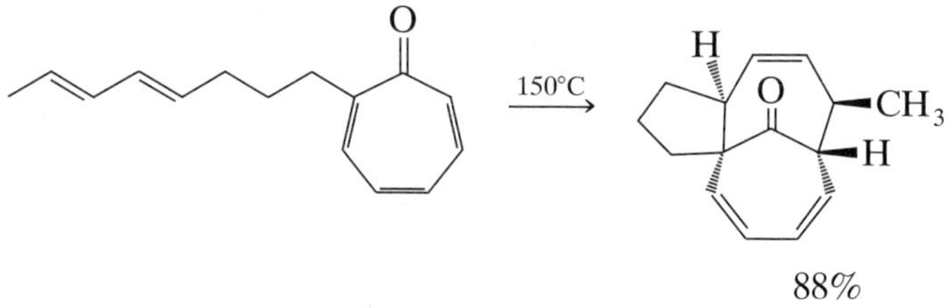

88%

Funk, R. L.; Bolton, G. L. *J. Am. Chem. Soc.* **1986**, *108*, 4655–4657.

d.

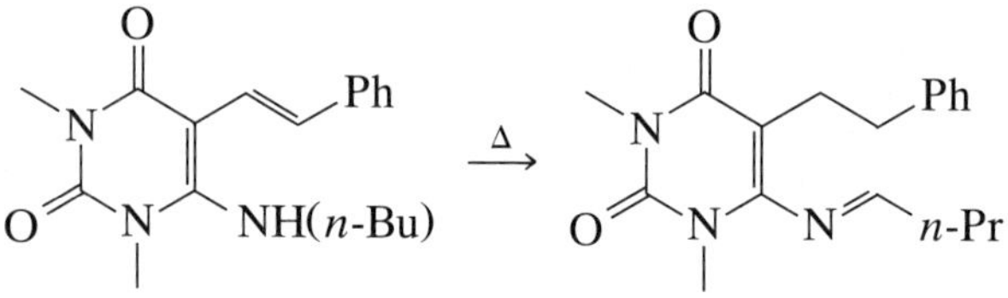

Matyus, P.; Zolyomi, G.; Eckhardt, G.; Wamhoff, H. *Chem. Ber.* **1986**, *119*, 943–949.

e.

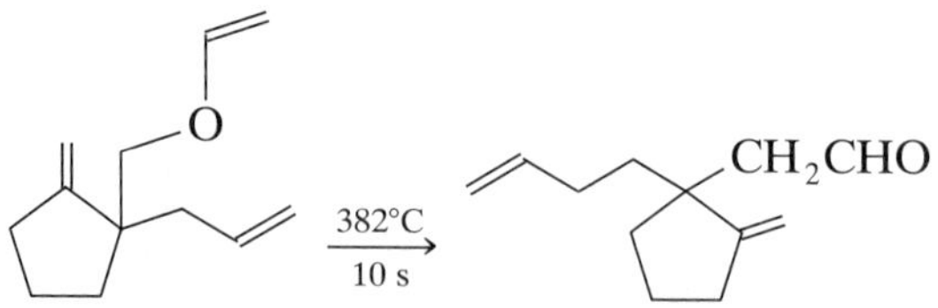

Ziegler, F. E.; Piwinski, J. J. *J. Am. Chem. Soc.* **1982**, *104*, 7181–7190.

f.

Alder, K.; Dortmann, H. A. *Chem. Ber.* **1952**, *85*, 556–565.

The following thermal transformation involves three pericyclic changes. **PROBLEM 6.19**
The first two are electrocyclic and the third is a sigmatropic rearrangement. Give structures for the two intermediates in the reaction.

PROBLEM 6.19
continued

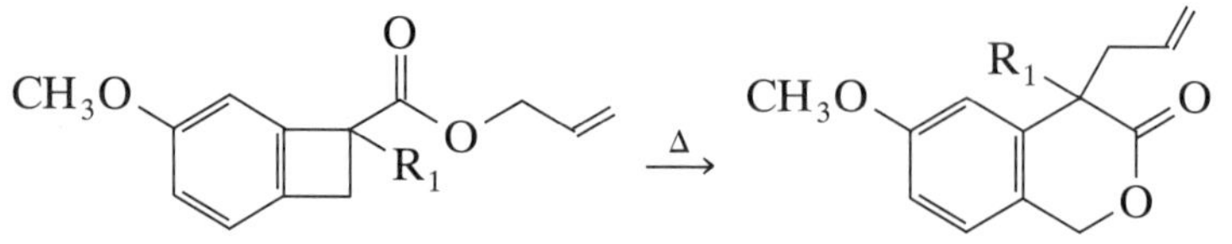

$$R_1 = \text{alkyl}$$

Shishido, K.; Shitara, E.; Fukumoto, K.; Kametani, T. *J. Am. Chem. Soc.* **1985**, *107*, 5810–5812.

PROBLEM 6.20   The following equation represents a thermal [4 + 4] cycloaddition, which occurs readily. Discuss orbital symmetry considerations in detail and propose a mechanism for the reaction.

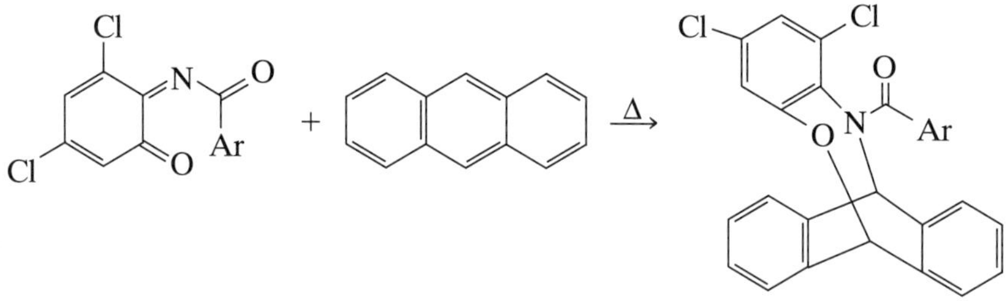

Heine, H. W.; Suriano, J. A.; Winkel, C.; Burik, A.; Taylor, C. M.; Williams, E. A. *J. Org. Chem.* **1989**, *54*, 5926–5930.

PROBLEM 6.21   Transformation of starting material to 6-17 involves a concerted electrocyclic transformation, followed by anion-promoted ring opening of the epoxide. The other transformations are one-step concerted processes. Explain the stereochemical preferences of each reaction. Why is the electrocyclic transformation of 6-17 to 6-19 favored when such reactions usually occur more readily in the opposite direction?

PROBLEM 6.21
continued

6-17

6-18   **6-19**

Coates, R. M.; Last, L. A. *J. Am. Chem. Soc.* **1983**, *105*, 7322–7326.

6. A MOLECULAR ORBITAL VIEW OF PERICYCLIC PROCESSES

Although electron counting and application of the selection rules provide a practical method of analyzing and predicting pericyclic reactions, a greater understanding of these reactions can be gained by analyzing the molecular orbitals involved. A quantitative mathematical approach is of interest for a detailed analysis of specific reactions, but qualitative analysis can be applied much more rapidly and with very good effect to a much larger number of systems. The theory and methods of the molecular orbital approach to pericyclic reactions are beyond the scope of this book. We simply provide some introductory material on how to derive molecular orbitals for the π systems involved in pericyclic reactions, along with a few illustrative examples of how molecular orbital theory looks at selected pericyclic reactions.

A. Orbitals

An orbital, whether atomic or molecular, represents the region of space where a particular electron is most likely to be found. Orbitals are derived by solving wave equations, which are based on the experimental finding that electrons, like photons, can behave like waves as well as like particles. The wave equations that can be written for the electrons associated with individual atoms are analogous to the equations that can be written to describe a standing wave in a vibrating string. This is a useful analogy because a wavefunction has amplitude and nodes, just like the standing waves in a vibrating string. For a given electron, the probability of finding an electron at a particular location can be obtained by squaring the amplitude given by the wavefunction. Plotting of these probabilities gives a visual representation of the orbital shape and electron density.

Hint 6.6   A node, where the amplitude and electron density are zero, separates an orbital into lobes that have different algebraic signs. An atomic *s* orbital has no nodes, whereas in an atomic *p* orbital a nodal plane divides the orbital into two lobes of differing sign.

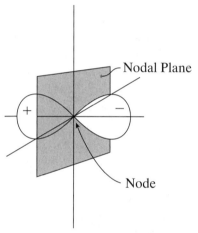

Atomic *p* orbital

In the diagram used in this chapter, the sign of the amplitude is represented by + and − signs within the lobes of the orbital. (An alternative representation uses shaded or unshaded lobes.)

B. Molecular Orbitals

Molecular orbitals (MOs) are derived mathematically by a linear combination of the wavefunctions for the atomic orbitals (AOs) of the individual atoms in a molecule. Usually, only the atomic orbitals of the valence electrons are considered, because these are the electrons involved in bonding. We can visualize the formation of MOs as proceeding from overlap of the AOs of the valence electrons.

Just as we can consider the *formation of molecules* from atoms in terms of the *interaction of atomic orbitals* to form molecular orbitals, we can consider *reactions of molecules*, both intramolecular and intermolecular, in terms of the *interaction of the molecular orbitals*. In the case of pericyclic reactions, a very useful simplification can be made on the basis of Huckel molecular orbital theory. We assume that, because the p orbitals that overlap to form π bonds are orthogonal (lie at right angles) to the σ bonds, the π bonding system can be considered independently of the single (σ) bonds in the reacting molecules and that the π bonding system is the most important factor in determining the chemical reactivity of conjugated polyenes and aromatic compounds.

Molecular orbitals can be classifed with regard to three elements:

Hint 6.7

1. The number of nodes
2. Symmetry with respect to a mirror plane (σ)
3. Symmetry with respect to rotation about a two-fold axis (C_2)

For each molecular orbital in a molecule, the classification according to these three characteristics is unique. In other words, for any molecule these three descriptors cannot be identical for two different molecular orbitals.

In molecular orbitals, as in atomic orbitals, *nodes* are regions of zero electron density that divide an orbital into lobes with amplitudes of opposite sign. When a node coincides with a nuclear position, there are no lobes depicted on that atom. In the following diagram, we see that the bonding π molecular orbital for ethylene has no nodes perpendicular to the bond axis, whereas the antibonding π^* orbital has one node perpendicular to the bond axis.

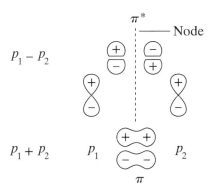

Electron distribution of atomic p orbitals and molecular orbitals (π and π^*).

The orbital representation is an approximation of the electron distribution.

The symmetry of a π system can be examined with respect to a *mirror plane* bisecting the π system and perpendicular to the σ bond molecular framework. (A mirror plane similarly bisects a fork or a spoon.) An orbital is symmetric (S) with respect to a mirror plane (usually called a σ plane) if it has the same sign on both sides, and antisymmetric (A) if the two sides have opposite sign. For example, in ethylene the bonding π orbital is symmetric with respect to the mirror plane, whereas the antibonding orbital is antisymmetric, as shown in Table 6.5. Note that antisymmetric does *not* mean the same as asymmetric.

The symmetry of a π system also can be classified with respect to *rotation about a C_2 axis bisecting the π system and lying in the plane defined by its σ bond framework.* A C_2 axis is defined operationally: rotation of 180° about a C_2 axis gives an arrangement indistinguishable from the original. (A twin-bladed fan or propeller is an everyday object with a C_2 axis. A flathead screwdriver has both a mirror plane and a C_2 axis.) The following diagram shows the C_2 axis for the antibonding π^* orbital in ethylene. When the orbitals are rotated 180° with respect to this axis, the $+$ and $-$ orbitals coincide with the $+$ and $-$ orbitals of the original arrangement.

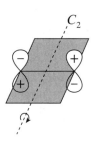

C. Generating and Analyzing π Molecular Orbitals

In analyzing the π molecular orbitals of reacting molecules, we need to determine the number of orbitals, their relative energy, and the number and placement of nodes. Once we have determined the number and relative energy of the orbitals, we can determine the electron configurations of the reactants and analyze the amplitudes of the orbitals that overlap in the course of the reaction.

TABLE 6.5 Symmetry Characteristics of Ethylene π Molecular Orbitals

Molecular orbital	Reflection through mirror plane (σ)	Rotation about the C_2 axis
π^* Nodes = 1	Antisymmetric (A)	Symmetric (S)
π Nodes = 0	Symmetric (S)	Antisymmetric (A)

The number of molecular orbitals is equal to the number of atomic orbitals that was used to construct them.

Hint 6.8

Example 6.18. *The π molecular orbitals of ethylene.*

For ethylene ($H_2C{=}CH_2$), the simplest π system, there are two ways that we can combine the two p atomic orbitals from the two carbon atoms so that we get two π molecular orbitals.

Basis sets and molecular orbitals for ethylene.

If the wavefunctions for the two p orbitals are added, we form a bonding orbital (π); if the wavefunctions are subtracted, we form an antibonding orbital (π^*). Molecular π orbitals are commonly represented by the atomic orbitals that combine to form them. In the accompanying diagram, the orbitals on the left are the atomic orbitals. These are used to form the molecular orbitals of ethylene, which appear on the right. The atomic orbitals that are combined to form the molecular orbitals are called the *basis set*.

Hint 6.9   When atomic orbitals are combined to give molecular orbitals, the molecular orbitals can be bonding, antibonding, or nonbonding. The number of bonding and antibonding orbitals is equal. If the number of atomic orbitals used is odd, there is a nonbonding orbital in addition to the bonding and antibonding orbitals.

Electrons in the bonding (π) orbital are lower in energy than electrons in the isolated p orbitals, and electrons in the antibonding (π^*) orbital are higher in energy than electrons in the isolated p orbitals, as illustrated here for ethylene.

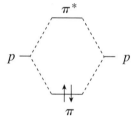

Energy levels of π molecular orbitals for ethylene.

The energy of a nonbonding orbital is comparable to that of the p orbitals of the isolated atoms.

Hint 6.10   For linear, conjugated polyenes, each molecular orbital has a unique energy.

The conjugated polyene systems taking part in pericyclic reactions are linear, so all the π molecular orbitals of the starting materials have a different energy. The distribution of energy levels is important because it affects the electronic configuration of the starting materials (see Example 6.19) and consequently the course of the reaction.

In symmetrical *cyclic conjugated* systems, such as benzene, there are sets of molecular orbitals with the same energy (degenerate orbitals). (The pattern of energy levels for these aromatic systems can be determined by Frost diagrams, which we will not consider; see Carey, F. A.; Sundberg, R. J. *Advanced Organic Chemistry, Part A: Structure and Mechanism*, 3rd ed; Plenum Press: New York, 1990; pp. 44–46.) The symbol Ψ is commonly used to designate the various molecular orbitals, which are distinguished by subscripts (i.e., in order of increasing energy, Ψ_1, Ψ_2, Ψ_3, ...).

We determine the electron configuration for molecular π electrons in the same way that we do for atomic electron configurations: we count the total number of electrons and then use these to fill each orbital with a maximum of two electrons, starting with the lowest energy orbital and filling the orbitals in order of increasing energy.

Hint 6.11

Example 6.19. *Energy levels for the allyl molecular orbitals and electron configurations of the allyl cation, radical, and anion.*

$$CH_2{=}CH{-}\overset{+}{C}H_2 \qquad CH_2{=}CH{-}\overset{\bullet}{C}H_2 \qquad CH_2{=}CH{-}\overset{-}{C}H_2$$

Allyl cation   Allyl radical   Allyl anion

The molecular orbitals of the allyl system are formed by the overlap of three atomic *p* orbitals. Because there is an odd number of atomic orbitals, one of the molecular orbitals is a nonbonding orbital, whose energy is comparable to that of the isolated *p* orbitals from which it was derived. Note that if there were degenerate molecular orbitals in the allyl system, the electronic configurations of various allyl species would be different. For example, if Ψ_2 and Ψ_3 for the allyl system had identical energy levels, the allyl anion would have two unpaired electrons.

The molecular orbital of lowest energy has no nodes, the next has one node, the next two nodes, and so on. The nodes are arranged symmetrically with respect to the center of a linear π system.

Hint 6.12

Nodes in the π molecular orbitals bisect the bonds forming the σ framework; for π systems derived from an odd number of atomic orbitals, nodes may also coincide with nuclei (see Example 6.20).

Hint 6.13

If the molecular orbitals are examined in order of increasing energy, the orbitals alternate in symmetry with regard to the mirror plane and C_2 axis of symmetry. The first orbital is symmetric (S) with respect to a mirror plane bisecting the linear system, the second orbital is antisymmetric (A) with respect to the same plane, and so on. Similarly, the first orbital is antisymmetric with respect to the C_2 axis, the second is symmetric with respect to C_2, and so on.

Example 6.20. *Molecular orbitals of the allyl system.*

		Nodes	Symmetry	
			Mirror (σ)	C_2
Ψ_3		2	S	A
Ψ_2		1	A	S
Ψ_1		0	S	A

PROBLEM 6.22   **Why would the following molecular orbital diagram for the allyl system be incorrect?**

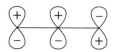

Explain why the following would not be the two lowest energy MOs for 1,3-butadiene.   **PROBLEM 6.23**

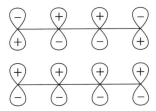

Write all the MOs for the four- and five-carbon π systems. Indicate the number of nodes and designate the symmetry of each orbital with respect to the mirror plane (σ) and the C_2 axis of rotation.   **PROBLEM 6.24**

D. HOMOs and LUMOs

In analyzing pericyclic reactions, two molecular orbitals are of particular interest: the π molecular orbital of highest energy that contains one or two electrons (the highest occupied molecular orbital, HOMO) and the molecular orbital of lowest energy that contains no electrons (the lowest unoccupied molecular orbital, LUMO). For electrocyclic reactions, where there is only one π system, the important orbital is the HOMO. When more than one π system is involved, as in cycloaddition, reactions are considered to occur through a transition state in which the HOMO of one component overlaps the LUMO of the other.

In photochemically activated reactions, the HOMO and LUMO are not the same as those for thermal reactions because absorption of light promotes an electron from the HOMO to the LUMO of the ground state. Thus, the LUMO of the ground state becomes the HOMO of the photochemically excited state.

Example 6.21. *HOMOs and LUMOs for the butadiene system.*

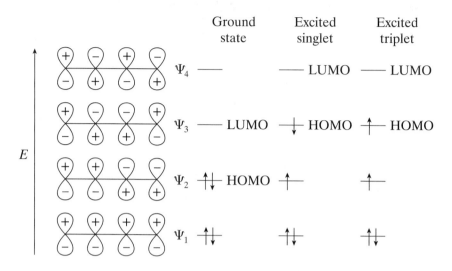

	Ground state	Excited singlet	Excited triplet

Ψ_4 — — LUMO — LUMO

Ψ_3 — LUMO — HOMO — HOMO

Ψ_2 — HOMO — —

Ψ_1 — — —

PROBLEM 6.25   **By using the orbitals drawn for Problem 6.24, indicate the electronic configuration for the pentadienyl cation, radical, and anion. Label the HOMO and LUMO for each case.**

Hint 6.14   For bond formation to occur, the overlapping (i.e., interacting) molecular orbitals must have amplitudes of like sign.

Bond formation during reactions is entirely analogous to the process whereby we form molecular orbitals from atomic orbitals in individual molecules. Interaction between orbitals of like sign results in constructive interference (i.e., bonding), whereas interaction between orbitals of opposing sign results in destructive interference.

E. Correlation Diagrams

In a pericyclic reaction, the pathway predicted by the selection rules is the one that allows maximum orbital overlap along the reaction pathway, including the transition state. Maximum orbital overlap corresponds to the path of minimum energy and is achieved if the orbitals involved are similar in energy and if the symmetry of the orbitals is maintained throughout the reaction path.

When pericyclic reactions are analyzed in terms of correlation diagrams, all of the π and σ molecular orbitals taking part in the reaction are analyzed in terms of their symmetry properties with respect to reflection in a mirror plane and rotation about a C_2 axis. The symmetry elements of importance are those that are found in both reactants and products *and* are preserved during the course of the reaction. If the symmetry properties of all of the orbitals remain unchanged (are conserved) throughout the reaction, then maximum orbital interaction is maintained throughout the course of the reaction and the reaction is *symmetry-allowed*. If the symmetry properties change, then the reaction is *symmetry-forbidden*.

Example 6.22. *Orbital correlation diagrams for the interconversion of butadiene and cyclobutene.*

Classifying the Relevant Orbitals

As in the electron-counting approach, we consider only the electrons and bonds that change in the reaction.

This is a 4π electron process. The bonds involved are the π bonds of the conjugated diene system and the π bond and the ring-forming σ bond of the cyclobutene. The electronic configurations and symmetry characteristics of the orbitals involved are as follows:

Symmetry Correlations between Bonding Orbitals of Starting Materials and Products

If we examine the classification of orbitals of starting material and product with respect to each of the two symmetry elements in turn, we see that the bonding orbitals, Ψ_1 and Ψ_2, for starting material and product share symmetry only with respect to the C_2 axis. (With respect to the σ plane, both bonding orbitals of the cyclobutene system are symmetric, but in the butadiene system, only Ψ_1 is symmetric with respect to the σ plane.)

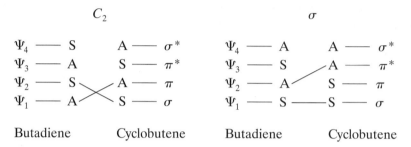

	C_2					σ		
Ψ_4 — S		A — σ^*		Ψ_4 — A		A — σ^*	
Ψ_3 — A		S — π^*		Ψ_3 — S		A — π^*	
Ψ_2 — S	A — π			Ψ_2 — A		S — π	
Ψ_1 — A	S — σ			Ψ_1 — S		S — σ	

Butadiene	Cyclobutene	Butadiene	Cyclobutene

This means that, during the course of a reaction, maximum bonding overlap can be maintained only if all of the intermediates along the reaction pathway also have C_2 symmetry; in other words, the whole process must be symmetrical with respect to C_2. If we look at symmetry correlations with respect to the σ plane, we see that to maintain symmetry with respect to the σ plane, the butadiene orbital Ψ_2 has to correlate with the π^* orbital. This means that in order to maintain symmetry with respect to the σ plane, the reaction would have to form the product in an excited state. The energy requirement for such a conversion would be very large, so that the conversion requiring the correlation of Ψ_2 with π^* would be *symmetry-forbidden.*

Symmetry Characteristics of the Reaction

As we saw in Section 2.B, the interconversion of butadiene and cyclobutene can occur through either a disrotatory or a conrotatory process. As the following diagram shows, the disrotatory opening is symmetrical with respect to a σ plane, whereas conrotatory opening is symmetrical with respect to the C_2 axis.

disrotatory opening   conrotatory opening

Because conrotatory ring opening proceeds so that intermediate structures have C_2 symmetry at all points along the reaction pathway linking starting material and product, it is logical that thermal ring opening of cyclobutene occurs by a conrotatory process. Furthermore, because the disrotatory process does not proceed along a pathway that maintains symmetry, it is not expected to occur.

Orbital Phase Correlations

An alternative way of analyzing the thermal ring-opening reaction is to look at the *phases* of the molecular orbitals involved. These can be visualized as follows:

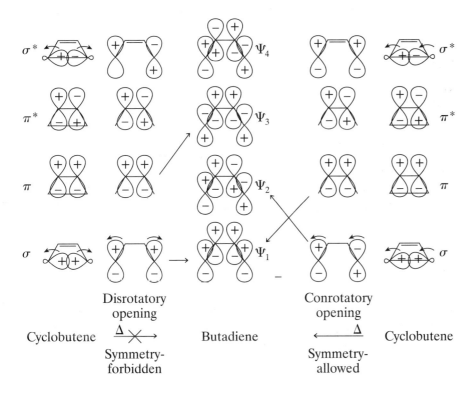

In conrotatory ring opening, the reoriented σ orbitals derived from cyclobutene look like part of the butadiene molecular orbitals Ψ_2 and Ψ_4. The orbitals derived from the double bond of cyclobutene look like part of the butadiene molecular orbitals Ψ_1 and Ψ_3. Because the signs of the cyclobutene orbitals can be correlated with bonding orbitals of butadiene by conrotatory opening, this mode of ring opening is allowed, whereas disrotatory opening, in which the signs of the cyclobutene orbitals must be correlated with an antibonding orbital, is forbidden.

F. Frontier Orbitals

The frontier orbital approach leads to the same predictions as those based on an analysis of correlation diagrams, but instead of considering all of the orbitals of the π system, only the frontier orbitals, the HOMOs and LUMOs are considered. A general discussion of frontier orbital theory is given in Fukui, K. *Acc. Chem. Res.* **1971**, 57–64. The choice of which HOMOs and LUMOs to combine requires an analysis that we will not undertake here. We will simply show that when these orbitals are combined, we can rationalize reactions observed experimentally in terms of the orbitals derived from this type of analysis.

Electrocyclic Reactions

The frontier orbital analysis of electrocyclic reactions focuses on the HOMO of the open-chain polyene.

Hint 6.15

In the frontier orbital approach to electrocyclic transformations, the course of reaction is determined by the amplitude of the lobes at the termini of the HOMO of the polyene. Rotation of bonds during reaction will occur in such a way that orbitals of like sign can overlap.

Example 6.23. *The frontier orbital approach to the thermal ring opening of cyclobutenes.*

The HOMO for butadiene is Ψ_2. In the conrotatory cyclizations shown here, rotation of the lobes at the ends of the HOMO leads to overlap of orbitals with like signs and production of a σ bond.

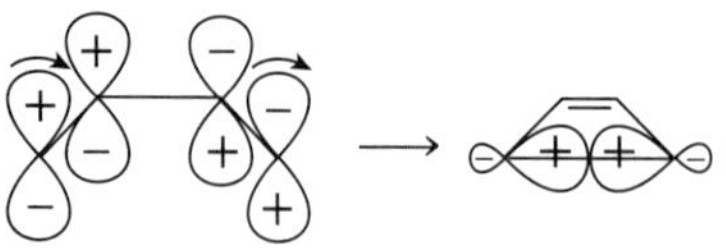

conrotatory 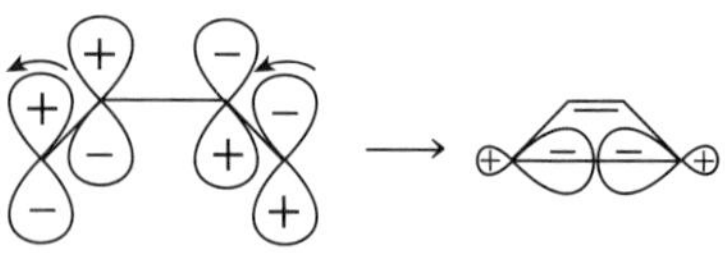   conrotatory

A disrotatory process, on the other hand, would lead to an antibonding interaction between the ends of the chain and would not lead to formation of the σ bond needed for cyclization.

The π bonding orbital of cyclobutene arises from reorganization of the butadiene π system. This process is considered in Woodward, R. B.; Hoffmann, R. *The Conservation of Orbital Symmetry*; Verlag Chemie: Weinheim, 1970; p. 38 ff. and references cited therein.

Cycloaddition

In cycloaddition reactions, frontier orbital analysis considers the interaction of the HOMO of one component and the LUMO of the other.

Example 6.24.   *Frontier orbital overlap in a [4 + 2] cycloaddition.*

In a [4 + 2] cycloaddition, we can consider overlap between the HOMO of the diene and the LUMO of the olefin.

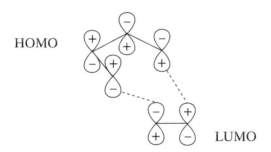

Overlap of lobes with like signs can occur by suprafacial–suprafacial addition or by an antarafacial–antarafacial process. The latter would not be expected for a [4 + 2] reaction because of geometric constraints.

Sigmatropic Rearrangements

Example 6.25.   *Frontier orbital analysis of sigmatropic [1,3] and [1,5] hydrogen shifts.*

We can think of a hydrogen shift as resulting from cleavage of the C—H sigma bond, followed by movement of a hydrogen atom across a π radical system to form a new sigma bond. (Alternative schemes in which a proton moves across a π anion system or a hydride moves across a π cation system would require more energy because these would require separating charges as well as breaking a carbon–hydrogen bond.) For a [1,3] shift, the interacting orbitals that are closest in energy are the $1s$ orbital of hydrogen and the Ψ_2 orbital of the allyl system. Each of these orbitals contains a single electron, unlike the HOMOs involved in electrocyclic reactions and cycloadditions, each of which contains a pair of electrons. In the case of the [1,5] hydrogen shift, the orbitals involved are the $1s$ orbital of hydrogen and the Ψ_3 orbital of the pentadienyl system. We can represent the orbital interactions as follows:

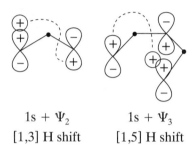

$$1s + \Psi_2$$
[1,3] H shift

$$1s + \Psi_3$$
[1,5] H shift

In the [1,3] shift, a total of 4π electrons are involved. A bonding interaction can be maintained only if the hydrogen moves to the opposite site of the π system (i.e., the rearrangement is antarfacial). Geometry is against this move. Either the hydrogen has to enter a region outside the π system in order to get from one side to the other, or the bond angles of the allyl system have to be strained severely in order for the hydrogen to move from one face to the other while maintaining bonding overlap. Both of these are very high-energy options, so the reaction is not observed.

In the [1,5] shift, 6π electrons are involved. In this case, bonding can be maintained throughout a suprafacial process, and this type of reaction occurs with ease.

Example 6.26.   *Orbitals involved in sigmatropic shifts of carbon.*

Frontier orbitals can help to explain the results seen when a sigmatropic shift involves migration of carbon rather than hydrogen. A [1,3] migration of carbon involves 4π electrons. We can think of the reaction as migration of a carbon radical across the π system of an allylic radical. In this case, overlap of the p orbital of the carbon radical with the allylic π system can be visualized as shown in the following diagram.

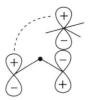

Because the *p* orbital has a node, the carbon atom can move across the π system. As bonding interaction between the + lobes of the *p* orbital and the π system decreases, interaction develops between the − lobes. In the process, the migrating carbon undergoes inversion analogous to that seen in an S_N2 process. In both cases, the bonds breaking and forming are at a 180° angle.

Example 6.27. *Orbitals involved in sigmatropic shifts in the exo-6-methyl bicyclo[3.1.0]hexenyl cation.*

In this system, as we saw previously, the methyl group remains *exo* as the migration of carbon-6 proceeds around the ring.

As we saw in Example 6.16, this is a [1,4] sigmatropic shift of the carbon labeled 1′ in the equation. (Note that the numbers shown in the equation are not those used to name the compound. Thus, the carbon labeled 1′ for classification of the sigmatropic shift is numbered 6 for the purpose of nomenclature.)

One way to consider the orbital interactions is to think of the process as migration of a carbocation across a butadiene π system. This is a reasonable choice because, in contrast to Example 6.25, the reacting species is already charged. If we consider the reaction as interaction between the LUMO of the carbocation (atomic *p* orbital) and the HOMO of the butadiene π system (Ψ_2), we can represent the interaction visually as follows:

In this scheme, inversion of configuration occurs at C-6, which means that the methyl group remains *exo*. If it is difficult to see that such a transformation leads to the methyl group remaining *exo*, molecular models can be of assistance.

For the actual structures studied see Hart, J.; Rodgers, T. R.; Griffiths, J. *J. Am. Chem. Soc.* **1969**, *91*, 754–756.

ANSWERS TO PROBLEMS

Problem 6.1

a. For the stereochemistry shown, the process must be disrotatory. This is a four-electron process and, thus, is symmetry-forbidden thermally. The term symmetry-forbidden does not mean the reaction is impossible, merely that so much energy is required that other processes usually occur instead.

This is an example of a symmetry-forbidden reaction that, nonetheless, actually takes place. Failure to conform to the selection rules usually is taken to mean that the reaction proceeds by a nonconcerted mechanism. The electronic effects of the substituents, R_1 and R_2, on the course of the reaction are complex.

b. In order to rotate the substituents into their positions in the product, a conrotatory mode is required. This is a 4π electron process and is thermally allowed. Note that the diene, and not the cyclobutene, is considered in giving the designation and stereochemistry.

Problem 6.1
continued

Note the unusual *trans* orientation of the ring carbons across the lower double bond in the product.

a. According to Table 6.1, a concerted thermal reaction of a 6π electron system requires a disrotatory mode of ring closing, which leads to a *cis* orientation of the hydrogens.

Problem 6.2

b. To determine the stereochemical consequences of the reaction, first rotate about the single bonds of the starting material to get a conformation in which the ends of the tetraene system are close enough to react. These rotations change the conformation of the molecule, but not the stereochemistry. Note, however, that you may *not* rotate about any of the double bonds to get the ends into position because that would change the stereochemistry of the tetraene. All of the double bonds have *cis* stereochemistry. Because this is an 8π electron system, the thermally allowed process will be conrotatory. Conrotatory rotation gives the product with the two methyl groups *trans*.

c. As in the previous example, the starting material must be placed in a conformation in which the ends of the π system are close enough to react.

Problem 6.2
continued

If it is not clear to you which electrons are necessary for the transformation, drawing arrows for the redistribution of electrons can be helpful.

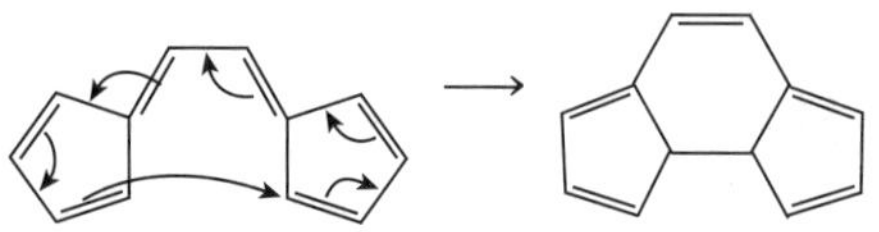

In order to get the product, all 12π electrons must be involved. Thus, the thermal reaction should be conrotatory, and the hydrogens will be *trans*.

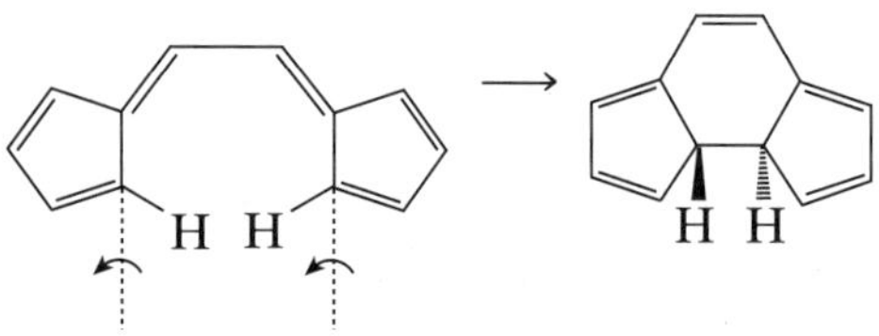

Problem 6.3

To obtain the stereochemistry observed in the product the long chain (carbon A) must move behind the plane of the page and hydrogen B (H_B) must move in front of the plane of the page. This is a disrotatory process.

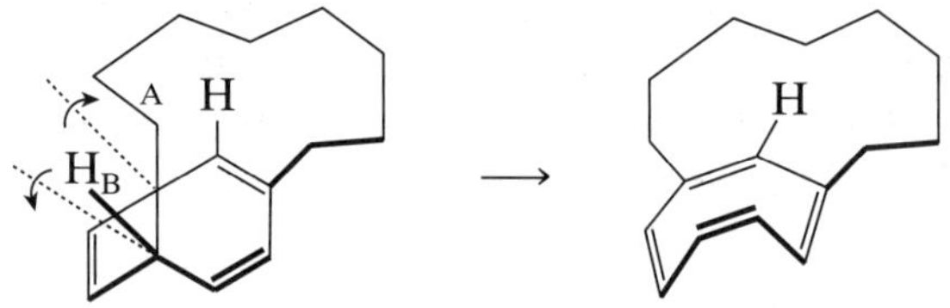

We can consider the change from starting material to product as involving either 4π electrons or 6π electrons, depending on whether we consider the process to be the ring opening of a four-membered ring or a six-membered ring. If it is a 4π-electron process, disrotatory opening can occur only under photochemical conditions. If the reaction is a 6π-electron process, its is symmetry allowed under thermal conditions.

Problem 6.3
continued

a. The thermal reaction will be disrotatory, leading to *trans*-methyl groups:

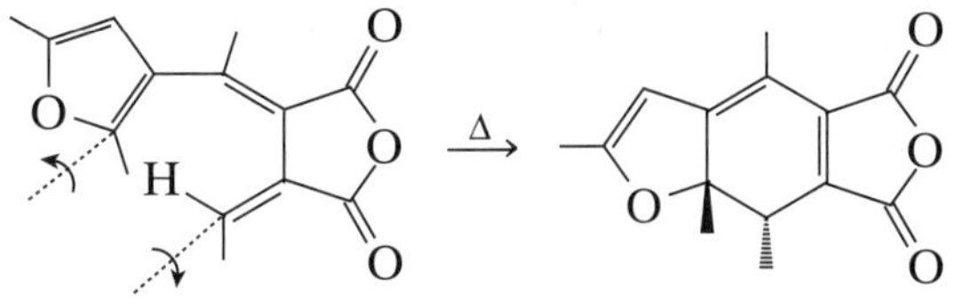

Problem 6.4

b. This is a 6π electron process. (Only one of the π bonds in the furan ring is involved in the reaction.) The photochemical reaction will be conrotatory (antarafacial), leading to *cis*-methyl groups:

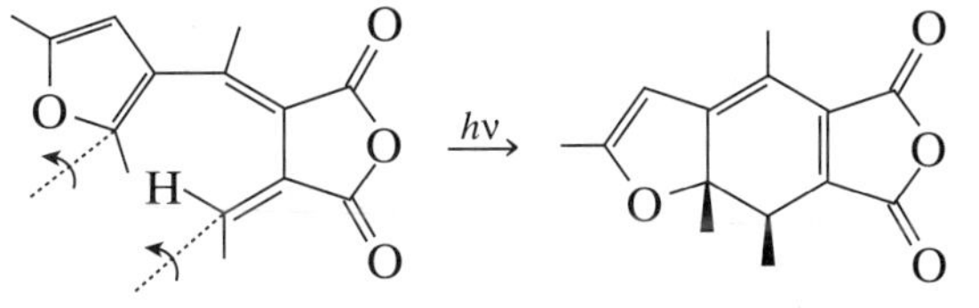

c. The photochemical ring opening of the thermal product will be conrotatory. (Remember, the total number of electrons counted in a process is that of the open-chain compound.) Thus, the product will be an isomer of the starting material for the original process.

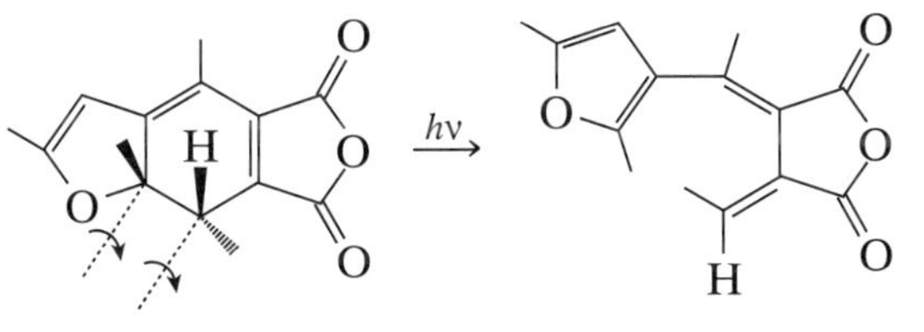

Problem 6.5
Ionization occurs to a 2π electron allylic cation. To be thermally allowed, therefore, the motion must be disrotatory. Only the disrotatory motion, which moves the hydrogens outward, will occur. This places the developing p orbitals opposite the bromide leaving group only for the second compound. The bromide ion then reacts with the cation to produce the product. In the first compound, the developing p orbitals will be opposite the fluoride group, which is such a poor leaving group that no reaction occurs.

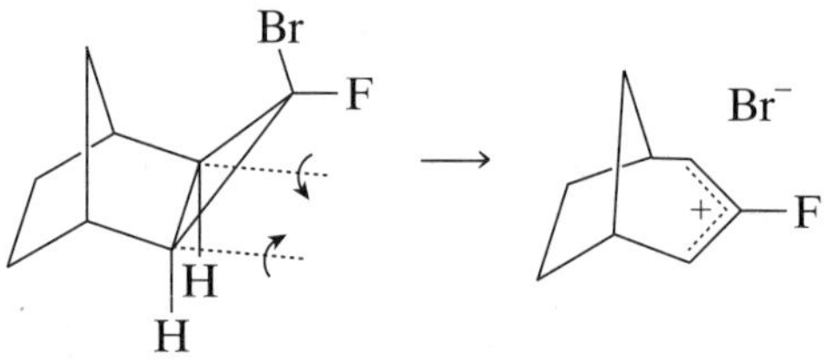

Problem 6.6
a. Both the cycloaddition to give the intermediate and the cycloreversion of the intermediate to give the product are [4 + 2] reactions. The electrons involved in each process are highlighted in the following structures:

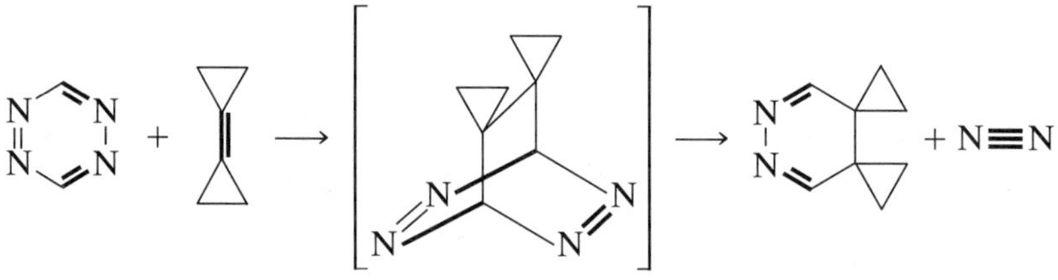

In the retrocycloaddition, only two of the π electrons of the N_2 product are used. The other two π electrons are perpendicular (orthogonal) to the first two and cannot take place in the reaction. Thus, by considering the products of the reaction, this is a $[\pi_s^4 + \pi_s^2]$ process. On the other hand, the reaction of the intermediate would be designated as a $[\sigma_s^2 + \sigma_s^2 + \pi_s^2]$ process.

b. This is an [8 + 8] cycloaddition, as becomes evident when arrows are used to show the flow of electrons leading to product. Because this is a 4*n* cycloaddition, the selection rules dictate a supra–antara cyclization, which is equivalent to a concerted *trans* addition. This is difficult to visualize. It is easier to visualize the supra–supra cyclization, which would lead to the isomer in which the hydrogen atoms are *trans*. From this we can conclude that because the product has the two hydrogens *cis*, the reaction must be supra–antara or antara–supra.

Problem 6.6
continued

c. This is a [6 + 4] addition. The π electrons of the carbonyl are not involved. The selection rules predict a supra–supra or antara–antara process, which results in the observed *cis* stereochemistry.

d. The singlet oxygen has all electrons paired and so reacts as the two-electron component of a [4 + 2] addition. In the more commonly encountered triplet state, oxygen has two unpaired electrons and reacts as a diradical.

a. $[\pi_s^4 + \pi_s^2]$. The phenyls of the diene and the carbonyls of the heterocycle are not involved in the ring-forming process, so the π electrons in these substituents are not counted. The two hydrogens, which are shown up (*cis*) in the product, also are *cis* in the starting material. In addition, the two phenyl groups of the diene end up on the same side of the six-membered ring. Both of these stereochemical consequences are the result of suprafacial processes. This can be demonstrated readily by using models.

Problem 6.7

b. $[\pi_s^2 + \pi_s^2]$. The best way to understand the stereochemistry of the process is to look at models. Placing the appropriate termini together in a model of the starting material shows that the stereochemistry of the product is produced when both components react suprafacially. The π electrons of the C=O group and the dimethoxy-substituted π bond are not counted because they are not involved in the reaction. This is a thermal reaction, so that a nonconcerted pathway is expected.

$[\pi^4 + \pi^4]$. The nitrone component contains four electrons: the two π electrons of the double bond and the two π electrons on the oxygen that overlap with them. The diene component also contains four electrons. The nitrone component contains three atoms and the diene component contains four atoms, so that a [4 + 3] adduct is formed. The authors cited propose that this adduct is formed by a diradical mechanism. For more on this reaction, see Problem 7.2.b.

Problem 6.8

Problem 6.9

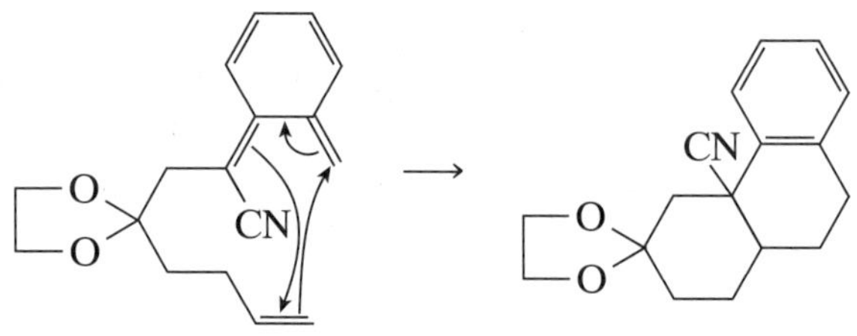

Problem 6.10   The following reaction shows the actual regiochemistry observed:

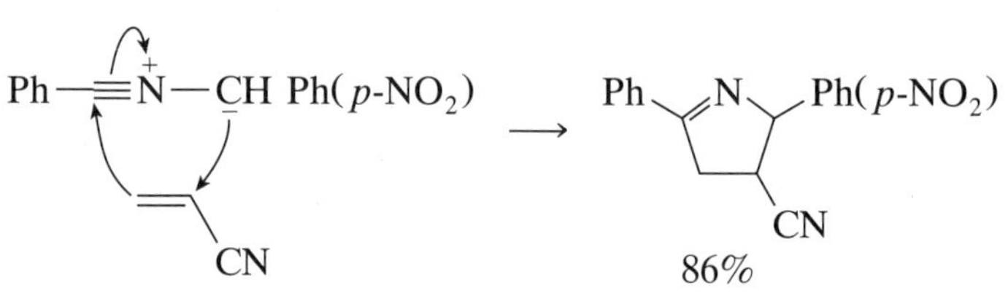

Because the double bond is polarized by the nitrile group, we might have expected the following reaction to predominate:

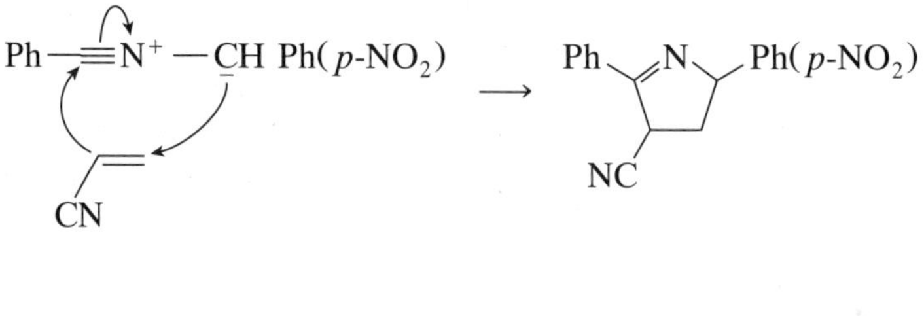

As indicated in the text, many of these reactions show high regioselectivity. However, it is not possible to predict this regiochemistry without molecular orbital calculations.

A possible mechanism is intermolecular proton transfer from one molecule of the starting ester–imine to another to give a 1,3-dipole, which then cycloadds to the acetylene.

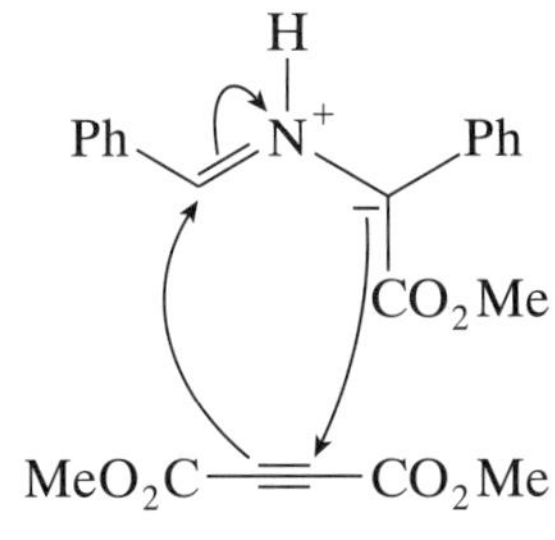

Problem 6.11

a. This is a [1,5] sigmatropic shift. The highlighted σ bond has been broken, the π system has shifted, and a new σ bond has been formed at the other end of the π system. Numbering atoms from each end of the original σ bond through the terminal atoms of the new σ bond gives the following:

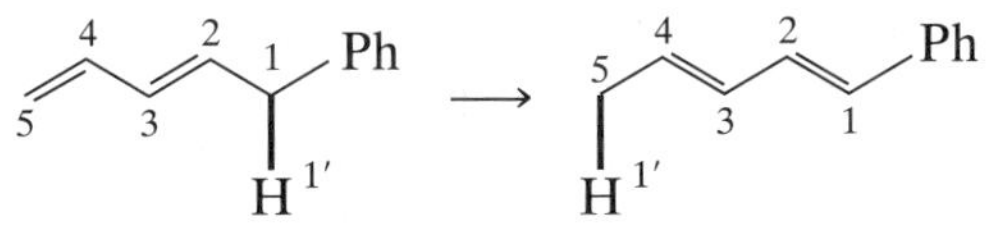

Problem 6.12

Thus, the hydrogen has moved from the 1 position to the 5 position on the chain. The 1 in the designation [1,5] indicates that at one end of the new σ bond is an atom (the hydrogen) that also was at one end of the old σ bond. The 5 indicates that the other end of the new σ bond is formed at the 5 position along the carbon chain, atom number 5.

b. This is a [3,3] sigmatropic shift. It is an example of the Claisen rearrangement, a [3,3] sigmatropic rearrangement of a vinyl ether.

Problem 6.12
continued

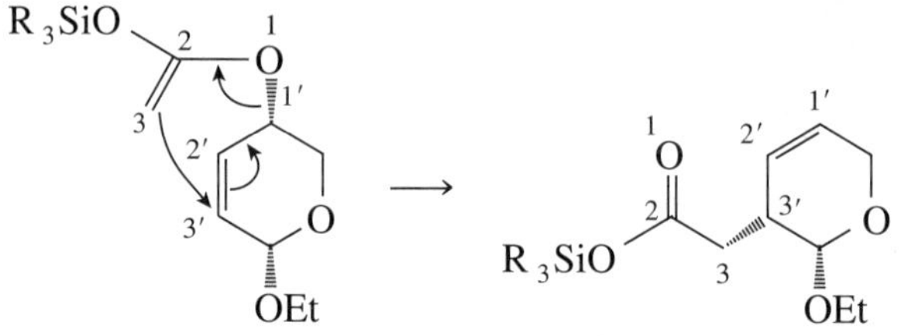

c. [2,3]

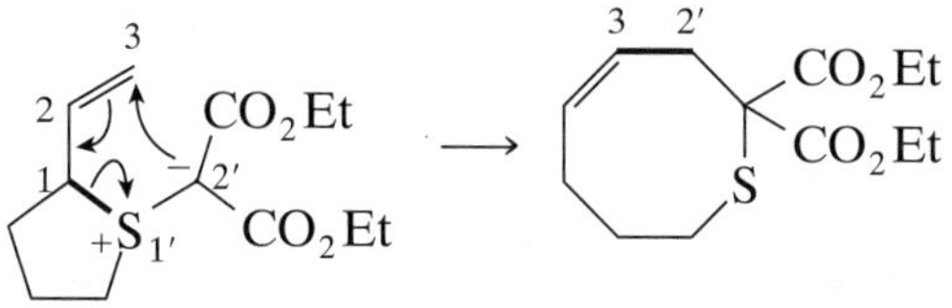

d. This is a [3,3] sigmatropic shift. It is an example of the Cope rearrangement, in which one of the carbons has been replaced by nitrogen. The σ bond broken and the new σ bond formed are highlighted in the structures.

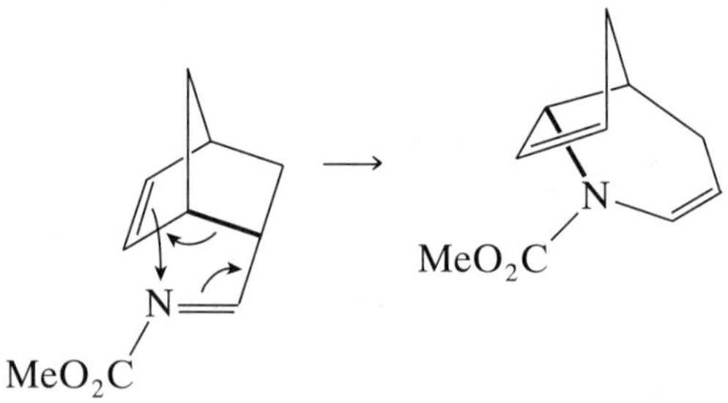

Rearrangement to the aromatic toluene requires a [1,3] sigmatropic shift of hydrogen.

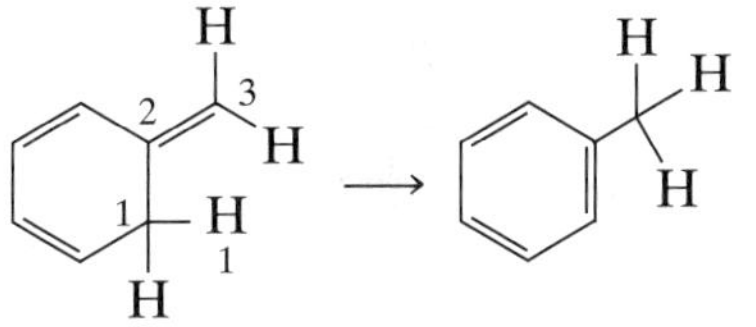

Problem 6.13

A suprafacial shift does not obey the selection rules and the geometry required makes an antarafacial shift very difficult, so that the compound is kinetically stable. The activation energy for rearrangement is large enough that the compound can be isolated even though it is thermodynamically unstable, just like diamond, which is kinetically stable even though it is thermodynamically unstable compared to graphite.

The photochemical transformation is a 6π electron $(4n + 2)$ electrocyclic ring opening. The selection rules predict a conrotatory process as illustrated:

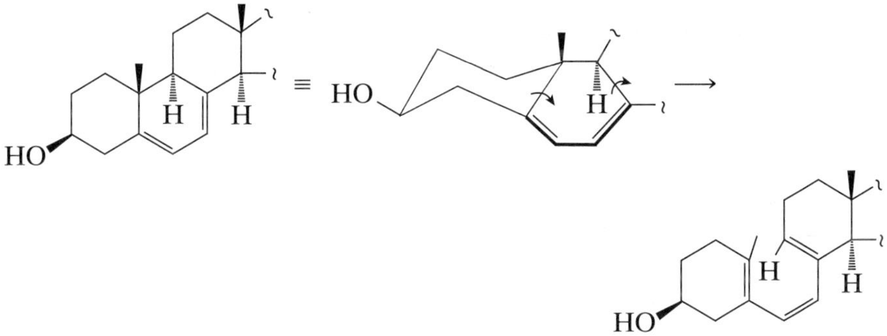

Problem 6.14

The alternative conrotatory opening would not be possible because it would result in the formation of *trans* double bonds in both cyclohexene rings. The second reaction is a thermal [1,7] sigmatropic hydrogen shift.

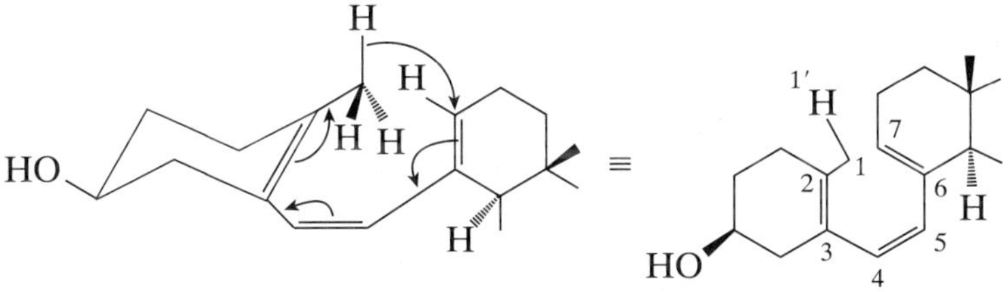

Problem 6.14
continued

The selection rules predict an antarafacial process. In this case, suprafacial and antarafacial processes would lead to the same product.

Problem 6.15

a. This rearrangement could take place by two different mechanisms. One is a [1,5] sigmatropic shift to give the product directly:

The other is two sequential [3,3] sigmatropic shifts:

The authors of the paper cited favor the [1,5] rearrangement because direct formation of the product is thermodynamically more favorable than proceeding through the intermediate from the first [3,3] sigmatropic shift.

b. This transformation was described by the authors cited as a [1,5] shift, followed by a "spontaneous" [1,7] shift.

6-20

The [1,5] shift would be a suprafacial reaction. The [1,7] shift must be antarafacial in the π component to be thermally allowed. The proton, located at carbon A in **6-20**, moves from the bottom side of the six-membered ring to the top side of the π system at the other end. In general, the antarafacial process, necessary for a thermal [1,7] sigmatropic shift, occurs with ease, especially when compared to the antarafacial process that would be required for a thermal [1,3] sigmatropic shift.

Problem 6.15
continued

Problem 6.16

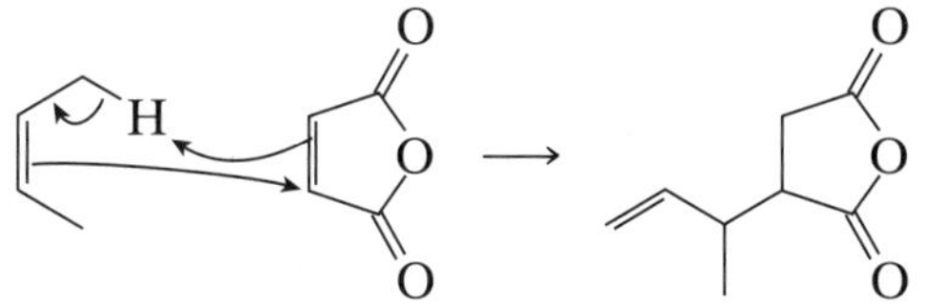

The mechanism can be written as an ene cyclization if we write the enol form of the keto group.

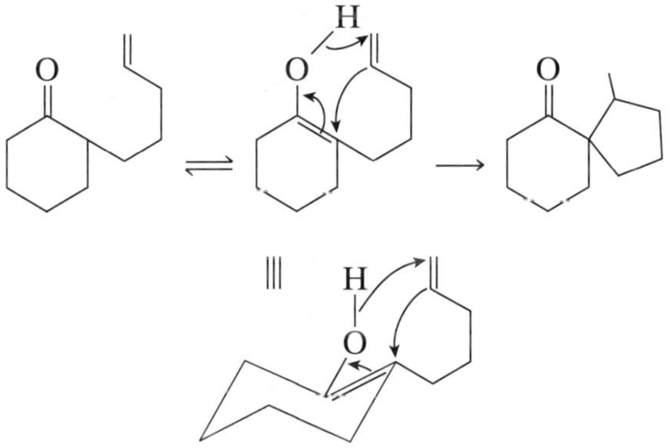

Problem 6.17

Problem 6.18   a. The major pathway, a $[\pi_s^2 + \pi_s^2 + \sigma_s^2]$ reaction, is an ene reaction.

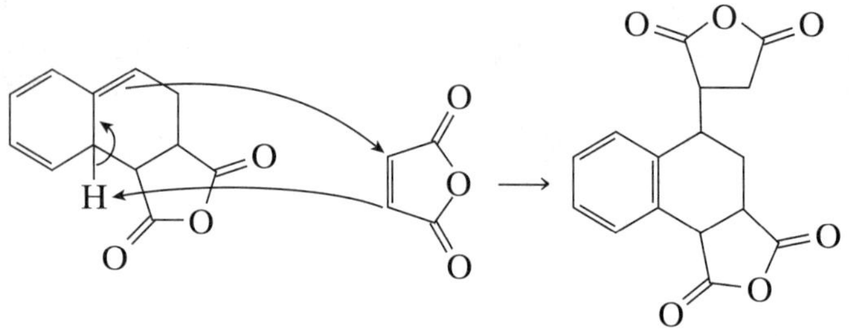

A stepwise mechanism also could be written for this reaction:

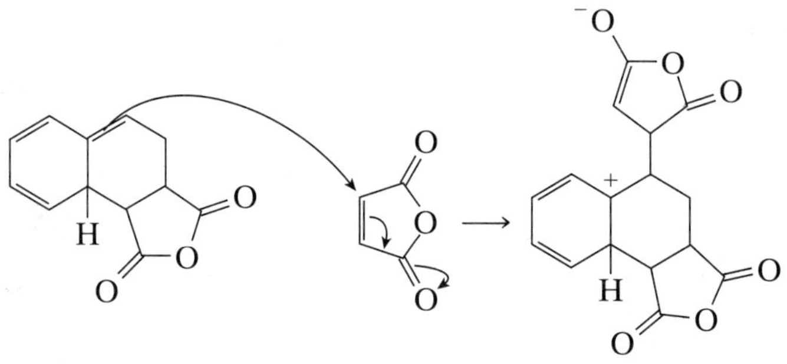

6-21

The intermediate formed, initially, **6-21**, is both an acid and a base. Appropriate intermolecular acid–base reactions, followed by tautomerization, can generate the product readily.

The minor product, formed in 11% yield, results from a Diels–Alder reaction, a $[\pi_s^4 + \pi_s^2]$ cycloaddition.

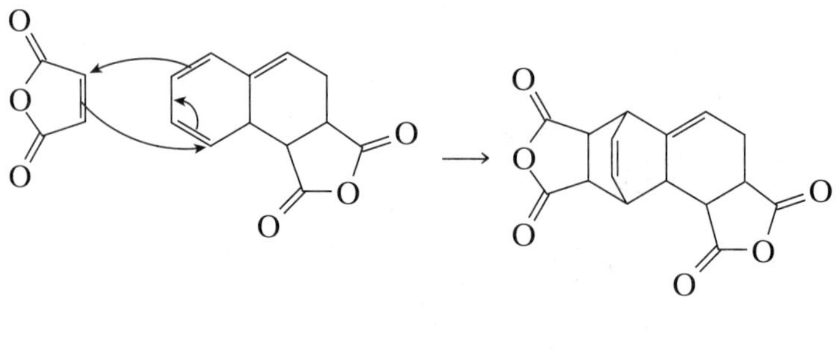

b. The first step is a $[\pi_s^4 + \pi_s^2]$ cycloaddition to give **6-22**.

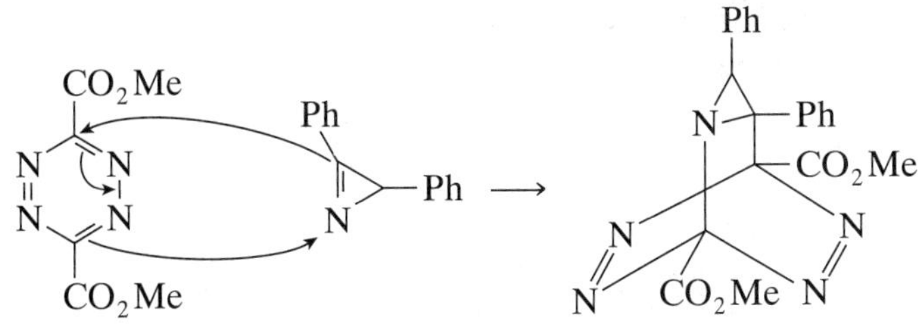

Problem 6.18
continued

6-22

The second step is a retrocycloaddition to give nitrogen and **6-23**, a $[\pi_s^4 + \pi_s^2]$ process. Finally, **6-23** undergoes a [1,5] sigmatropic shift to form the final product. In **6-23**, the bonds involved in the sigmatropic shift are highlighted.

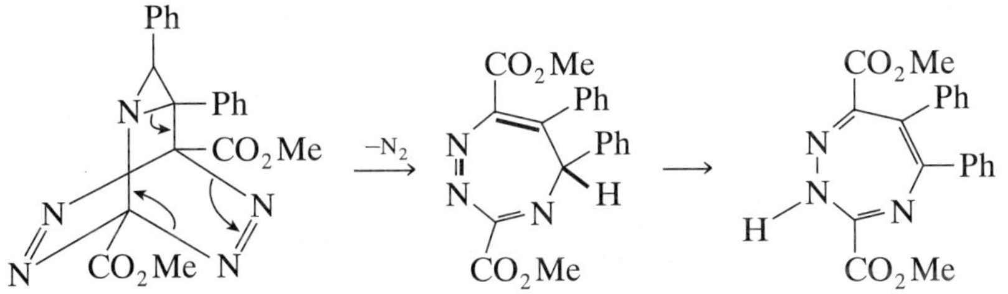

6-22   **6-23**

Another possible decomposition of **6-22**, $[\pi_s^4 + \pi_s^2]$, could give **6-24**, which then could open in a 6π electron electrocyclic reaction to give **6-23**.

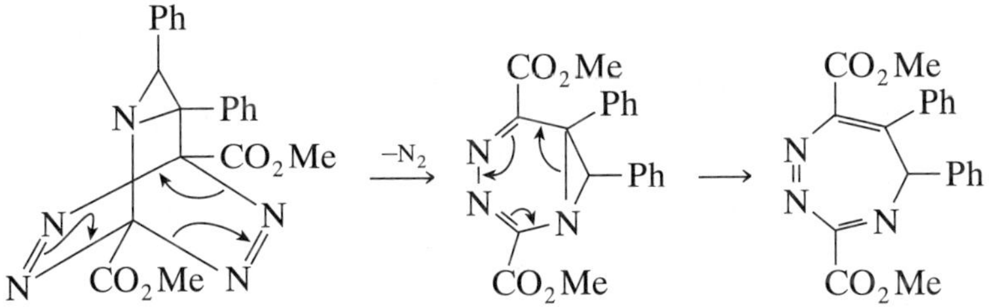

6-22   **6-24**   **6-23**

The direct opening of **6-22** to **6-23** appears more favorable, on the basis of immediate relief of the strain of the three-membered ring.

c. This reaction is a $[\pi_s^6 + \pi_s^4]$ intramolecular cycloaddition. The paper cited noted that the reaction is "periselective"; that is, none of the $[4 + 2]$ cycloadduct is produced.

The two alternative $[\pi_a^6 + \pi_a^4]$ modes of reaction are extremely unlikely, because they give products with very high strain energy.

d. The product can be explained as the result of two consecutive $[1,5]$ sigmatropic shifts.

e. The product can be produced by two consecutive [3,3] sigmatropic shifts.

Problem 6.18
continued

The first reaction is a Cope rearrangement, and the second reaction, rearrangement of a vinyl ether, is a Claisen rearrangement.

f. Step a looks like a [1,3] sigmatropic shift. However, because a concerted shift would have to be a antarafacial-suprafacial reaction, which is unfavorable sterically, the mechanism probably is not concerted. A radical chain reaction, initiated by a small amount of peroxide or oxygen, is possible.

Step b is the ene reaction.

Steps c and d are electrocyclic ring opening ($[\pi_a^4]$) processes. Because there are only hydrogens on the sp^3-hybridized carbons of the cyclobutene rings, there is no way to observe the stereochemical consequences of ring opening. Therefore, the arrows shown indicate only the rearrangement of electrons.

Step c:

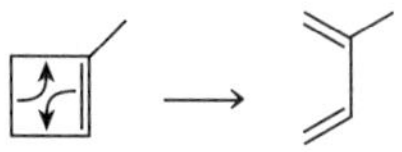

Step d:

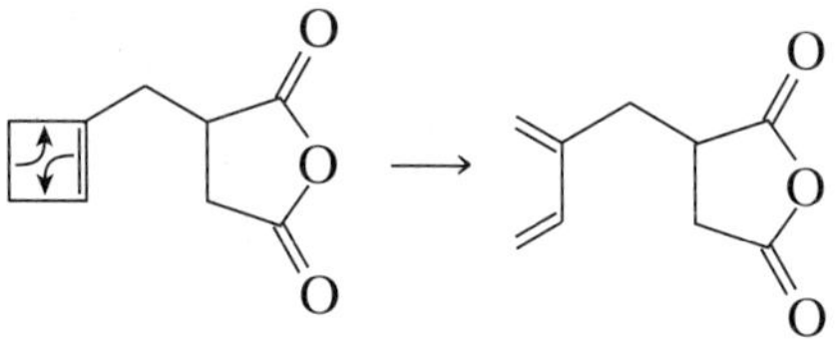

Steps e and f are $[\pi_s^4 + \pi_s^2]$ cycloadditions.

Problem 6.19

In the following equations, R is the allyl group. The first step, electrocyclic ring opening of the four-membered ring, is very common for benzocyclobutenes. There are two possible products for this ring opening (each from a possible conrotatory mode), but only **6-25** has the correct orientation to continue the reaction. The next two steps are both [3,3] sigmatropic shifts.

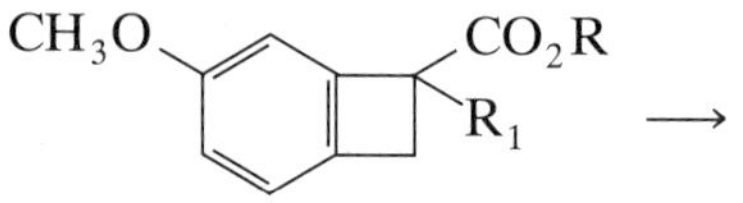

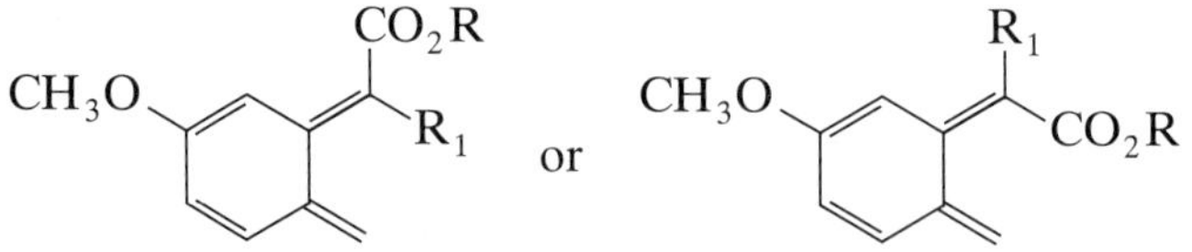

6-25

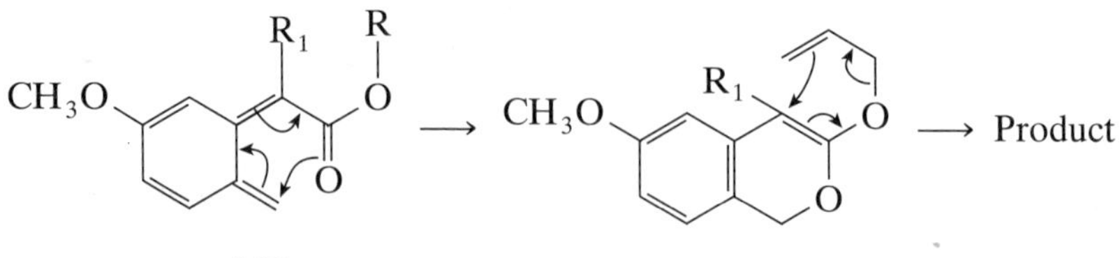

6-25   Product

The fact that the reaction occurs readily suggests that it is not concerted, because a $[\pi_s^4 + \pi_s^4]$ reaction is not a thermally allowed process (see Table 6.2). Whereas the stereochemistry of the product shows that anthracene has reacted suprafacially, the presence of oxygen and nitrogen at the ends of the other 4π electron component precludes a determination of the stereochemistry of addition at this component. However, antarafacial reaction of this component is unlikely because it would require a large amount of twist in the transition state. Thus, this transformation probably proceeds through an intermediate, which is likely to be charged. A positive charge could be stabilized by conjugation with two aromatic rings of the anthracene component. A negative charge could be stabilized by N and/or O, as well as by the electron-withdrawing chlorines. Therefore, reaction of anthracene on either oxygen or nitrogen gives a stabilized intermediate:

Problem 6.20

Problem 6.21

a. The first ring opening is a conrotatory process typical of a 4π electron thermal reaction. That is, in **6-26** there are an anion and a π bond giving a total of four electrons:

Formation of **6-19** follows from conrotatory ring closure of the heptadiene:

Notice that the conrotatory motion moves the hydrogens up and the carbons down. Ring closure to form **6-19** is favored because of the ring strain associated with a *trans* double bond in a seven-membered ring.

The isobenzofuran acts as a 4π electron component in a $[4 + 2]$ cycloaddition to trap the intermediate. As usual, both components must react suprafacially. Thus, because the fusion between the six- and seven-membered rings in **6-18** is *trans*, it must be the *trans* double bond in **6-16** that reacts in this trapping reaction.

Problem 6.21
continued

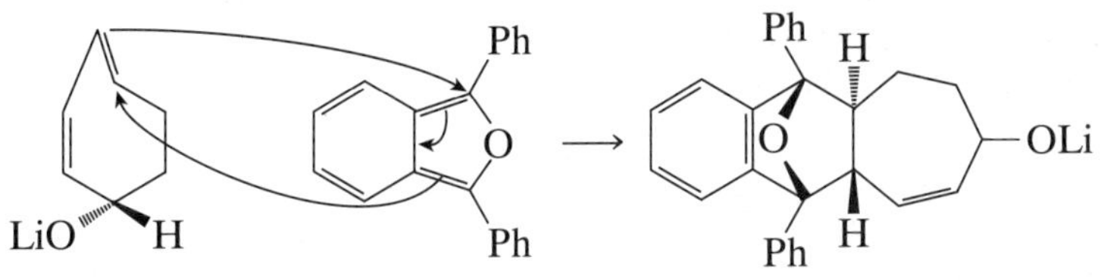

6-18

The single node should be located symmetrically, that is, at the center of the system. Because the node is located incorrectly, the orbital is neither symmetric nor antisymmetric.

Problem 6.22

The lowest energy orbital shown has no nodes, but the next higher one has two nodes. There should be an MO of intermediate energy with one node. In addition, the orbitals do not alternate in symmetry. Both of the orbitals have the same symmetry with respect to the σ plane (S) and the C_2 axis (A).

Problem 6.23

Problem 6.24

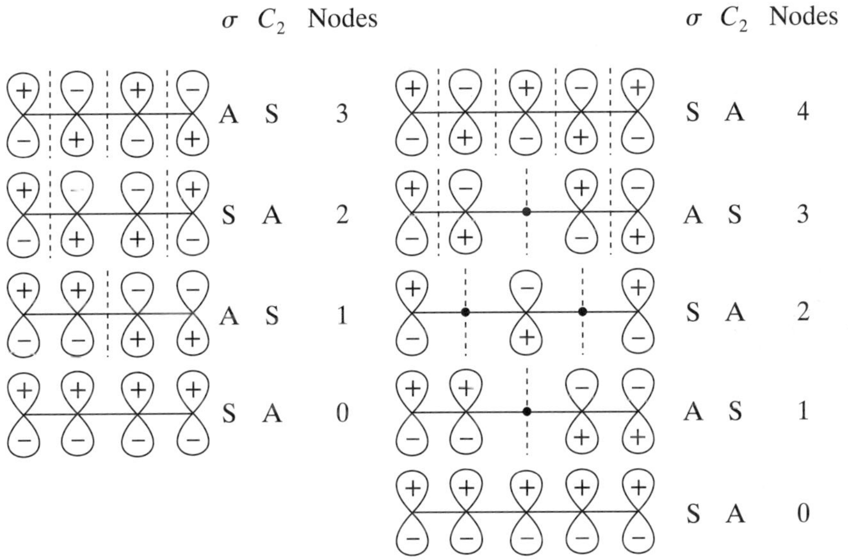

Problem 6.24
continued

Some of the nodes coincide with nuclei in the five-carbon π system because the nodes must be placed symmetrically with respect to the mirror plane that bisects the system.

Problem 6.25

The pentadienyl cation has 4π electrons; the radical and the anion have 5π and 6π electrons, respectively. To obtain the electronic configuration, fill in the MOs in order, starting with the orbital of lowest energy.

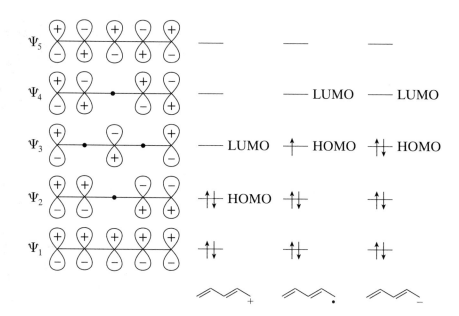

7

Additional Problems

This chapter includes additional problems related to the material in Chapters 3–6. Some of the mechanisms are mixed; for example, there might be a pericyclic reaction followed by a hydrolysis in either acid or base. If a reaction appears to be pericyclic, be sure to determine whether the reaction is symmetry-allowed or symmetry-forbidden under the reaction conditions. If it is symmetry-forbidden, a nonconcerted reaction pathway through radical or charged intermediates will be the most likely mechanism.

Choose one principle or mechanism from each of Chapters 3–6, and find a reaction from the recent (past 5 years) literature that illustrates that principle or mechanism. Write a detailed step-by-step mechanism for each reaction. Journals to look at include *J. Org. Chem., Tetrahedron Lett., Synthesis,* and *Synthetic Comm.* Sometimes a synthetic sequence can be found that answers the entire question!

PROBLEM 7.1

PROBLEM 7.2   **Show how the following transformations could occur.**

a.

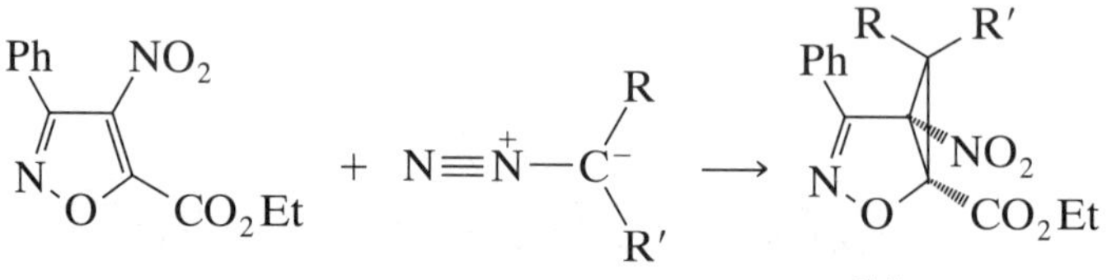

7-1

Nesi, R.; Giomi, D.; Papelao, S.; Bracci, S. *J. Org. Chem.* **1989**, *54*, 706–708.

b.

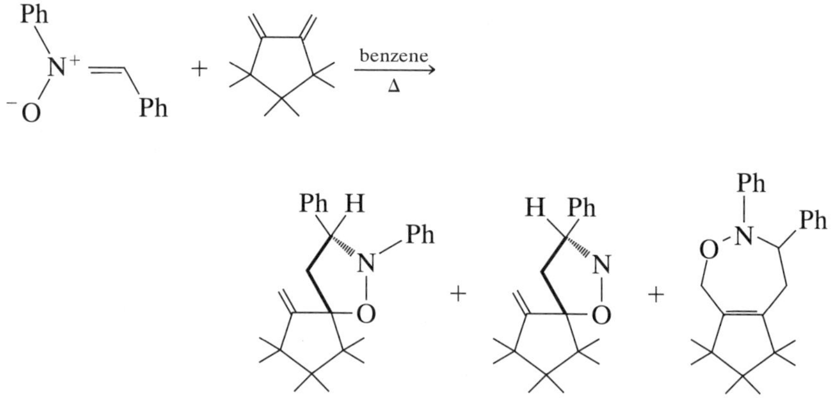

7-2   **7-3**   **7-4**

Baran, J.; Mayr, H. *J. Am. Chem. Soc.* **1987**, *109*, 6519–6521.

c.

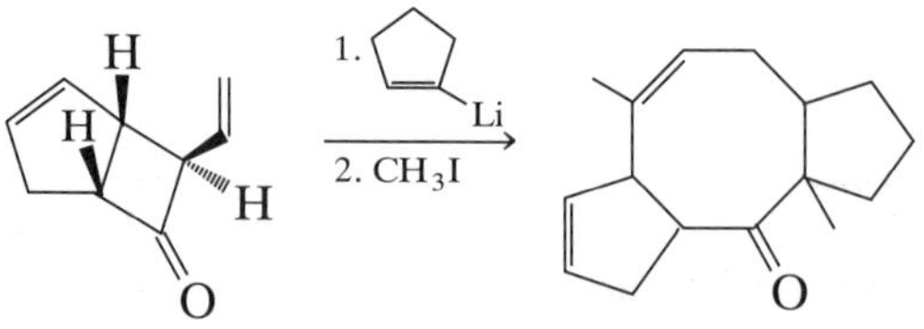

7-5   **7-6**

Paquette, L. A.; Andrews, D. R.; Springer, J. P. *J. Org. Chem.* **1983**, *48*, 1147–1149.

What is the designation for the sigmatropic shift in the transformation **PROBLEM 7.3**
of 7-7 to 7-8 in the following reaction sequence?

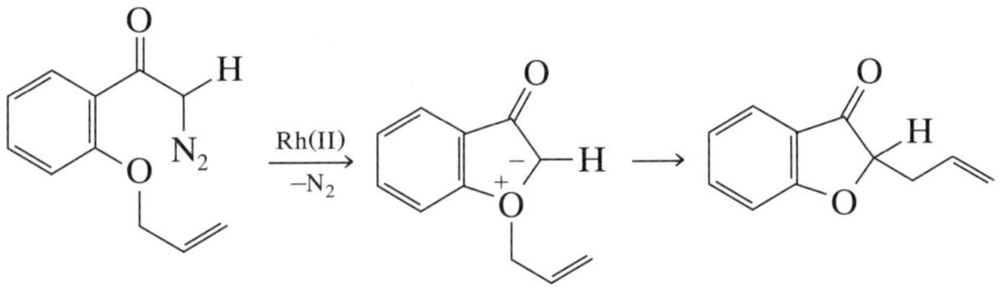

7-7

7-8

Pirrung, M. C.; Werner, J. A. *J. Am. Chem. Soc.* **1986**, *108*, 6060–6062.

Write a step-by-step mechanism for the following transformation. Other **PROBLEM 7.4**
products for this reaction are covered by Problem 4.17.b.

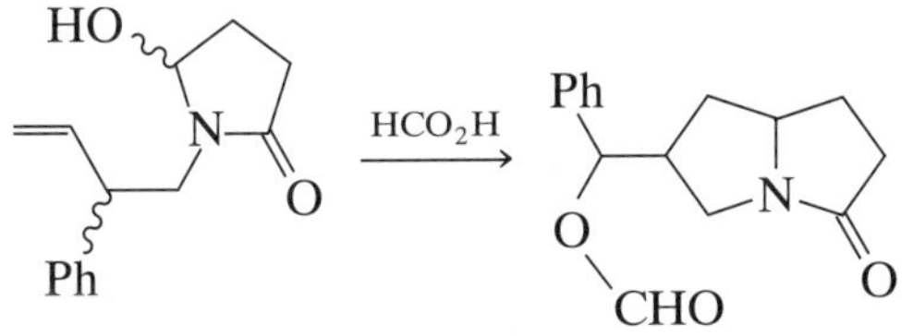

Ent, H.; de Koning, H.; Speckamp, W. N. *J. Org. Chem.* **1986**, *51*, 1687–1691.

Problem 5.13 showed the transformation of dibenzyl sulfide to a num- **PROBLEM 7.5**
ber of products by reaction with *n*-butyllithium, followed by treatment
with methyl iodide. Some tetramethylenediamine was included in
the reaction mixture to coordinate the lithium. Another product, 7-9,
is produced by this reaction. Write a reasonable mechanism for its
formation.

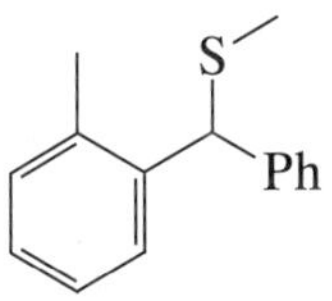

7-9

PROBLEM 7.6   By using mechanistic principles as well as the molecular formulas, determine structures for 7-10 and 7-11.

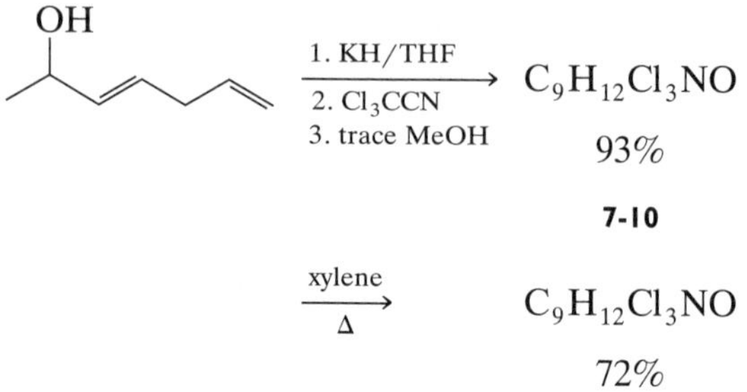

$$\xrightarrow[\text{2. Cl}_3\text{CCN}]{\text{1. KH/THF}} \text{C}_9\text{H}_{12}\text{Cl}_3\text{NO}$$
3. trace MeOH   93%

7-10

$$\xrightarrow[\Delta]{\text{xylene}} \text{C}_9\text{H}_{12}\text{Cl}_3\text{NO}$$
72%

7-11

Roush, W. R.; Straub, J. A.; Brown, R. J. *J. Org. Chem.* **1987**, *52*, 5127–5136.

PROBLEM 7.7   a. Write a step-by-step mechanism for the formation of 7-14.
b. Propose a reasonable structure for the isomeric product, 7-15.

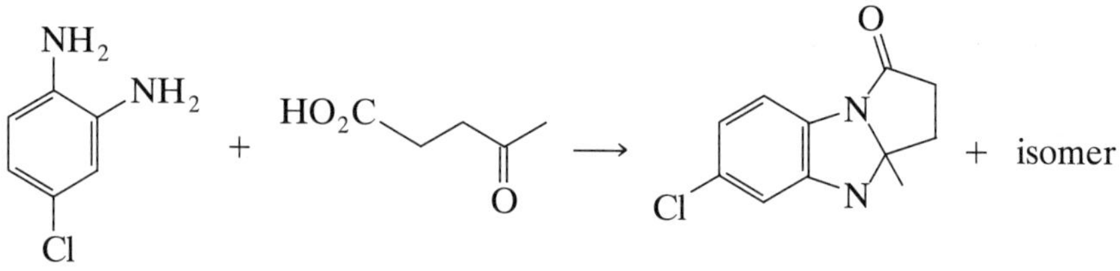

7-12   **7-13**   **7-14**   **7-15**

The reaction is run in refluxing toluene with azeotropic removal of water.

Chimirri, A.; Grasso, S.; Monforte, P.; Romeo, G.; Zappala, M. *Heterocycles* **1988**, *27*, 93–100.

PROBLEM 7.8   Write a mechanism for the following reaction, a reduction–rearrangement in which zinc is a reducing agent. The first steps of a mechanism could involve protonation of the nitrogen and reaction with zinc.

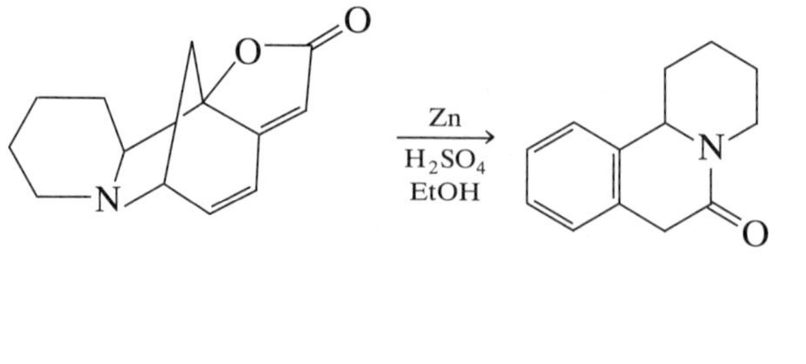

$$\xrightarrow[\substack{\text{H}_2\text{SO}_4 \\ \text{EtOH}}]{\text{Zn}}$$

Write mechanisms for the formation of both products derived from the following degradation of the fungal neurotoxin, verrucosidin. **PROBLEM 7.9**

Ganguli, M.; Burka, L. T.; Harris, T. M. *J. Org. Chem.* **1984**, *49*, 3762–3766.

Propose mechanisms for the steps shown in the following synthetic sequence. The overall process involves reduction by the transition metal reagents, as well as cyclization. **PROBLEM 7.10**

Zarecki, P. A.; Fang, H.; Ribiero, A. A. *J. Org. Chem.* **1998**, *63*, 4779.

ANSWERS TO PROBLEMS

Problem 7.2   a. There are several reasons why it is unlikely that loss of nitrogen from the diazo compound, followed by direct addition of the carbene to the double bond, is the mechanism for the reaction. (1) Even though the carbene has two alkyl substituents, it is still quite electrophilic and would not be expected to react with one of the double bonds to the exclusion of the other. (2) There is no catalyst or ultraviolet light to aid in the generation of the carbene.

A more probable mechanism could be 1,3-dipolar addition to the electrophilic double bond to give **7-19**, followed by loss of nitrogen from the intermediate to give the product.

7-19

The loss of nitrogen might occur with the intermediate formation of ions or radicals. In the heterolytic process, a negative charge develops on carbon because of resonance stabilization by the nitro group and the nitrogen in the oxazole ring.

Problem 7.2
continued

b. The formation of **7-2** and **7-3** can result from 1,3-dipolar cycloaddition of the nitrone to one double bond. This is a $[\pi_s^4 + \pi_s^2]$, thermally allowed reaction. The four-electron component is the nitrone, and the two-electron component is the alkene.

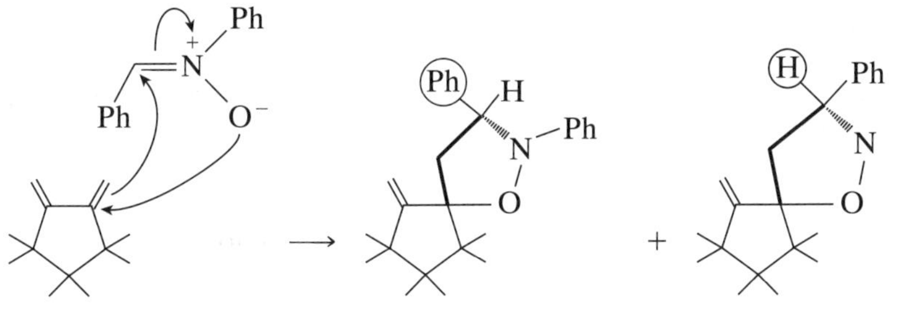

	7-2		7-3

Representing the reaction in this way does not clearly indicate how the two isomers arise. A better appreciation of what is going on is obtained by considering the orientation required for the molecular orbitals of the two components to interact in a pericyclic process. The most favorable orientation has the faces of the π systems parallel to one another, as shown in the following structures.

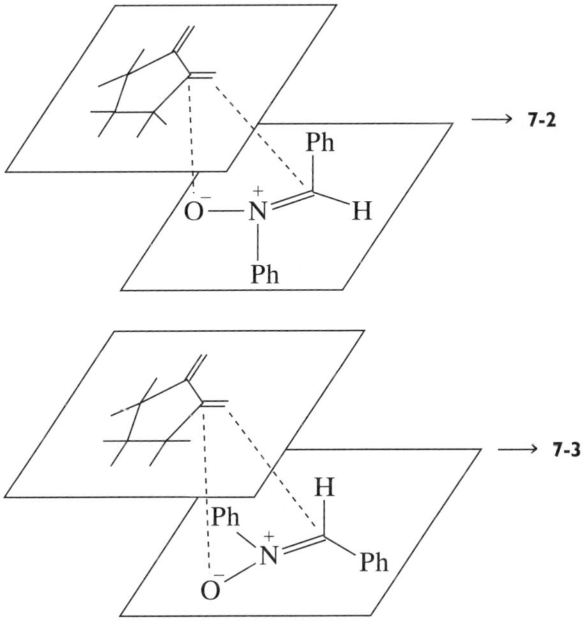

Problem 7.2
continued

The formation of the stereoisomeric products **7-2** and **7-3** is analogous to the formation of *exo* and *endo* adducts in the Diels–Alder reaction. One way to keep track of the stereochemical relationships is to note that in **7-2** the circled phenyl group and the *exo*-methylene group in the cyclopentane ring are on the same side of the oxazole ring (*cis*), whereas in **7-3** it is the circled hydrogen that is *cis* to the *exo*-methylene group.

Concerted formation of **7-4** would be a $[\pi_s^4 + \pi_s^4]$ cycloaddition, which is not thermally allowed. Thus, formation of this product probably involves a diradical intermediate:

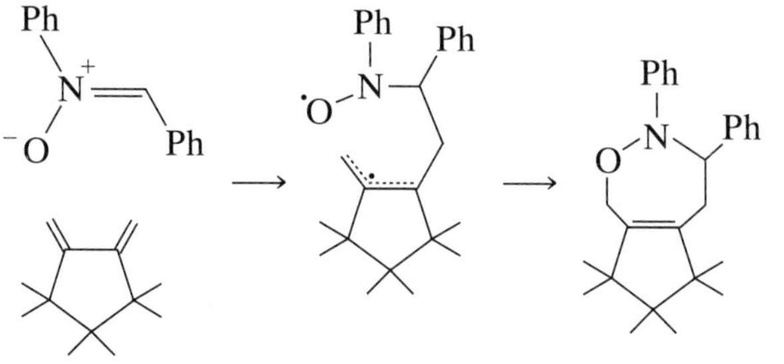

7-4

Another possibility is formation of an intermediate ion pair, **7-20**. However, in a nonpolar solvent like benzene, radical intermediates, which require less solvation than ions, seem more likely.

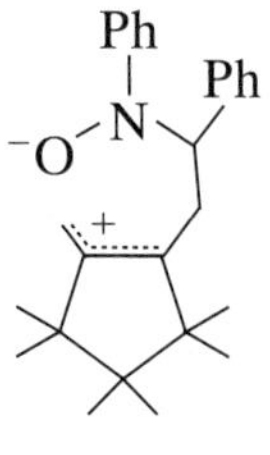

7-20

Formation of **7-2** and **7-3** also is possible through these intermediates.

The following reaction mechanism was ruled out by the authors cited: [4 + 2] cycloaddition of the diene with the nitrone to give tertiary amine oxide, **7-21**, which then thermally rearranges to the product. (Thermal rearrangement of a tertiary amine oxide to an alkylated hydroxylamine is called a Meisenheimer rearrangement.)

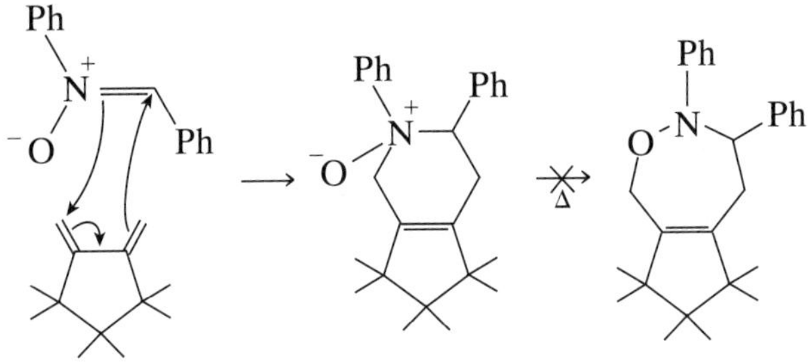

Problem 7.2
continued

When **7-21**, synthesized by an alternate route, was subjected to the reaction conditions, it gave **7-22**, not **7-4**. Therefore, **7-21** cannot be an intermediate in the mechanism to **7-4**.

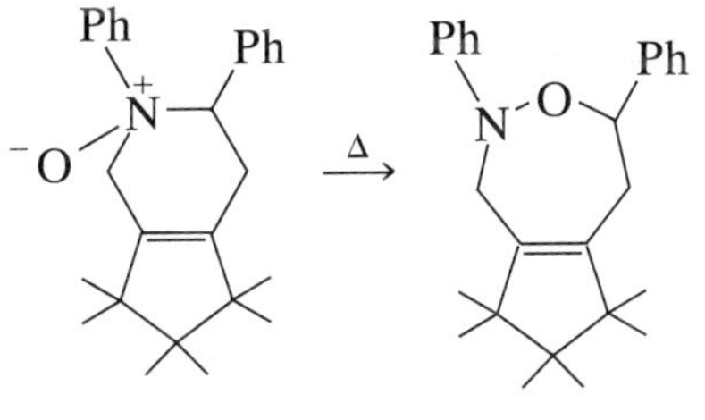

c. The lithium reagent must react with the carbonyl carbon of **7-5** to give **7-23** as an intermediate. Because neither of the double bonds in the starting material is conjugated with the carbonyl, 1,4-addition of the lithium reagent is not possible.

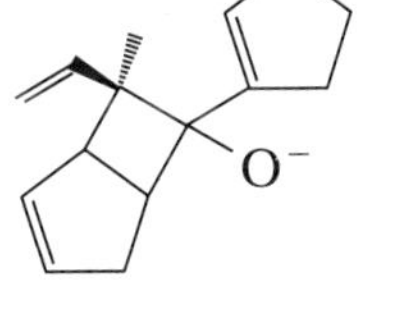

7-23

Compare the structures of the starting material **7-5**, intermediate **7-23**, and the product **7-6**. Note the following:

(1) The double bond in the five-membered ring of the starting ketone, **7-5**, is retained in the product.

(2) The other two double bonds in **7-23** have rearranged. The allyl double bond and the double bond in the new cyclopentenyl group are situated suitably for a [3,3] sigmatropic shift.

(3) There is one methyl group in the starting material and two methyl groups in the product. The second methyl must be introduced by an S_N2 reaction of methyl iodide. One of the methyls in the product is bonded to a carbon α to the carbonyl group. Thus, the nucleophile in the S_N2 reaction most likely is an enolate ion.

Ring opening to allyl anion **7-25** would be less likely, because **7-25** would be considerably less stable than enolate anion **7-24**, which is resonance-stabilized.

The usual numbering scheme shows that this is a [2,3] sigmatropic shift involving a total of six electrons.

Another potential mechanism would be a [1,2] sigmatropic shift of the allyl group. However, if concerted, this would be a $[\pi_s^2 + \sigma_s^2]$ reaction, which is not thermally allowed. For purposes of illustration *only*, suppose that a [1,2] sigmatropic shift takes place. There are two imaginable ways to write the flow of electrons for this reaction. The usual arrow-pushing technique cannot be correct, because it leaves a pentavalent, 10-electron carbon with a negative charge, which is not possible.

The method of writing electron flow, as we have been doing it in this book, *actually means the shift of a single electron from one atom to another*. That is, in the preceding mechanism, the two electrons that were on C-2' are shared with oxygen in the product. C-2' now is perceived as having just one of those original two electrons; that is, a single electron has shifted to the oxygen.

Problem 7.3
continued

The other way to write the flow of electrons for a hypothetical [1,2] shift would be transfer of two electrons from carbon-2′ to the oxygen, as illustrated in the following. But, as just reemphasized, reaction mechanisms are not written in this way.

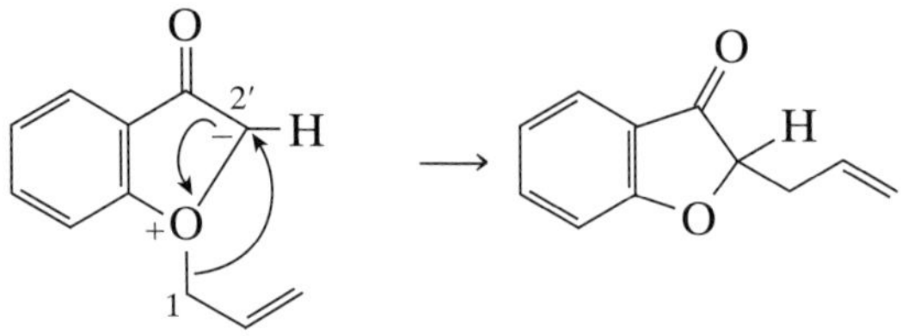

For further reading on this concept of the single-electron shift, see Pross, A. "The Single Electron Shift as a Fundamental Process in Organic Chemistry: The Relationship between Polar and Electron-Transfer Pathways." *Acc. Chem. Res.* **1985**, *18*, 212–219.

Problem 7.4

Protonation of the hydroxyl group followed by loss of water gives an intermediate, **7-26**, which can undergo a [3,3] sigmatropic shift to **7-27**.

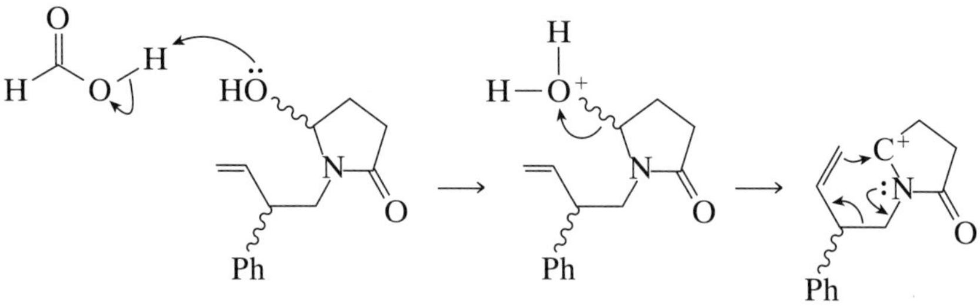

7-26

As in the case of an analogous radical cyclization (see Chapter 5, Section 6.B), the *exo* ionic cyclization of **7-27** is favored here. The reasons for this preference are subtle. With other substitution patterns, there might be steric and/or electronic factors that would favor cyclization to the six-membered ring.

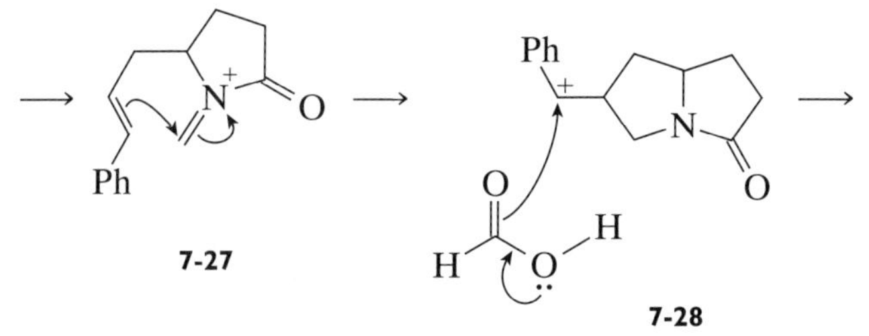

Problem 7.4
continued

7-27   **7-28**

In the nucleophilic reaction of the carboxylic acid with the cation, **7-28**, the carbonyl oxygen, not the C—O oxygen, acts as the nucleophile because only then is the resulting intermediate stabilized by resonance.

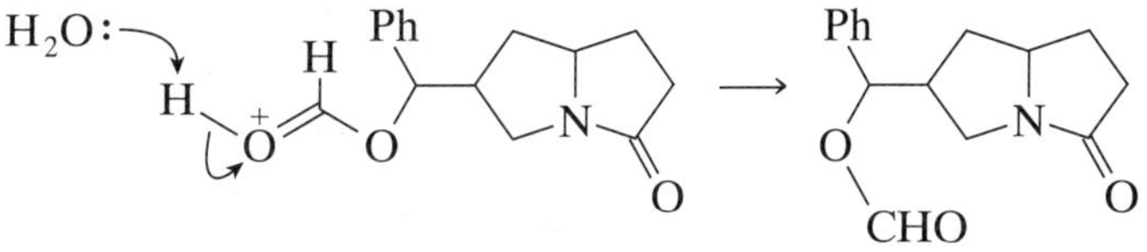

The first step in the reaction is removal of the most acidic proton in dibenzyl sulfide, **7-29**, by *n*-butyllithium to give anion **7-30**. This anion then undergoes a [2,3] sigmatropic rearrangement to **7-31**. This rearrangement utilizes a σ bond, the anionic electrons, and one π bond of one of the aromatic rings.

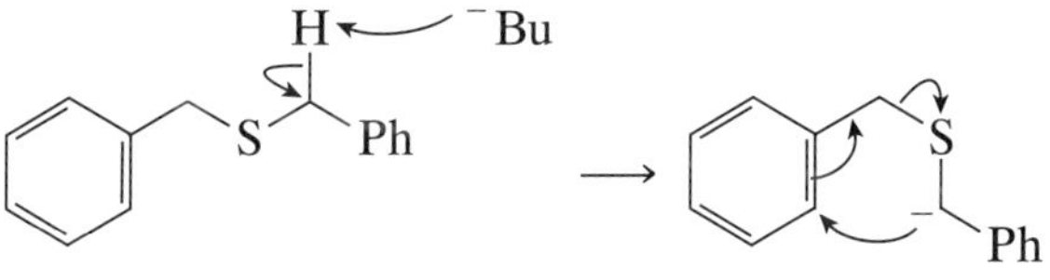

Problem 7.5

7-29   **7-30**

The intermediate, **7-31**, can be converted to the more stable benzylic anion, **7-32**, by a tautomerization.

The base, which removes the proton from **7-31**, could be butyllithium, a molecule of anion **7-32**, or the anion from tetramethylenediamine (TMEDA). Anion **7-32** then can remove a proton from a molecule of **7-31** or from a molecule of TMEDA to produce **7-33**. It is unlikely that **7-32** picks up a proton from butane because of the high pK_a value of butane relative to the amine or **7-31**. (Compare the following pK_a's from Appendix C: diisopropylamine, 36 or 39; ethane, 50; toluene, 43; and ammonia, 41.)

A concerted [1,3] thermal shift of hydrogen to convert **7-31** to **7-32** is not a reasonable mechanism because such shifts are ruled out by the selection rules for pericyclic reactions (see Chapter 6).

The last step is an S_N2 reaction of the nucleophilic sulfur anion with methyl iodide to give the methylated sulfur.

The molecular formula of **7-10** indicates that trichloroacetonitrile has
reacted with the starting material. By using Hint 1.2, three rings and/or π
bonds are calculated. Because the starting alcohol has two π bonds and
the nitrile has two π bonds, a nucleophilic reaction with the nitrile is
likely. Potassium hydride should remove the most acidic proton from the
starting alcohol, the proton on the oxygen.

There are two possible condensation reactions for the nitrile. One is
nucleophilic reaction of the oxyanion with the carbon of the nitrile
functional group. This carbon is activated by nitrogen and by the strongly
electron-withdrawing trichloromethyl group. The other possibility, S_N2
displacement of chloride, is ruled out because there are three chlorines
and four π bonds in the product. After the nucleophilic reaction, a trace
of methanol is needed to form a neutral product by protonation of the
anion.

The second reaction is an isomerization. That is, the molecular formula of

Problem 7.6

7-10

The second reaction is an isomerization. That is, the molecular formula of
7-11 is the same as that of **7-10**. Because xylene is an unreactive aromatic
hydrocarbon, a reasonable assumption is that it functions only as the
solvent. Therefore, the first thing to look for is a concerted reaction. A
[3,3] sigmatropic reaction is possible.

7-10   **7-11**

Problem 7.7

a. Two types of condensations occur. One is reaction of an amine with a carboxylic acid, and the other is reaction of an amine with a ketone. The latter should occur more readily (and is written first in the following mechanism) because the resonance effect of the hydroxyl of the carboxylic acid group reduces the electrophilicity of its carbonyl group compared to that of a ketone. Two rings are formed in this reaction. Either the smaller five-membered ring or the larger eight-membered one could be formed first. Entropy factors would favor more rapid formation of the five-membered ring.

The first step is protonation of the ketone carbonyl group by the carboxyl group. This might be either an intermolecular or an intramolecular reaction.

Then the nitrogen of the aniline can react, as a nucleophile, at the electrophilic carbon of the protonated carbonyl group:

Subsequently, the positively charged nitrogen undergoes deprotonation, followed by protonation of the hydroxyl group.

Problem 7.7
continued

Loss of a molecule of water leads to an iminium ion, **7-35**, which is susceptible to nucleophilic reaction at carbon:

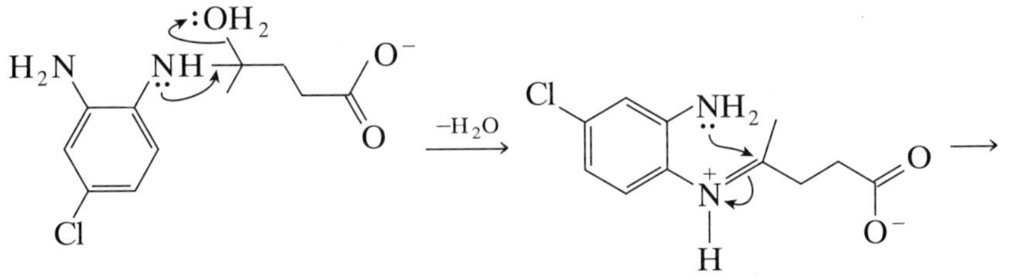

After deprotonation of the positively charged nitrogen, one of the nitrogens acts as a nucleophile with the electrophilic carbon of the neutral carboxy group. In the structures that have been written up to this point, the carboxylate anion has been indicated. However, this anion and the neutral carboxylic acid can be written interchangeably because they both would be present under the reaction conditions. (As stated previously, entropy favors the initial formation of the five-membered ring because of the low probability for productive collision of the ends of an eight-atom chain.)

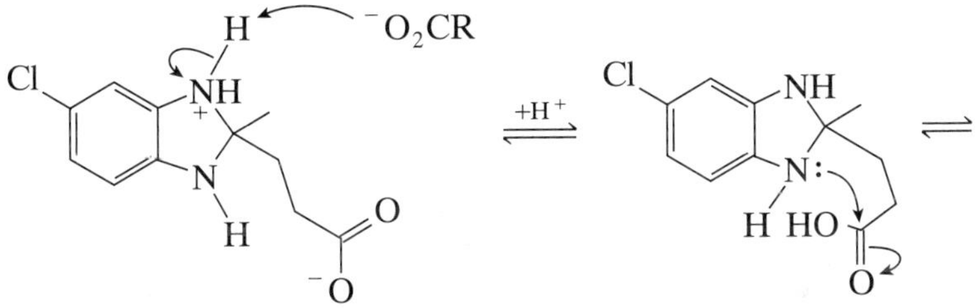

Protonation of oxygen and deprotonation of nitrogen occur.

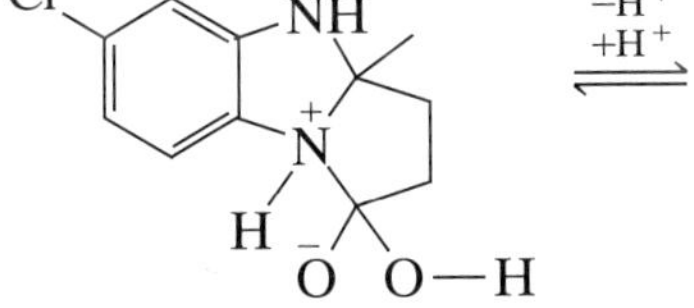

Then, protonation of one of the hydroxyl groups, elimination of water, and deprotonation give the product.

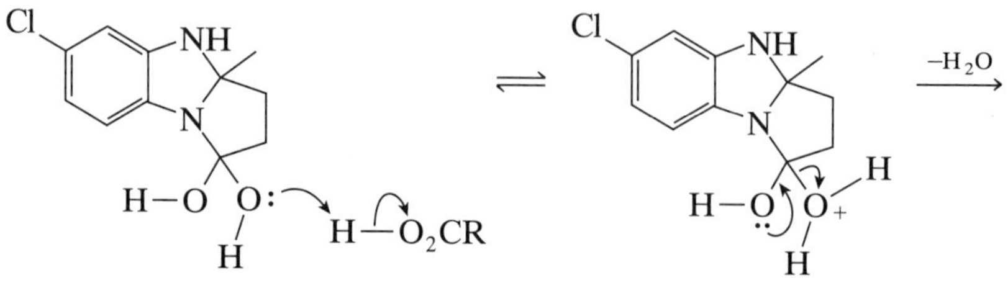

During the formation of **7-14**, water is lost in two separate steps. Both steps are reversible, but the iminium ion **7-35** is hydrolyzed much more readily than the amide group in **7-14**. Azeotropic removal of water by reflux in toluene drives both reactions toward completion.

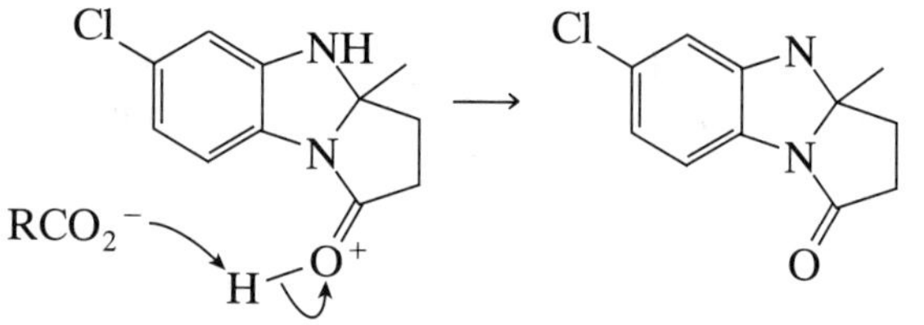

7-14

Additional Comments

(1) The following equilibrium, which undoubtedly occurs under the reaction conditions, is unproductive. That is, it is not a reaction of the amine that is on a pathway to the product.

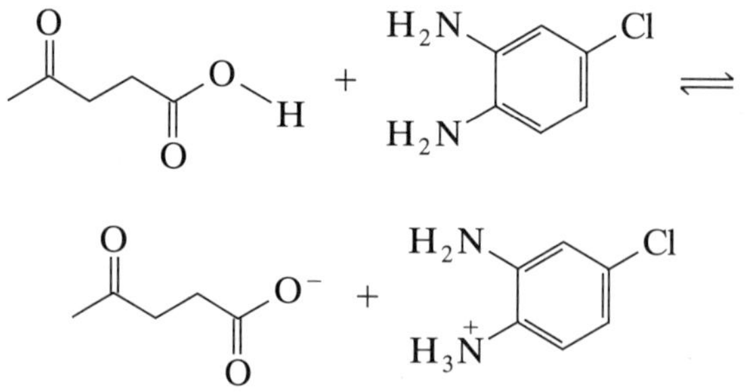

Problem 7.7
continued

(2) There are several possible mechanisms for the loss of water from the protonated aminol, **7-34**. A partial structure representing the possibility depicted earlier is as follows:

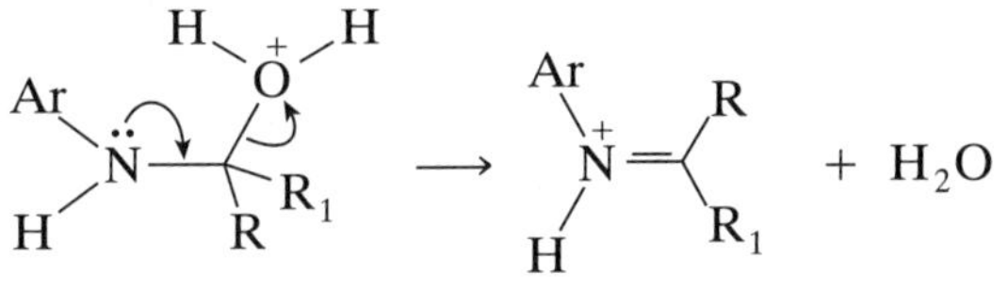

7-34   7-35

Another possibility is the simultaneous loss of a proton from the nitrogen:

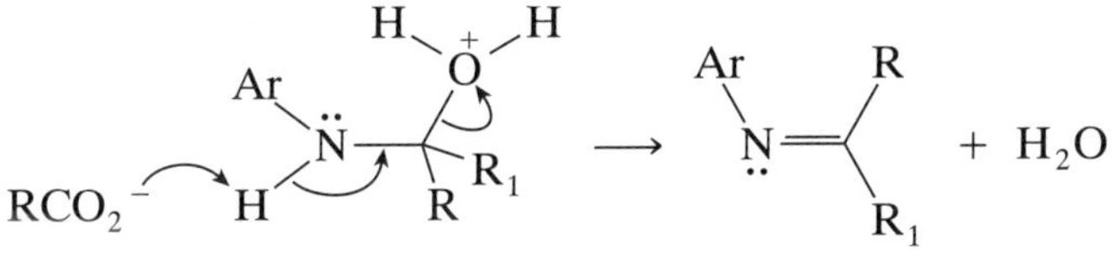

Finally, the proton might be removed from the nitrogen before elimination of water takes place:

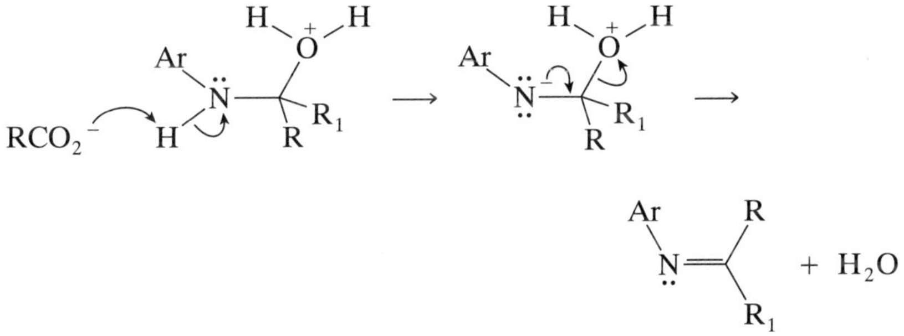

This last mechanism is an example of an E_{1CB} elimination, which is not common unless the intermediate anion is especially stabilized by strongly electron-withdrawing groups. Although the anion produced in this reaction is resonance-stabilized by the ring, it probably is not stable enough to be produced without the driving force of simultaneous formation of the double bond. Also, the bases usually used in E_{1CB} reactions are much stronger than the carboxylate anion.

(3) Estimations of the basicity of carboxylate ion **7-37**, formed from **7-13**, and the starting aniline, **7-12**, suggest that either one may be used as the base in the mechanisms written.

7-37

From Appendix C, the pK_a of *p*-chloroanilinium ion is 4.0, and that of acetic acid is 4.76. This means that the acetate ion is more basic than *p*-chloroaniline, but by less than a factor of 10. However, the basicity of **7-12** would be enhanced by the second amino group, and the basicity of **7-37** would be decreased by the keto group. Thus, the basicities of these two bases may be approximately the same.

(4) The anion **7-38**, which is formed by removal of a proton from one of the amino groups, would not function as a base in this reaction because the equilibrium to form **7-38** is completely on the side of starting material. This can be calculated from values given in Appendix C. The hydronium ion, H_3O^+, ($pK_a = -1.7$) is at least 10^{20} times more acidic than the substituted aniline (pK_a of *m*-chloroaniline is 26.7).

7-38

b. Formation of the isomeric product, **7-15**.

Problem 7.7
continued

If the other nitrogen in **7-36** reacts with the carboxyl carbon, the following isomeric product would be formed:

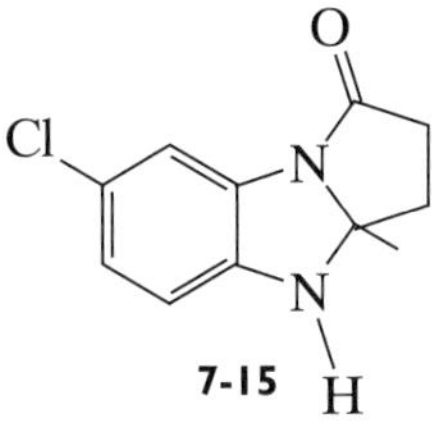

The nitrogens in the starting amine are not identical, so that **7-15** also could be produced by the nitrogen *meta* to chlorine, acting as the original nucleophile. The mechanism then would be written just as it was in part a. See the paper cited for experimental evidence that supports just one of these two possibilities.

Protonation of the nitrogen is followed by reduction by zinc. This is a two-electron reduction, in which zinc is oxidized to Zn^{2+}. The remainder of the mechanism involves acid-catalyzed reactions.

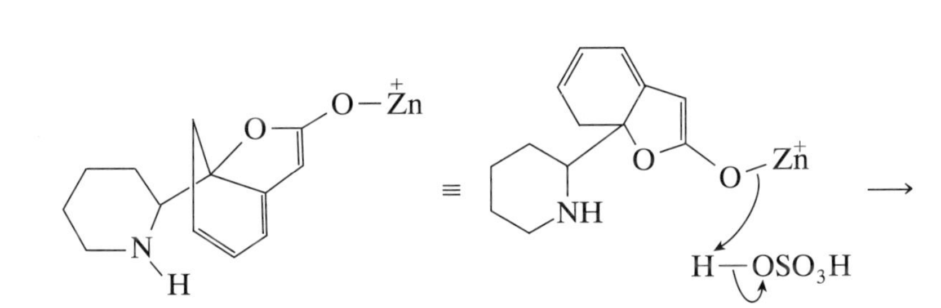

Problem 7.8

Problem 7.8
continued

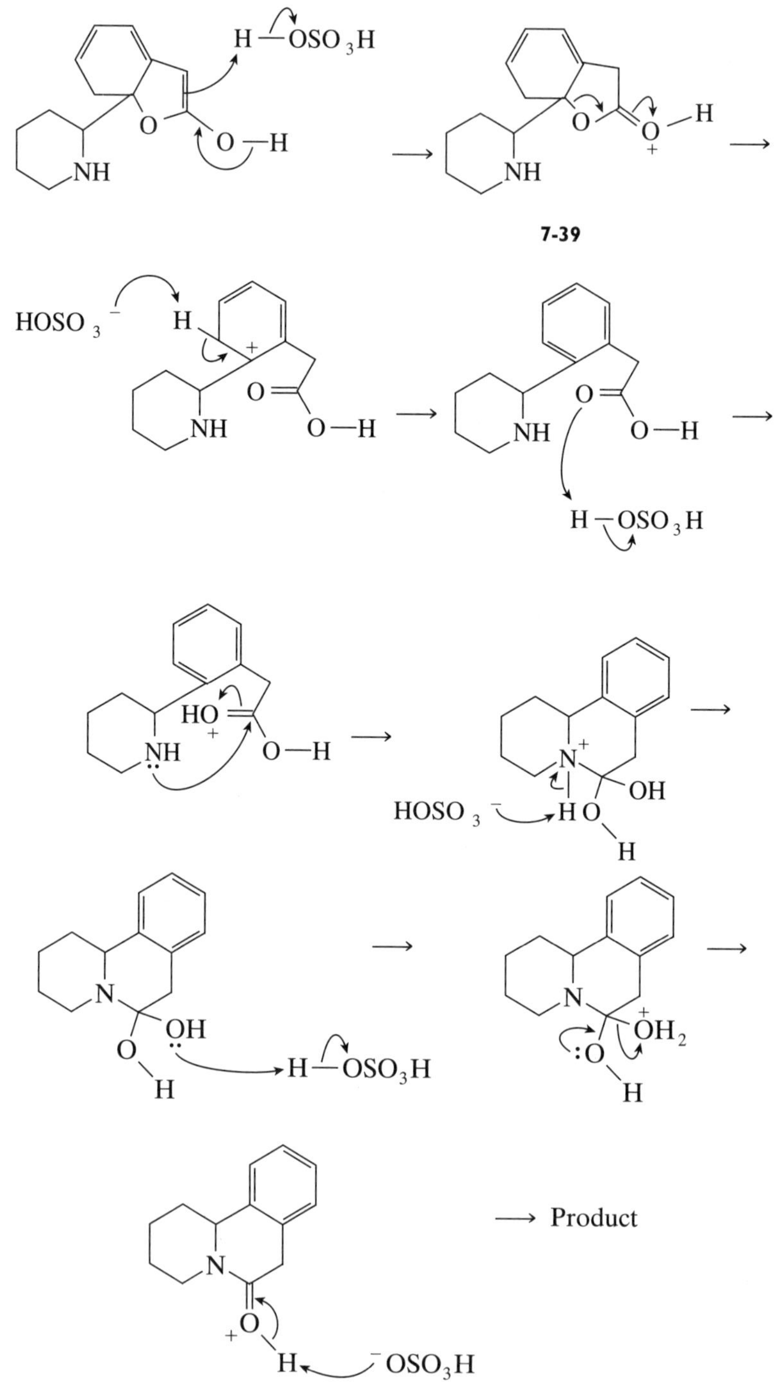

7-39

$\longrightarrow$ Product

Problem 7.8 *continued*

An alternative route from **7-39** involves formation of the remaining six-membered ring before the five-membered ring opens:

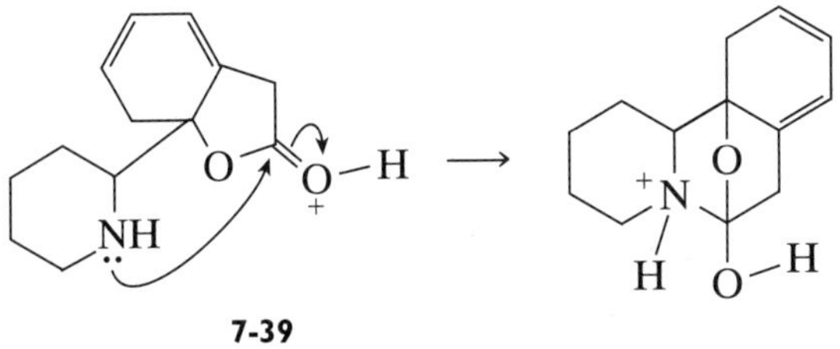

7-39

Subsequent acid-catalyzed steps will also give the product.

If you failed to use zinc in this reaction, a reduction would not occur. The product of acid-catalyzed reactions of the starting material then would contain two less hydrogens, as in **7-40**.

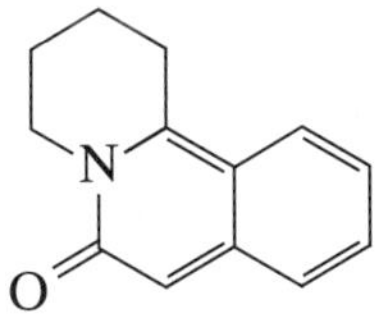

7-40

Problem 7.9

Application of Hint 2.15 may help to organize your thoughts about how the products might form. Formation of the first product involves the loss of one carbon. It appears most likely that this would be CO_2 or CO. It is probably CO_2, because CO usually is lost only in strong acids or, under thermal conditions, in reverse Diels–Alder reactions or free radical decarbonylation of aldehydes.

Once the atoms in the starting material, **7-16**, have been numbered, the carbons attached to thc R group and the methoxy group in product **7-17** can be numbered.

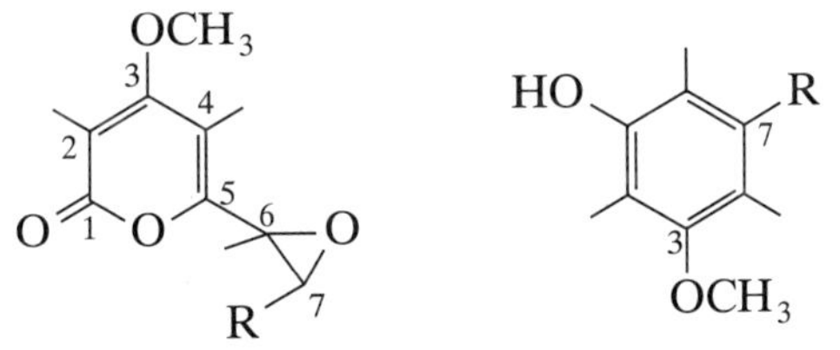

7-16 **7-17**

Problem 7.9
continued

There are two possible numbering schemes for the remaining atoms in **7-17**:

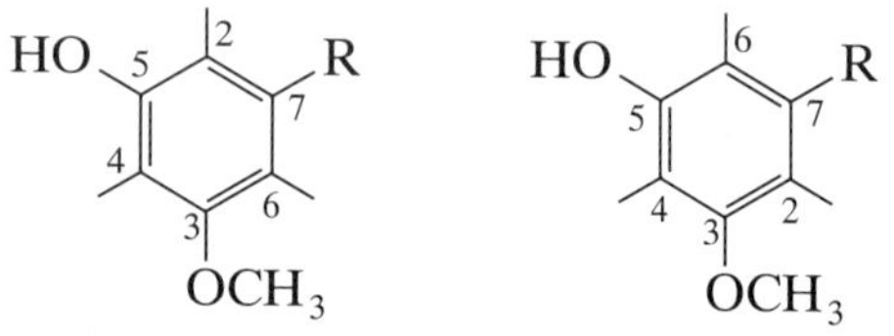

7-41   **7-42**

Getting to **7-42** involves much less bond making and breaking than getting to **7-41**. Thus, application of Hint 2.17 suggests that **7-42** is a better guide to what is occurring than **7-41**. That is, C-2 becomes bonded to C-7, and C-1 is lost. All other carbons remain attached to the same atoms.

Because the reaction conditions are basic, the next thing to do is to search for protons in the molecule that are acidic and for carbons susceptible to reaction with a nucleophile. At this point, we will look at possible nucleophilic reactions and will return to acidic protons later in the discussion. Two obvious electrophilic carbons are those of the epoxide. Moreover, if the C-2—C-7 bond is formed by nucleophilic reaction of an anion at C-2 on carbon C-7, ring opening of the epoxide would also occur in this step. This ring opening would be a driving force for the reaction. Thus, we should search for steps that would lead to an anion at C-2. This is not likely to be obvious from initial inspection of the product. In addition to the epoxide carbons, electrophilic positions in **7-16** include the carbonyl carbon (C-1) and the positions β and δ to it (C-3 and C-5).

One other comment should be made before we start writing a mechanism. The acidities of methanol ($pK_a = 15.5$) and water ($pK_a = 15.7$) are practically identical, so that methoxide or hydroxide ion as base and methanol or water as acid can be used interchangeably in the mechanism.

$$^-OH + HOCH_3 \rightleftharpoons HOH + {}^-OCH_3$$

We need to start trying the various nucleophilic reactions suggested previously. We will start by adding hydroxide ion to the carbonyl group. Then a ring-opening reaction can occur.

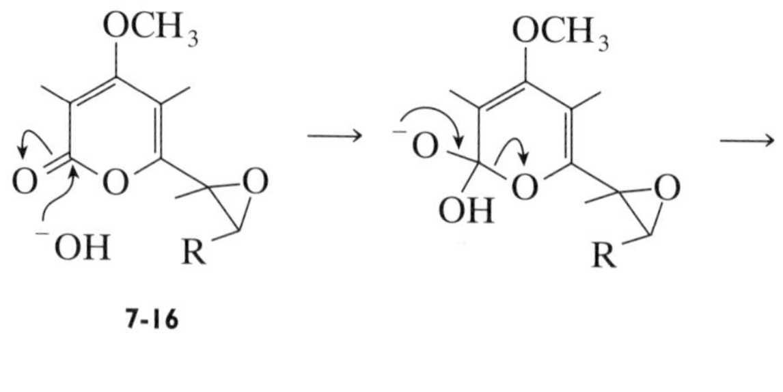

7-16

This ring-opening reaction leads to **7-43**, a resonance-stabilized anion:

Problem 7.9
continued

7-43-1   **7-43-2**

7-43-3   **7-44**

In the basic medium, the carboxylic acid will be converted rapidly to the carboxylate anion, **7-44**. Rotation in this intermediate brings C-2 and C-7 close enough to react with each other. After ring opening of the epoxide, the resulting oxyanion can remove a proton from solvent to give the corresponding alcohol.

7-44   MeO—H

The relationship of the carboxylate ion to the ketone carbonyl makes it possible to write a mechanism analogous to the decarboxylation of acetoacetic acid. This suggests that decarboxylation should occur readily.

Problem 7.9
continued

The resulting anion then undergoes base-promoted elimination of water to give the anion of a phenol, **7-45**. This adds a proton, upon workup, to give the neutral product, **7-17**. The driving force for this elimination is the formation of an aromatic system. Hence, this reaction occurs with facility, even though the leaving group is a poor one.

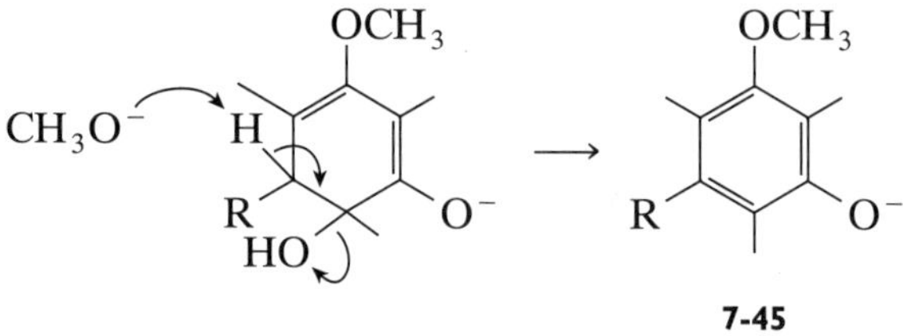

7-45

Another possible mechanism for transforming **7-44** to product involves a different sequence of reactions. Thus, **7-44** might pick up a proton to give **7-46**.

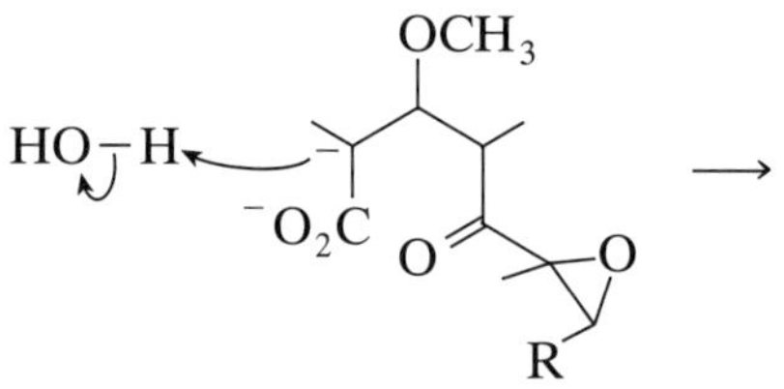

7-44

This intermediate can then decarboxylate to give an anion, which can then ring close with the epoxide.

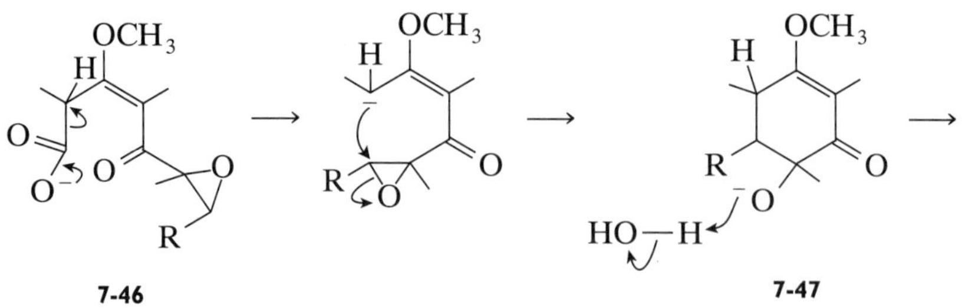

7-46 **7-47**

Protonation of **7-47** by water or methanol gives an intermediate that can undergo base-promoted elimination of water and tautomerization to give product **7-17**.

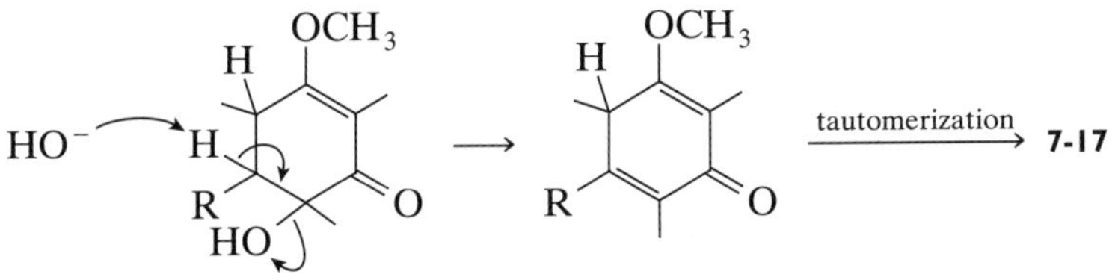

Problem 7.9
continued

Another possible mechanism for the formation of **7-17** begins with nucleophilic addition of hydroxide to C-5 of **7-16**. The resulting anion reacts with the epoxide ring, which opens to give **7-48**.

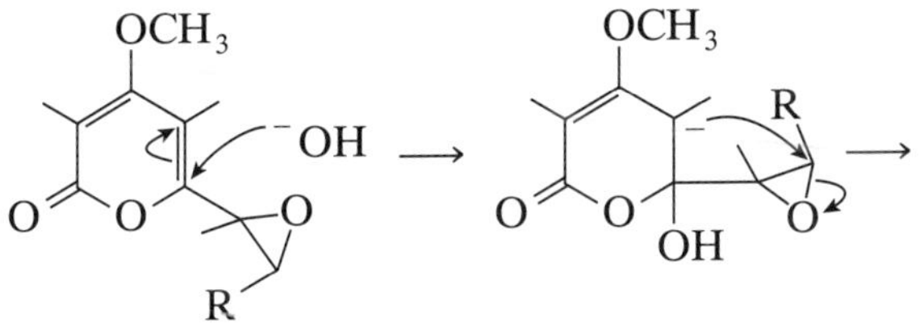

7-16

Intermediate **7-48** picks up a proton on one oxygen and loses a proton from another oxygen.

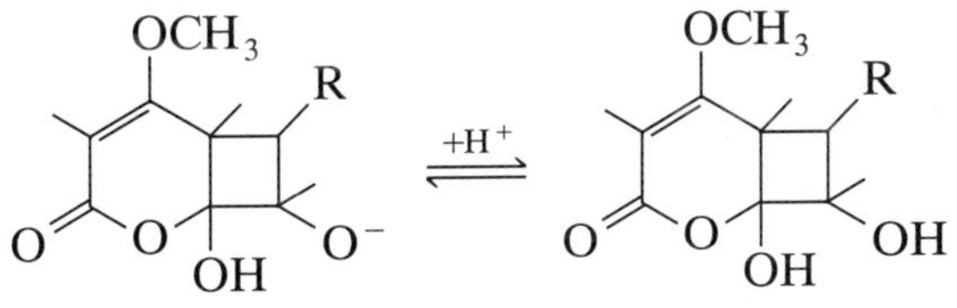

7-48

The six-membered ring of the new alkoxide ion, **7-49**, then opens with decarboxylation, to give an anion, **7-50**. Cyclization then occurs by reaction of the anion with the carbonyl group to give **7-51**.

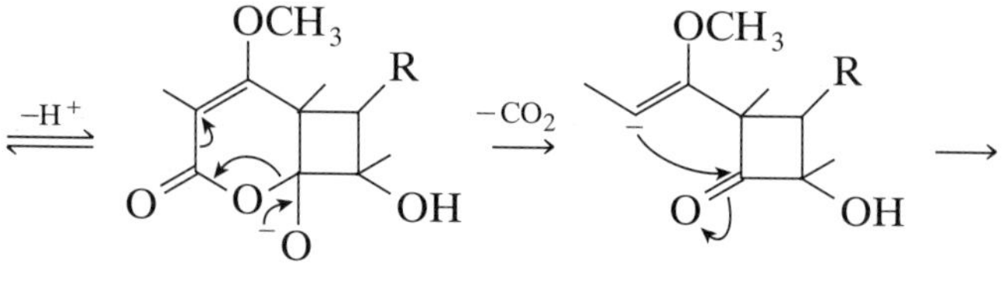

7-49   **7-50**

The remaining steps from **7-51** to product are protonation, electrocyclic ring opening, and elimination of water.

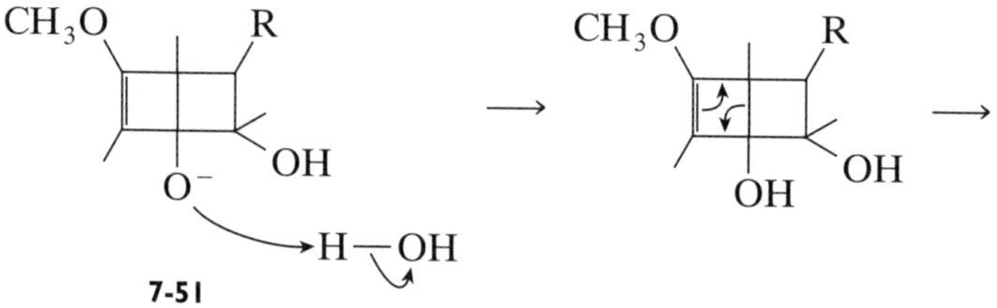

7-51

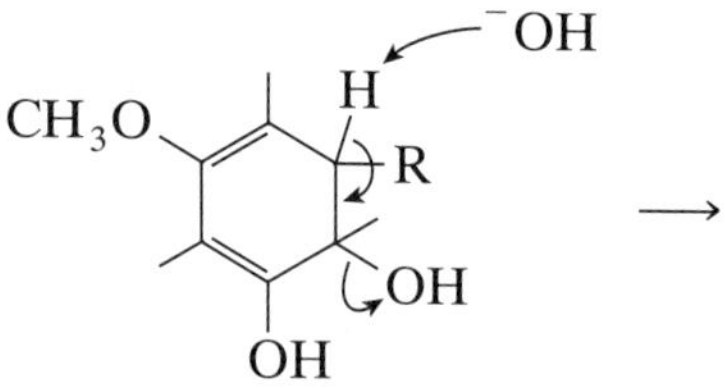

$\longrightarrow$ **7-17**

This mechanism has several disadvantages when compared to those given previously. Intermediate carbanion **7-50** is relatively unstable because its negative charge is localized. In contrast, the carbanions of the other mechanisms are stabilized by delocalization. Another intermediate of this mechanism, the bicyclo[2.2.0]hexenyl system, is not very stable because of high strain energy.

Problem 7.9
continued

Mechanism for the formation of **7-18**.

Use of Hint 2.15 to analyze the relationship between **7-16** and **7-18** gives the following:

7-16   **7-18**

This numbering scheme in the product gives the fewest changes in bond making and bond breaking and shows that the only new connection is between C-6 and C-1. C-7 and the R group attached to it are not present in the product. The paper cited shows the following mechanism for this transformation:

7-16

7-52

Problem 7.9
continued

The phenolate then can pick up a proton during acidic workup to give the product phenol.

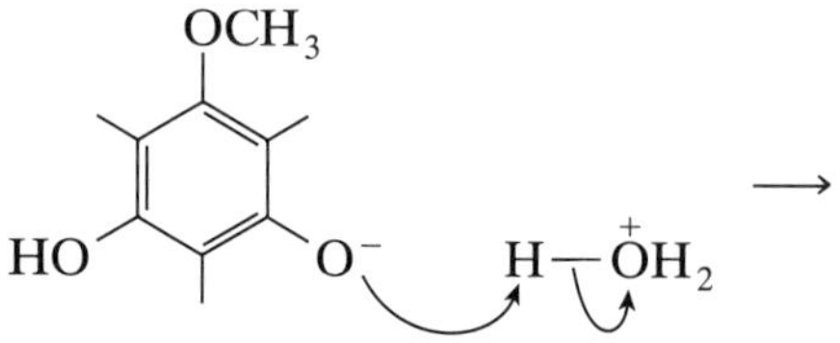

7-18

An alternative mechanism for the formation of intermediate **7-52** might be nucleophilic addition of hydroxide to C-5 of **7-16**, followed by ring opening and subsequent modification of the epoxide:

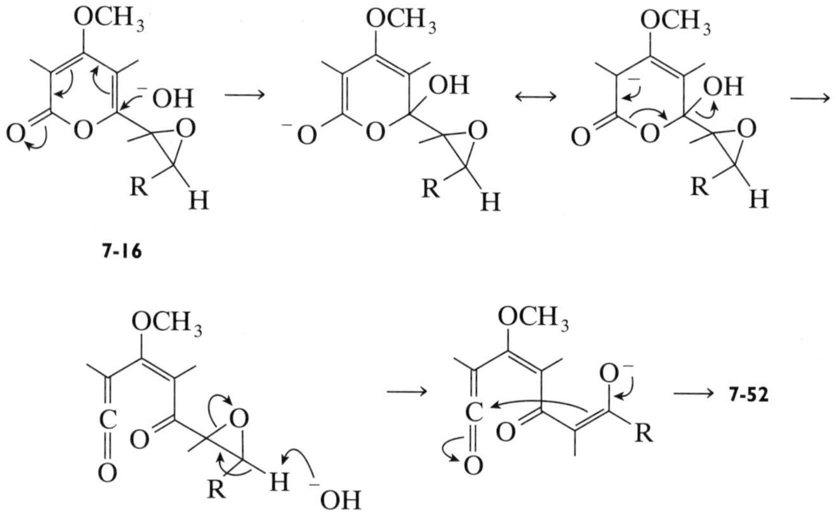

7-16

7-52

The following reaction was proposed by a student as part of a mechanism for the reaction of **7-16**.

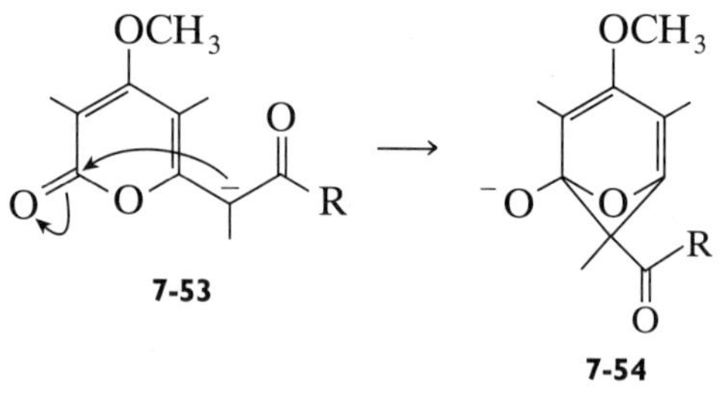

7-53

7-54

Problem 7.9
continued

This mechanistic step suffers from three problems. First, anion **7-53** is attacking itself! This anion is a resonance hybrid; one of the resonance forms, **7-53-2**, shows that there already is considerable electron density at the carbonyl group:

Second, all of the atoms involved in resonance stabilization must be coplanar. Thus, the exocyclic carbanion is not in a position geometrically favorable for reaction with the carbonyl group. Third, in **7-54** there is a double bond at a bridgehead that involves considerable strain energy.

Another student wrote the following as a first step in the reaction:

Anion **7-55** is not stabilized by resonance in any way. The acidity of the proton on the epoxide ring, relative to that of normal hydrocarbons, is enhanced by the higher s character in the C—H bond and by the adjacent oxygen. Nonetheless, neither of these factors enhances its acidity enough to bring it close to the pK_a's of methanol or water. The mechanism given in the paper cited starts with removal of this proton, but makes it part of a concerted reaction in which the epoxide ring is opened. In this case, the release of strain energy and the stability of the resulting resonance-stabilized anion provide considerable driving force.

Problem 7.10

In analyzing the first cyclization, it may be useful to ask ourselves what kind of intermediate is involved. We note that 2 mol of the manganese(III) compound and 1 mol of the copper(II) compound are used in the reaction. Manganese has stable oxidation states from +2 to +7, so that manganese(III) could function as either an oxidizing or a reducing agent. Copper is stable in the +1 and +2 states, so that the copper(II) compound would be expected to function as an oxidizing agent. By comparing the formulas of the starting material and product, we can see that the product has two less hydrogen atoms than the starting material. How is the hydrogen lost? As a first step, we could conceive of loss of a proton, a hydrogen atom, or a hydride ion. In acidic solution, loss of a proton to form a carbanion is not likely to occur. Unassisted loss of a hydride ion is an unlikely process, but loss of hydride by transfer to one of the transition metals (especially manganese) is a possibility. However, there are no positions in the starting material where transfer of a hydride could lead to a reasonably stable carbocation. On the other hand, abstraction of a hydrogen atom from the β-ketoester group leads to a radical that is stabilized by resonance with both the keto and the ester groups. The authors of the article cited suggest that this radical (**7-56**) is an intermediate in the cyclization. This suggestion seems reasonable, especially because it is known that heating manganese(III) acetate produces the radical $\cdot CH_2CO_2H$, which could abstract a hydrogen atom from the starting material.

7-56-1

7-56-2   **7-56-3**

The radical **7-56** could then cyclize as shown to form the ring system of the tricyclic intermediate.

Problem 7.10
continued

Notice that although there are four asymmetric centers in the molecule, one isomer is formed in 58% yield. This points to a concerted cyclization step in which the stereochemistry incorporated in the C=C double bonds is transferred to the developing asymmetric centers in the product. (This phenomenon is analogous to the well-known stereospecific formation of the steroid skeleton by cationic cyclizations of similar olefinic systems with defined stereochemistry at the C=C double bonds.)

Although a radical intermediate can be written for this transition-metal-mediated reaction, this is not a radical chain reaction because the transition metal reagents are employed in stoichiometric amounts. In addition, if the reaction were to proceed by a chain reaction, one might reasonably expect to obtain a different product (**7-59**) as a result of hydrogen abstraction.

Problem 7.10
continued

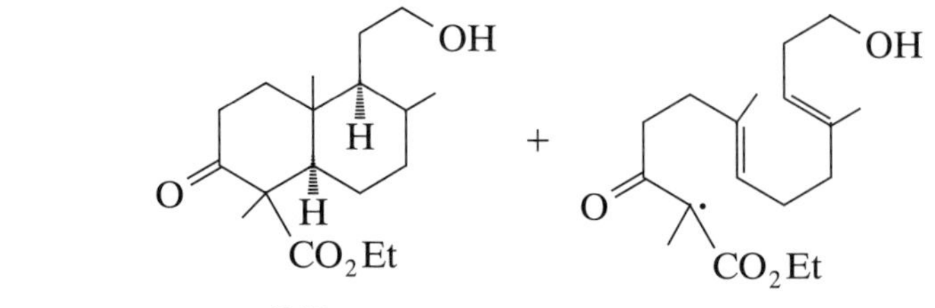

7-59   **7-56**

How can we get from the cyclized radical **7-57** to the product? One possibility is to transfer an electron to the copper(II) acetate to form a tertiary carbocation, **7-60**, which then loses a proton to form the exocyclic olefin.

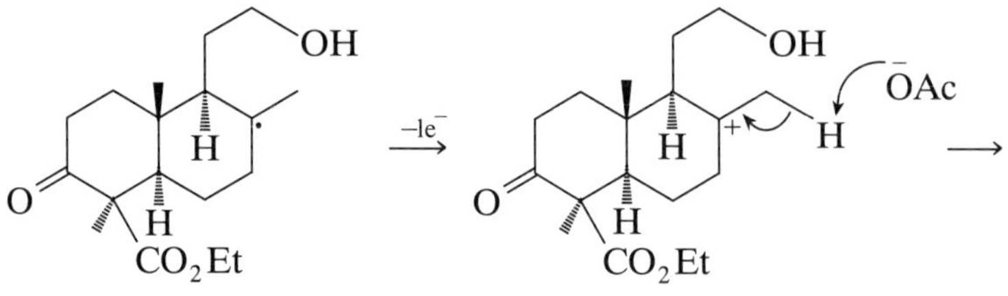

7-57   **7-60**

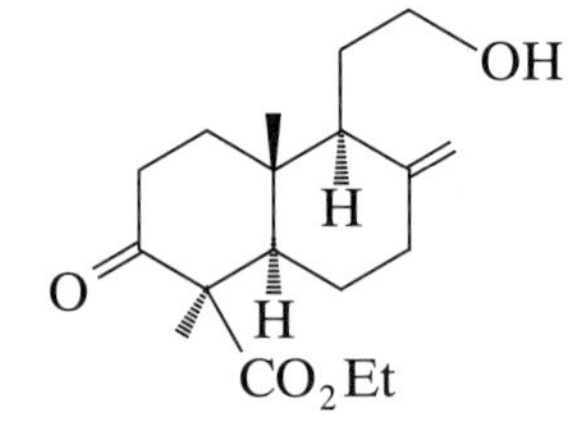

7-58

The second step of the synthetic sequence is relatively straightforward. The exocyclic olefin is protonated in trifluoroacetic acid to form the tertiary carbocation, which cyclizes by an intramolecular S_N1 reaction to give the final product.

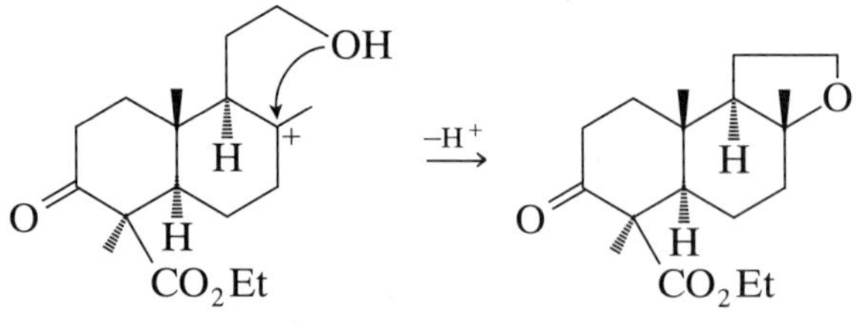

7-60   **7-61**

If the carbocation **7-60** is an intermediate in the first reaction, why does it not immediately cyclize to form the final product? To answer this, it may help to look at the reaction conditions. The first cyclization is carried out in acetic acid, whereas the final cyclization uses trifluoroacetic acid. To form the exocyclic olefin, the loss of a proton from **7-60** to form the exocyclic methylene group in **7-58** must be a rapid process compared to reaction with a nucleophile, even by the intramolecular hydroxy group. Once **7-58** is formed, acetic acid is not a strong enough acid to form an appreciable concentration of the tertiary carbocation **7-60**. However, in trifluoroacetic acid (a much stronger acid), **7-58** can be protonated to reform **7-60** as many times as necessary to give the hydroxy group time to react, and the reaction can proceed to the more stable product, **7-61**.

Problem 7.10
continued

Lewis Structures of Common Functional Groups

$-\overset{\textstyle\vert}{\underset{\textstyle\vert}{C}}-$	aliphatic carbon	$\overset{\displaystyle :O:}{\underset{\displaystyle \parallel}{-C}}-\ddot{N}H_2$	amide
$\overset{\backslash}{\underset{/}{C}}=\overset{/}{\underset{\backslash}{C}}$	alkene	$:\dot{\ddot{O}}-\dot{\ddot{O}}:$	molecular oxygen (triplet)
$-C\equiv C-$	alkyne	$-:\ddot{O}:-H$	hydroxy
$\overset{\backslash}{\underset{/}{C}}=\ddot{O}$	carbonyl	$-\ddot{O}-$	ether
$\overset{\displaystyle :O:}{\underset{\displaystyle \parallel}{-C}}-\ddot{O}H$	carboxylic acid	$-\ddot{O}-\ddot{O}-H$	hydroperoxy

(continues)

453

454

(*continued*)

$-\ddot{\text{O}}-\ddot{\text{O}}-$	peroxy	$-\ddot{\text{N}}-\ddot{\text{N}}=\ddot{\text{O}}$	nitrosamine
$\text{R}-\ddot{\underset{..}{\text{X}}}:$	halide (X = F, Cl, Br, I)	$\text{R}-\ddot{\text{N}}=\ddot{\text{N}}-\text{R}$	azo
$\text{R}_3\text{N}:$	amine (tertiary)	$\overset{\backslash}{\underset{/}{}}\text{C}=\overset{+}{\text{N}}=\ddot{\underset{..}{\text{N}}}:^{-}$	diazo
$\overset{\backslash}{\underset{/}{}}\text{C}=\ddot{\text{N}}\overset{\backslash}{}$	imine	$-\ddot{\text{N}}=\overset{+}{\text{N}}=\ddot{\underset{..}{\text{N}}}:^{-}$	azide
$\overset{\backslash}{\underset{/}{}}\text{C}=\text{C}\overset{\text{H}}{\underset{\ddot{\text{N}}\text{R}_2}{}}$	enamine	$-\ddot{\underset{..}{\text{S}}}-$	sulfide
$-\text{C}\equiv\text{N}:$	nitrile	$-\ddot{\underset{..}{\text{S}}}-\text{H}$	thiol
$-\text{N}\equiv\text{C}:$	isontrile	$-\ddot{\underset{..}{\text{S}}}-\ddot{\underset{..}{\text{S}}}-$	disulfide
$\text{R}_3\overset{+}{\text{N}}-\ddot{\underset{..}{\text{O}}}:^{-}$	amine oxide	$-\overset{:\text{O}:}{\underset{||}{\text{S}}}-$	sulfoxide
$\overset{\backslash}{\underset{/}{}}\text{C}=\ddot{\text{N}}-\ddot{\underset{..}{\text{O}}}-\text{H}$	oxime	$-\overset{:\text{O}:}{\underset{:\text{O}:}{\overset{||}{\underset{||}{\text{S}}}}}-$	sulfone
$-\ddot{\text{N}}=\ddot{\text{O}}$	nitroso	$\text{R}_3\text{P}:$	phosphine
$-\text{N}\overset{\ddot{\text{O}}}{\underset{\ddot{\underset{..}{\text{O}}}:}{}}$	nitro	$\text{R}_3\text{P}=\ddot{\text{O}}$	phosphine oxide
$-\ddot{\text{N}}\text{H}\ddot{\text{N}}\text{H}_2$	hydrazine	$(\text{RO})_3\text{P}:$	phosphite ester
$\overset{\backslash}{\underset{/}{}}\text{C}=\ddot{\text{N}}-\ddot{\text{N}}\text{H}_2$	hydrazone	$(\text{RO})_3\text{P}=\ddot{\text{O}}$	phosphate ester

B

Symbols and Abbreviations Used in Chemical Notation

Symbol	Meaning
$\longrightarrow$	Reaction
$\rightleftharpoons$ or $\rightleftharpoons$	Equilibrium
$\longleftrightarrow$	Resonance structures
⌣↗	Movement of an electron pair
⌣↗ or ⌣	Movement of an unpaired electron
↑ or ↓	An electron (in molecular orbital diagrams)
: or |	Electron pair
·	Unpaired electron
⬡ (circle inside)	Circle inside a polygon denotes aromaticity
⬡ (dotted)	Dotted line indicates delocalized π system. It may or may not be aromatic.
Me	CH_3-, methyl group
Et	CH_3CH_2-, ethyl group

(*continues*)

(*continued*)

Symbol	Meaning
i-Pr	$(CH_3)_2CH-$, isopropyl group
t-Bu	$(CH_3)_3C-$, *t*-butyl group

Ph or φ   , phenyl group

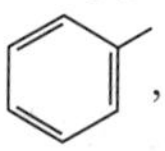

Ar   Aryl, any aromatic residue, e.g.,   ,   ,

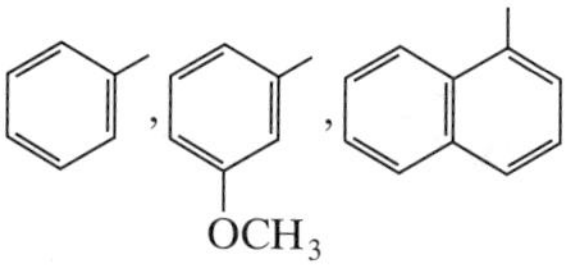

OCH_3

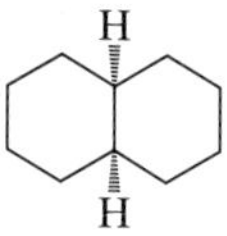

Dotted wedge indicates that the bond is directed below the plane of the page. The structure shown has a *cis* ring junction.

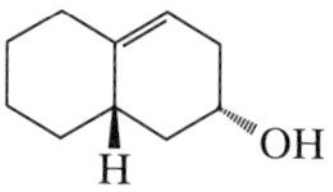

Solid wedge indicates that the bond is directed toward the viewer (above the plane of the page). The hydrogen and hydroxy groups are *trans* to one another.

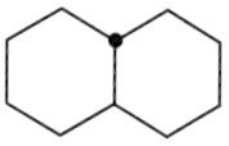

Solid circle indicates a hydrogen atom above the plane of the page. The ring junction is *trans*.

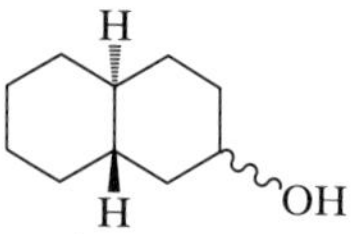

Wavy bond indicates that the stereochemistry is unspecified.

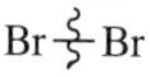

Br ξ Br   Wavy line across bond indicates homolytic cleavage.

*   Excited state

C

Relative Acidities of Common Organic and Inorganic Substances[a]

Acid	Solvent	pK_a	Conjugate base	Reference[h]
HI	Extrapolated[b]	−10	I⁻	1
HBr	Aqueous H_2SO_4	−9	Br⁻	2
HCl	Aqueous H_2SO_4	−8	Cl⁻	2
$(CH_3)_2\overset{+}{S}H$	Aqueous H_2SO_4	−6.99	$(CH_3)_2S$	3
Ph$\overset{+}{O}$(H)(CH₃)	Aqueous H_2SO_4	−6.5	Ph—OCH₃	4
Ph(C=$\overset{+}{O}$H)OEt	Aqueous H_2SO_4	−6.2	$PhCO_2Et$	5
CF_3SO_3H	Estimate[c]	−5.1(−5.9)	$CF_3SO_3^-$	3, 6
$HClO_4$	Estimate[c]	−5.0	ClO_4^-	6
FSO_3H	Estimate c	−4.8(−6.4)	FSO_3^-	3, 6

(continues)

(*continued*)

Acid	Solvent	pK_a	Conjugate base	Reference[h]
Ph–C(OH)=$\overset{+}{O}$H (protonated benzoic acid)	Aqueous H$_2$SO$_4$	−4.7	PhCO$_2$H	5
Ph–C(CH$_3$)=$\overset{+}{O}$H (protonated acetophenone)	Aqueous H$_2$SO$_4$	−4.3	Ph–C(=O)–CH$_3$	5
PhCH=$\overset{+}{O}$H	Aqueous H$_2$SO$_4$	−3.9	PhCH=O	5
(CH$_3$)$_2$$\overset{+}{O}$H	Aqueous H$_2$SO$_4$	−3.8	(CH$_3$)$_2$O	7
Ph–C(NH$_2$)=$\overset{+}{S}$H (protonated thiobenzamide)	Aqueous H$_2$SO$_4$	−3.20	Ph–C(=S)–NH$_2$	8
(CH$_3$)$_2$C=$\overset{+}{O}$H	Aqueous H$_2$SO$_4$	−2.85	(CH$_3$)$_2$C=O	3
PhSO$_3$H	Estimate[c]	−2.8	PhSO$_3^-$	6
H$_2$SO$_4$	Estimate[c]	−2.8	HSO$_4^-$	6
protonated pyran ($\overset{+}{O}$–H)	Aqueous H$_2$SO$_4$	−2.8	pyran (O)	7
CH$_3$–C(NH$_2$)=$\overset{+}{S}$H (protonated thioacetamide)	Aqueous H$_2$SO$_4$	−2.51	CH$_3$–C(=S)–NH$_2$	8
CH$_3$$\overset{+}{O}H_2$	Aqueous H$_2$SO$_4$	−2.5	CH$_3$OH	9
CH$_3$SO$_3$H	Extrapolated[d]	−1.9	CH$_3$SO$_3^-$	1
Ph–C(NH$_2$)=$\overset{+}{O}$H (protonated benzamide)	Aqueous H$_2$SO$_4$	−1.74	Ph–C(=O)–NH$_2$	10
H$_3$O$^+$	Estimate[c]	−1.7	H$_2$O	6
Ph–C(NHCH$_3$)=$\overset{+}{O}$H	Aqueous H$_2$SO$_4$	−1.7	Ph–C(=O)–NHCH$_3$	10
CH$_3$–$\overset{+}{S}$(OH)–CH$_3$ (protonated DMSO)	Aqueous H$_2$SO$_4$	−1.5	CH$_3$–S(=O)–CH$_3$	8
HNO$_3$	Estimate[c]	−1.3	NO$_3^-$	6
NH$_2$–C(NH$_2$)=$\overset{+}{S}$H (protonated thiourea)	Aqueous H$_2$SO$_4$	−1.26	NH$_2$–C(=O)–NH$_2$	11

(*continues*)

(continued)

Acid	Solvent	pK$_a$	Conjugate base	Reference[h]
CH$_3$C(OH$^+$)NH$_2$ (protonated acetamide)	Aqueous H$_2$SO$_4$	−0.6	CH$_3$C(O)NH$_2$	10
CF$_3$CO$_2$H	H$_2$O	−0.6	CF$_3$CO$_2^-$	1
Cl$_3$CCO$_2$H	H$_2$O	−0.5	Cl$_3$CCO$_2^-$	1
HC(OH$^+$)NH$_2$ (protonated formamide)	Acetic acid[e]	−0.48	HC(O)NH$_2$	12
CH$_3$C(OH$^+$)N(CH$_3$)$_2$	H$_2$O, CH$_3$NO$_2$	0.1	CH$_3$C(O)N(CH$_3$)$_2$	13
NH$_2$C(OH$^+$)NH$_2$ (protonated urea)	Acetic acid[e]	0.5	NH$_2$C(O)NH$_2$	12
(Ph$_2$)NH$_2^+$	H$_2$O	0.8	(Ph)$_2$NH	14
PhSO$_2$H	H$_2$O	1.2	PhSO$_2^-$	1
HO$_2$CCO$_2$H	H$_2$O	1.25	HO$_2$CCO$_2^-$	1
Cl$_2$CHCO$_2$H	H$_2$O	1.35	Cl$_2$CHCO$_2^-$	1
PhCH=N$^+$HOH	H$_2$O	2.0	PhCH=NOH	1
H$_3$PO$_4$	H$_2$O	2.1	H$_2$PO$_4^-$	14
CH$_3$SO$_2$H	H$_2$O	2.3	CH$_3$SO$_2^-$	1
$^+$NH$_3$CH$_2$CO$_2$H	H$_2$O	2.35	$^+$NH$_3$CH$_2$CO$_2^-$	1
FCH$_2$CO$_2$H	H$_2$O	2.6	FCH$_2$CO$_2^-$	1
ClCH$_2$CO$_2$H	H$_2$O	2.86	ClCH$_2$CO$_2^-$	1
HF	H$_2$O	3.2	F$^-$	2
HNO$_2$	H$_2$O	3.4	NO$_2^-$	6
CH$_3$COSH	H$_2$O	3.4	CH$_3$COS$^-$	15
O$_2$N–C$_6$H$_4$–CO$_2$H	H$_2$O	3.44	O$_2$N–C$_6$H$_4$–CO$_2^-$	1
H$_2$CO$_3$	H$_2$O	3.7	HCO$_3^-$	16
HCO$_2$H	H$_2$O	3.75	HCO$_2^-$	17
HOCH$_2$CO$_2$H	H$_2$O	3.8	HOCH$_2$CO$_2^-$	17
Cl–C$_6$H$_4$–$^+$NH$_3$	H$_2$O	4.0	Cl–C$_6$H$_4$–NH$_2$	18
O$_2$N–C$_6$H$_3$(NO$_2$)–OH	H$_2$O	4.1	O$_2$N–C$_6$H$_3$(NO$_2$)–O$^-$	17

(continues)

(*continued*)

Acid	Solvent	pK_a	Conjugate base	Reference[h]
$PhCO_2H$	H_2O	4.2	$PhCO_2^-$	17
	H_2O	4.6		18
CH_3CO_2H	H_2O	4.76	$CH_3CO_2^-$	6
$PhCH_2\overset{+}{N}H_2OH$	H_2O	4.9	$PhCH_2NHOH$	1
	H_2O	4.92		14
$Ph\overset{+}{N}H(CH_3)_2$	H_2O	5.1	$PhN(CH_3)_2$	19
	H_2O	5.2		19
$CH_3\overset{+}{N}H_2OH$	H_2O	6.0	CH_3NHOH	1
$^+NH_3OH$	H_2O	6.0	NH_2OH	1
	H_2O	6.5		20
H_2S	H_2O	7.0	HS^-	14
	H_2O	7.2		17
$^+NH_3OH$	H_2O	8.0	NH_2OH	45
	H_2O	8.3		17
	H_2O	8.3		17
$CH_3COCH_2COCH_3$	H_2O	9.0	$CH_3COCHCOCH_3$	21
NH_4^+	H_2O	9.2	NH_3	9
	H_2O	9.6		22
$^+NH_3CH_2CO_2^-$	H_2O	9.8	$NH_2CH_2CO_2^-$	14
	H_2O	10.0		17
CH_3NO_2	H_2O	10.0	$^-CH_2NO_2$	2
HCO_3^-	H_2O	10.2	CO_3^{2-}	14
$PhSH$	DMSO	10.3	PhS^-	23
CH_3CH_2SH	H_2O	10.6	$CH_3CH_2S^-$	1

(*continues*)

(*continued*)

Acid	Solvent	pK_a	Conjugate base	Reference[h]
cyclohexyl-$\overset{+}{N}H_3$	H_2O	10.7	cyclohexyl-NH_2	19
$(CH_3CH_2)_3\overset{+}{N}H$	H_2O	10.8	$(CH_3CH_2)_3N$	19
cyclohexanone-$CO_2CH_2CH_3$	H_2O	10.9	cyclohexanone-$\overset{-}{C}O_2CH_2CH_3$	17
$CH_2{=}C(OH)CH_3$	H_2O	11.0	$CH_2{=}C(O^-)CH_3$	24
$PhCO_2H$	DMSO	11.0	$PhCO_2^-$	2
$(CH_3CH_2)_2\overset{+}{N}H_2$	H_2O	11.0	$(CH_3CH_2)_2NH$	19
$CH_2(CN)_2$	DMSO	11.1	$\overset{-}{C}H(CN)_2$	25
piperidinium $\overset{+}{N}H_2$	H_2O	11.1	piperidine NH	19
$CH_2(CN)_2$	H_2O	11.4	$\overset{-}{C}H(CN)_2$	26
$HOOH$	H_2O	11.6	HOO^-	27
indanone ${=}O$	H_2O	12.2	indanone anion ${=}O$	28
$PhCH_2NO_2$	DMSO	12.3	$Ph\overset{-}{C}HNO_2$	2
CH_3CO_2H	DMSO	12.3	$CH_3CO_2^-$	2
CF_3CH_2OH	H_2O	12.4	$CF_3CH_2O^-$	29
$CH_2(SO_2CH_3)_2$	H_2O	12.5	$\overset{-}{C}H(SO_2CH_3)_2$	47
$NCCH_2CO_2CH_3$	DMSO	12.8	$NC\overset{-}{C}HCO_2CH_3$	30
$CH_2(COCH_3)_2$	DMSO	13.4	$\overset{-}{C}H(COCH_3)_2$	25
$\overset{+}{N}H_2{=}C(NH_2)NH_2$	H_2O	13.4	$HN{=}C(NH_2)NH_2$	1
$CH_3COCH_2CO_2CH_2CH_3$	DMSO	14.2	$CH_3CO\overset{-}{C}HCO_2CH_2CH_3$	30
imidazole $N{\cdots}NH$	H_2O	14.5	imidazolide $N{\cdots}N^-$	31
CH_3OH	H_2O	15.5	CH_3O^-	29
H_2O	H_2O^f	15.7	HO^-	32
CH_3CH_2OH	H_2O	15.9	$CH_3CH_2O^-$	32
$O_2N{-}$aryl(NO_2)$-NH_2$	DMSO	15.9	$O_2N{-}$aryl(NO_2)$-\overset{-}{N}H$	33

(*continues*)

(*continued*)

Acid	Solvent	pK_a	Conjugate base	Reference[h]
Ph–CH₂–CO–CH₃ (PhCH₂COCH₃)	H₂O	15.9	Ph–CH⁻–CO–CH₃	28
CH_3CHO	H_2O	16.5	$\bar{C}H_2CHO$	34
$(CH_3)_2CHNO_2$	DMSO	16.9	$(CH_3)_2\bar{C}NO_2$	31
$(CH_3)_2CHOH$	H_2O	17.1	$(CH_3)_2CHO^-$	32
CH_3NO_2	DMSO	17.2	$\bar{C}H_2NO_2$	35
$(CH_3)_3COH$	H_2O	18	$(CH_3)_3CO^-$	32
(cyclopentadiene)	DMSO	18.1	(cyclopentadienyl anion)	25
CH_3CSNH_2	DMSO	18.5	$CH_3C\bar{S}NH$	36
CH_3COCH_3	H_2O	19.2	$\bar{C}H_2COCH_3$	24
O_2N–C₆H₄–NH_2	DMSO	20.9	O_2N–C₆H₄–$\bar{N}H$	33
$PhCH_2CN$	DMSO	21.9	$Ph\bar{C}HCN$	35
Ph_2NH	$H_2O/DMSO$	22.4	$Ph_2\bar{N}$	37
(fluorene)	DMSO	22.6	(fluorenyl anion)	35
Ph_2NH	DMSO	23.5	Ph_2N^-	33
$CHCl_3$	H_2O^g	24	$^-CCl_3$	46
CH_3COPh	DMSO	24.7	$\bar{C}H_2COPh$	35
$HC{\equiv}CH$		25	$HC{\equiv}C^-$	38
CH_3CONH_2	DMSO	25.5	$CH_3CO\bar{N}H$	36
Ph–CO–CH(CH₃)₂ (PhCOCH(CH₃)₂)	DMSO	26.3	Ph–CO–$\bar{C}$(CH₃)₂	35
CH_3COCH_3	DMSO	26.5	$\bar{C}H_2COCH_3$	35
3-Cl–C₆H₄–NH_2	DMSO	26.7	3-Cl–C₆H₄–$\bar{N}H$	33
NH_2CONH_2	DMSO	26.9	$NH_2CO\bar{N}H$	3
$CH_3CH_2COCH_2CH_3$	DMSO	27.1	$CH_3CH_2CO\bar{C}HCH_3$	35
$Ph–C{\equiv}CH$	DMSO	28.8	$Ph–C{\equiv}C^-$	35
CH_3SO_2Ph	DMSO	29.0	$\bar{C}H_2SO_2Ph$	35
Cl–C₆H₄–NH_2	DMSO	29.4	Cl–C₆H₄–$\bar{N}H$	35

(*continues*)

(*continued*)

Acid	Solvent	pK_a	Conjugate base	Reference[h]
CH_3CO_2Et	DMSO	30.5	$\bar{C}H_2CO_2Et$	30
$(Ph)_3CH$	DMSO	30.6	Ph_3C^-	35
$PhNH_2$	DMSO	30.7	$Ph\bar{N}H$	33
(dithiane)—H	DMSO	30.6	(dithiane)⁻	2
(dithiane)—Ph	DMSO	30.7	(dithiane)=Ph	2
$CH_3SO_2CH_3$	DMSO	31.1	$\bar{C}H_2SO_2CH_3$	35
CH_3CN	DMSO	31.2	$\bar{C}H_2CN$	25
$(Ph)_2CH_2$	DMSO	32.3	$(Ph)_2\bar{C}H$	25
$CH_3CON(CH_2CH_3)_2$	DMSO	34.5	$\bar{C}H_2CON(CH_2CH_3)_2$	30
$[(CH_3)_2CH]_2NH$	THF	35.7	$[(CH_3)_2CH]_2\bar{N}$	39
$[(CH_3)_2CH]_2NH$	THF	39	$[(CH_3)_2CH]_2\bar{N}$	40
NH_3	35	41	$\bar{N}H_2$	25
$PhCH_3$	DMSO	43	$Ph\bar{C}H_2$	25
(benzene)	CHA	43	(phenyl)⁻	41
$CH_2{=}CH_2$	35	44	$CH_2{=}\bar{C}H$	42
$CH_3CH{=}CH_2$	35	47.1–48.0	$\bar{C}H_2CH{=}CH_2$	43
CH_3CH_3	35	Approximately 50	$\bar{C}H_2CH_3$	44
CH_4	35	58 ± 5	$\bar{C}H_3$	43

[a] Abbreviations: DMSO, dimethyl sulfoxide; THF, tetrahydrofuran; CHA, cyclohexylamine. Most acidities were measured at 25°C. Some are extrapolated values; some are values from kinetic studies. Errors in some cases are several pK units. The farther the pK value is from 0–14, the larger the errors because of estimates and assumptions made when water is not the solvent. Values of pK's for the same substance in different solvents differ because of differences in solvation. Although the acids' actual structures are listed in this Appendix, not all references do this. Thus, you may find lists of the pK_a values for organic amines that refer to the pK_a of the protonated amine rather than the amine itself. A good rule of thumb is that if the pK_a value given for an amine is less than 15, it must be the pK_a of the protonated amine rather than the amine itself.

[b] Calculated from vapor pressure over a concentrated aqueous solution extrapolated to infinite dilution.

[c] Estimated from model kinetic studies, extrapolated to aqueous media.

[d] Highly concentrated solutions extrapolated to dilute aqueous media.

[e] Titrated in acetic acid and corrected to H_2O at 20°C.

[f] Corrected from 14 because H_2O concentration is 55 mol/liter.

[g] Acidities of very weak acids are measured and/or calculated by a variety of indirect methods and may contain large errors.

[h] References: 1. Stewart, R. *The Proton: Applications to Organic Chemistry*; Academic Press: New York, 1985. 2. Bordwell, F. G. *Acc. Chem. Res.* **1988**, *21*, 456–463. 3. Perdoncin, G.; Scorrano, G. *J. Am. Chem. Soc.* **1977**, *99*, 6983–6986. 4. Arnett, E. M.; Wu, C. Y. *Chem. Ind.* **1959**, 1488. 5. Edward, J. T.; Wong, S. C. *J. Am. Chem. Soc.* **1977**, *99*, 4229–4232. 6. Guthrie, J. P. *Can. J. Chem.* **1978**, *56*, 2342–2354. 7. Arnett, E. M.; Wu, C. Y. *J. Am. Chem. Soc.* **1960**, *82*, 4999–5000. 8. Lemetais, P.; Charpentier, J.-M. *J. Chem. Res. (Suppl.)* **1981**, 282–283. 9. Deno, N. C.; Turner, J. O. *J. Org. Chem.* **1966**, *31*, 1969–1970. 10. Yates, K.; Stevens, J. B. *Can. J. Chem.* **1965**, *43*, 529–537. 11. Janssen, M. J. *Recl. Trav.*

Chim. Pays-Bas **1962**, *81*, 650–660. 12. Huisgen, R.; Brade, H. *Chem. Ber*. **1957**, *90*, 1432–1436. 13. Adelman, R. L. *J. Org. Chem*. **1964**, *29*, 1837–1844. 14. *CRC Handbook of Chemistry and Physics*; Weast, R. C., Ed.; CRC Press: Boca Raton, FL, 1982–1983. 15. Kreevoy, M. M.; Eichinger, B. E.; Stary, F. E.; Katz, E. A.; Sellstedt, J. H. *J. Org. Chem*. **1964**, *29*, 1641–1642. 16. Bell, R. P. *The Proton in Chemistry*, 2nd ed.; Cornell Univ. Press: Ithaca, NY; 1973. 17. Bell, R. P.; Higginson, W. C. E. *Proc. R. Soc. (London)* **1949**, *197A*, 141–159. 18. Biggs, A. E.; Robinson, R. A. *J. Chem. Soc*. **1961**, 388–393. 19. Perrin, D. D. *Dissociation Constants of Organic Bases in Aqueous Solution*; Butterworths; London; 1965. 20. Liotta, C. L.; Perdue, E. M.; Hopkins, H. P., Jr. *J. Am. Chem. Soc*. **1974**, *96*, 7981–7985. 21. Pearson, R. G.; Dillon, R. L. *J. Am. Chem. Soc*. **1953**, *75*, 2439–2443. 22. Pine, S. H. *Organic Chemistry*, 5th ed.; McGraw-Hill: New York; 1987. 23. Bordwell, F. G.; Hughes, D. J. *J. Am. Chem. Soc*. **1985**, *107*, 4737–4744. 24. Chiang, Y.; Kresge, A. J.; Tang, Y. S.; Wirz, J. *J. Am. Chem. Soc*. **1984**, *106*, 460–462. 25. Bordwell, F. G.; Bartness, J. E.; Drucker, G. E.; Margolin, Z.; Matthews, W. S. *J. Am. Chem. Soc*. **1975**, *97*, 3226–3227. 26. Hojatti, M.; Kresge, A. J.; Wang, W. H. *J. Am. Chem. Soc*. **1987**, *109*, 4023–4028. 27. Everett, A. J.; Minkoff, G. J. *Trans. Faraday Soc*. **1953**, *49*, 410–414. 28. Ross, A. M.; Whalen, D. L.; Eldin, S.; Pollack, R. M. *J. Am. Chem. Soc*. **1988**, *110*, 1981–1982. 29. Ballinger, P.; Long, F. A. *J. Am. Chem. Soc*. **1959**, *81*, 1050–1053. 30. Bordwell, F. G.; Fried, H. E. *J. Org. Chem*. **1981**, *46*, 4327–4331. 31. Walba, H.; Isensee, R. W. *J. Am. Chem. Soc*. **1956**, *21*, 702–704. 32. Murto, J. *Acta Chem. Scand*. **1964**, *18*, 1043–1053. 33. Bordwell, F. G.; Algrim, D. J. *J. Am. Chem. Soc*. **1988**, *110*, 2964–2968. 34. Guthrie, J. P. *Can. J. Chem*. **1979**, *57*, 1177–1185. 35. Matthews, W. S.; Bares, J. E.; Bartmess, J. E.; Bordwell, F. G.; Cornforth, F. J.; Drucker, G. E.; Margolin, Z.; McCallum, R. J.; McCollum, G. J.; Vanier, N. R. *J. Am. Chem. Soc*. **1975**, *97*, 7006–7014. 36. Bordwell, F. G.; Algrim, D. J. *J. Org. Chem*. **1976**, *41*, 2507–2508. 37. Dolman, D.; Stewart, R. *Can. J. Chem*. **1967**, *45*, 911–925, 925–928. 38. Cram, D. J. *Fundamentals of Carbanion Chemistry*; Academic Press: New York; 1965. 39. Fraser, R. T.; Mansour, T. S. *J. Org. Chem*. **1984**, *49*, 3442–3443. 40. Chevrot, C.; Perichon, J. *Bull. Soc. Chim. Fr*. **1977**, 421–427. 41. Streitwieser, A., Jr.; Scannon, P. J.; Neimeyer, H. H. *J. Am. Chem. Soc*. **1972**, *94*, 7936–7937. 42. Maskornick, M. J.; Streitwieser, A., Jr. *Tetrahedron Lett*. **1972**, 1625–1628. 43. Juan, B.; Schwar, J.; Breslow, R. *J. Am. Chem. Soc*. **1980**, *102*, 5741–5748. 44. Streitwieser, A., Jr.; Heathcock, C. H. *Introduction to Organic Chemistry*, 3rd ed.; Macmillan: New York; 1985. 45. Bissot, T. C.; Parry, R. W.; Campbell, D. H. *J. Am. Chem. Soc*. **1957**, *79*, 796–800. 46. Margolin, Z.; Long, F. A. *J. Am. Chem. Soc*. **1973**, *95*, 2757–2762. 47. Hine, J., Philips, J. C.; Maxwell, J. I. *J. Org. Chem*. **1970**, *35*, 3943.

Acetal, hydrolysis and formation, 216–218

Acetic acid, esterification with methanol in strong acid, 71

Acetyl chloride, hydrolysis in water, 71–72

Acid–base equilibrium
  equilibrium constant calculation, 36–37, 60–61
  tautomers, 30

Acid catalysis
  acetic acid esterification in strong acid, 71
  carbonyl compounds
    1,4-addition, 218–220, 258–260
    hydrolysis of carboxylic acid derivatives
      amide hydrolysis, 213–215, 252–255
      ester hydrolysis, 215, 255–258
      steps, 213
    hydrolysis and formation of acetals, ketals, and orthoesters, 216–218
  nitrile hydrolysis, 256–258
  phenylhydrazone synthesis, 128–129
  zinc as reducing agent, 421, 437–439

Acidity, *see* pK_a

Acrolein, electrophilic addition of hydrochloric acid, 219, 258

Acylium ion, intermediate in electrophilic aromatic substitution, 263–264

Addition–elimination mechanism, nucleophilic substitution at aromatic carbons, 116–118, 164–166

Addition reactions, *see* Carbene; Cycloaddition reactions; Electrophilic addition; Nucleophilic addition; Radical

1,4-Additions
  carbon nucleophile to carbonyl compounds, 132–141, 177–179
  electrophilic, 218–220, 258–260

AIBN, *see* Azobis(isobutyronitrile)

Aldol condensation, carbon nucleophile addition to carbonyl compounds, 130–131, 141, 174–177, 338–339

Alkyllithium reagents, addition reactions
  aldehydes and ketones, 124
  carboxylic acid derivatives, 125–126

Amide hydrolysis, leaving groups, 83–84

Anions, representation, 13

Aromaticity
  antiaromatic compounds, 28
  aromatic carbocycles, 26–27
  aromatic heterocycles, 27–28
  classification of compounds, 29, 54–55

Arrows
  bond-making and bond-breaking, 66–69, 91–92
  electron density redistribution, 66–67, 427–428
  radical reaction representation, 283–284

Aryne mechanism, nucleophilic substitution at aromatic carbons, 118–120, 166–168, 188–191

Atom numbering, writing reaction mechanisms, 85–86, 88, 97–98, 144–145, 160–161, 201, 242–243, 440

Azide, intramolecular 1,3-dipolar cycloaddition, 364–365

Azobis(isobutyronitrile) (AIBN), decomposition, 285–286, 316–318, 335–336

Baeyer–Villager rearrangement, electron-deficient oxygen, 236, 238–239, 270–271
Balancing equations
  atoms, 64–65
  charges, 65–69, 90–92
  criteria in organic chemistry, 64
Baldwin rules, radical cyclization, 296–297
Barton nitrite photolysis, 298, 320–321
Basicity
  comparison with nucleophilicity, 37
  determination with resonance structures, 33, 35, 57–59
  leaving group ability, inverse relationship to base strength, 83
  solvent effects, 87
BDE, *see* Bond dissociation energy
Beckmann rearrangement, electrophilic nitrogen, 233–235, 267–269
Benzilic acid rearrangement, 142–144, 179–180
Benzoic acid, Birch reduction, 311
Benzophenone oxime, Beckmann rearrangement, 234
BHT, *see* Butylated hydroxytoluene
Bicyclo[3.1.0]hexane system, geometric constraint to disrotatory ring opening, 354–355
Bimolecular elimination, *see* E2 elimination
Birch reduction
  benzoic acid, 311
  mechanism, 311–312, 329–330
  solvent, 310
Bond dissociation energy (BDE)
  determining feasibility of radical reactions, 292–294, 316–317
  table of values, 293
Bond number
  carbon, 4
  estimation, 2–3, 41–44, 50–54
  hydrogen, 4
  nitrogen, 4–5
  oxygen, 5
  phosphorous, 5
  radicals, 11–12
  sulfur, 5

Bronsted acid, 32
Bronsted base, 32
Butadiene, cyclobutene interconversion and correlation diagrams, 389–392
Butene, bromination, 209–210
Butylated hydroxytoluene (BHT), radical inhibition, 290–291
t-Butyl hypochlorite, radical chain halogenation
  energetics, 292–294, 316–317
  mechanism, 288–290

Carbene
  addition reactions
    butene, addition of singlet dichlorocarbene, 228
    stereospecificity, 227–229
  carbenoid, 225, 230
  formation
    alkyl halides in base, 225–226
    diazo compounds as starting compounds, 227
    Simmons–Smith reagent, 226–227
  insertion reactions, 230
  reactivity, 224
  rearrangements, 230–231
  singlet carbene, 225
  substitution reactions, 229–230, 265–267
  triplet carbene, 225
Carbocation, *see also* Electrophilic addition
  fates, 199–200
  formation
    alkyl halide reaction with Lewis acid, 199
    electrophile addition to π bond, 197–199
    ionization, 196–197
  rearrangement
    alkyl shift, 201–203, 241–244
    dienone–phenol rearrangement, 204–205
    hydride shift, 201
    overview of pathways, 200
    pinacol rearrangement, 206–208, 244–249
    stability, 200

resonance stabilization in electrophilic aromatic substitution, 220–222, 260–262
  stability, factors affecting, 195–196
    hybridization, 196
    hyperconjugation, 195–196
    inductive effects, 196
    resonance effects, 196
Carbon, bond number, 4
Chain process, *see* Radical
Chemical notation, symbols and abbreviations, 455–456
2-Chlorobutane, reaction with aluminum trichloride, 199
Claisen rearrangement, 344, 403, 410–411
Conrotatory process, electrocyclic transformations, 347–348, 397–399, 405, 414
Cope rearrangement, 344, 366–367, 404, 410–411
Correlation diagrams, pericyclic reaction analysis
  classification of relevant orbitals, 389
  orbital phase correlations, 391–392
  principle, 388
  symmetry characteristics of reaction, 390–391
  symmetry correlations between bonding orbitals of reactants and products, 389–390
C_2 symmetry, molecular orbital theory, 382, 386–387, 415
Curtius rearrangement, electrophilic nitrogen, 235–236
Cycloaddition reactions
  allyl cations and 1,3-dipoles, 362–365, 402–403
  atom number in classification, 359–360, 402
  electron number in classification, 355–357, 400–401
  frontier orbital theory, 393
  orbital symmetry, 378, 413
  overview, 344, 355
  selection rules, 360–361
  stereochemistry
    allowed stereochemistry, 361
    antarafacial process, 356–357
    classification of reactions, 400–401

exo:endo ratio in Diels–Alder reaction, 361–362
suprafacial process, 356–357
Cyclobutene, butadiene interconversion and correlation diagrams, 389–392
Cyclooctatriene, thermal cyclization, 350–351
Cyclopentanone, Baeyer–Villager oxidation, 238–239, 270–271
Cyclopropyl cation, *see* Electrocyclic transformations
Cyclopropyl tosylates, solvolysis, 353–354

Dewar benzene, geometry, 24
Di-*t*-butyl nitroxide
  radical inhibition, 292
  stability, 284
Diazoacetophenone, Wolff rearrangement, 231
Diels–Alder reaction, 344, 355–358, 361–362, 374, 408, 424, 440
Dienone–phenol rearrangement, 204–205, 262
Dinitrobenzene, radical trapping, 23, 49–50, 291
1,4-Dinitrotoluene, radical inhibition, 291
1,3-Dipolar cycloaddition, 362–365, 402–403, 422–423
Dipole
  direction, 18, 45
  relative dipoles in common bonds, 17
Disrotatory process, electrocyclic transformations, 349, 354–355, 396–399
Driving forces, chemical reactions
  leaving groups, 83–84
  overview, 82–83
  small stable molecule formation, 84

E_{1CB} elimination, 436
E2 elimination
  aldol condensation, 131, 175, 338–339
  concerted process, 120
  leaving groups, 121–122
  stereochemistry, 120–121
Ei elimination
  pyrolytic elimination from a sulfoxide, 122

stereochemical restrictions, 122–123, 168
  transition states, 122
Electrocyclic transformations
  cyclooctatriene, thermal cyclization, 350–351
  cyclopropyl cation reactions, 353–355, 400
  frontier orbital theory, 392–393
  intermediates, 412
  overview, 343–344, 346
  selection rules, 346–347, 350
  stereochemistry
    conrotatory process, 347–348, 397–399, 405, 414
    disrotatory process, 349, 354–355, 396–399
    effect of reaction conditions, 399
Electronegativity
  periodic table, trends, 16
  polarity of bonds, 17
  relative values of elements, 16–17
Electrophile, *see also specific electrophiles*
  common types, 39
  definition, 37, 195
  identification of centers, 41, 61–62
Electrophilic addition
  1,4-addition, 218–220, 258–260
  regiospecificity, 208–209
  steps, 208
  stereochemistry
    anti addition, 209–210
    nonstereospecific addition, 211–212
    syn addition, 210–211
  temperature effects, 212, 249–252
Electrophilic substitution
  intermediate carbocations and resonance stabilization, 220–222, 260–262
  mechanisms, 224, 262–265
  metal-catalyzed intramolecular reaction, 222–223
  nitrenium ion intermediate, 273, 280
  substituent influence in aromatic substitution, 221–222
  toluene, electrophilic substitution by sulfur trioxide, 220–221

Elimination, *see* E_{1CB} elimination; E2 elimination; Ei elimination
Elimination–addition mechanism, nucleophilic substitution at aromatic carbons, 118–120, 166–168, 188–191
Ene reactions
  intramolecular reactions, 374–375, 407
  overview, 344, 373–374
Equilibrium constant, calculation, 36–37, 60–61
Ester, hydrolysis in acid, 215, 255–258
Ethyl 2-chloroethyl sulfide, neighboring group effect in hydrolysis, 111

Favorskii rearrangement, 141–144, 179–180
Formal charge
  calculation, 6–8, 10, 12, 41–44
  dimethyl sulfoxide, 10–11
Frontier orbital theory, pericyclic reactions
  cycloaddition reactions, 393
  electrocyclic reactions, 392–393
  overview, 345, 392
  sigmatropic rearrangements, 393–396

Grignard reagents, addition reactions
  aldehydes and ketones, 124, 127, 168–170
  esters, 127
  nitriles, 125–126, 128

Hexanamide, Hofmann rearrangement to pentylamine, 235, 237–238
Highest occupied molecular orbital (HOMO), pericyclic reaction analysis, 387–388, 394–395, 416
Hofmann rearrangement, electrophilic nitrogen, 235–238
HOMO, *see* Highest occupied molecular orbital
Homolytic bond cleavage, radical formation, 284–285, 316, 341
Hückel's rule
  aromatic carbocycles, 26–27
  aromatic heterocycles, 27–28
Hybrid orbitals, representation, 14–16, 44–45

Hydrogen
  bond number, 4
  sigmatropic shifts, 103, 344,
    366–371, 403–406, 410–411
Hydrogen abstraction
  incorporation in mechanisms,
    449–450
  radical formation, 285–286
  rates of abstraction and radical sta-
    bility, 286

Indene, chlorination, 210–211
Intermediate
  resonance stabilization, 25–26, 50,
    35–36, 59–60, 82, 100, 448
  stability required in mechanism
    writing, 79–81
  tautomer stabilization, 430
Intramolecular elimination, *see* Ei
    elimination

Ketal, hydrolysis and formation,
    216–218

Leaving group
  ability
    common groups, 83, 108
    inverse relationship to base
      strength, 83
    solvent effects, 107
  amide hydrolysis reactions, 83–84
  E2 elimination, 121–122
  S_N2 reactions, 107–108
Lewis acid
  alkyl halide reactions, 199
  carbonyl compound reactions,
    198–199
  definition, 195
Lewis structure, *see also* Resonance
    structure
  acetaldehyde, 6
  bond number estimation, 2–5
  common functional groups, table,
    453–454
  dimethyl sulfoxide, 10–11
  drawing, 3–6, 8–9, 12, 26, 41–44
  formal charge calculation, 6–8
Lone pairs, representation, 12–13
Lowest unoccupied molecular orbital
    (LUMO), pericyclic reaction
    analysis, 387–388, 394–395, 416

LUMO, *see* Lowest unoccupied
    molecular orbital

Meisenheimer rearrangement, 425
Methyl acetate, hydrolysis in strong
    base, 70
3-Methyl-2-cyclohexen-1-one, hy-
    bridization and geometry of
    atoms, 15–16
exo-6-Methylbicyclo[3.1.0]hexenyl
    cation, sigmatropic shifts,
    372–373
Michael reaction, carbon nucleophile
    addition to carbonyl compounds,
    132–133
Moebius–Huckel theory, pericyclic
    reactions, 345
Molecular orbital theory, pericyclic
    reaction analysis
  C_2 symmetry, 382, 386–387, 415
  correlation diagrams
    classification of relevant orbitals,
      389
    orbital phase correlations,
      391–392
    principle, 388
    symmetry characteristics of reac-
      tion, 390–391
    symmetry correlations between
      bonding orbitals of reactants
      and products, 389–390
  frontier orbital theory, 345, 392–396
  highest occupied molecular orbital,
    387–388, 394–395, 416
  lowest unoccupied molecular or-
    bital, 387–388, 394–395, 416
  mirror plane, 381–382
  nodes, 380–381, 383, 385, 387, 415
  π orbitals
    allyl system, 386, 415
    basis set, 384
    bonding system in chemical reac-
      tivity, 381
    energy levels, 384–387, 415
    ethylene, 383–384
    types, 384
  wavefunctions, 379–380, 383

Naphthalene, resonance structures,
    19–21
Neighboring group effect, nucle-
    ophilic substitutions, 111

Nitrene
  features, 232
  synthesis, 232
Nitrenium ion
  features, 232, 272–273
  intermediate in electrophilic substi-
    tution, 273, 280
  synthesis, 233
p-Nitroanisole, resonance structures,
    21–22
Nitrogen
  bond number, 4–5
  elctron defficient, rearrangements
    Beckmann rearrangement,
      233–235, 267–269
    Curtius rearrangement, 235–236
    Hofmann rearrangement,
      235–238
    Schmidt rearrangement, 235–236,
      238, 269–270
  positively-charged species, 81–82
  valence shell accommodation of
    electrons, 80–81
Notation, symbols and abbreviations,
    455–456
Nucleophile, *see also specific nucle-
    ophiles*
  common types, 38
  definition, 37
  hydroxide ion, 70
  identification of centers, 41, 61–62
  nucleophilicity
    comparison with basicity, 37
    ranking of nucleophiles, 37–38,
      40
    solvent dependence, 39–40
    substrate structure effects, 38–39
Nucleophilic addition
  addition followed by rearrange-
    ment, 144–146, 181–183
  carbonyl compounds
    carbon nucleophiles, reactions
      with carbonyl compounds
    1,4-additions, 132–141,
      177–179
    aldol condensation, 130–131,
      141, 174–177
    Michael reaction, 132–133
    nitrogen-containing nucleophiles,
      reactions with aldehydes and
      ketones

overview, 128
phenylhydrazone formation mechanism, 128–129, 170
steps in mechanism, 129, 170–174
organometallic reagents to aldehydes and ketones, 124, 127, 168–170
carboxylic acid derivatives, 125
esters, 127
nitriles, 125–126
overview, 123–124
reversibility of additions, 123
combination addition and substitution reactions, 146–148
overview, 105–106
Nucleophilic substitution
aromatic carbon substitution
addition–elimination mechanism, 116–118, 164–166
elimination–addition mechanism, 118–120, 166–168
carbonyl group substitution
ester hydrolysis in base, 112–114
examples of steps in substitution, 114–116, 157–164
resonance-stabilized intermediates, 159, 163–164
sp^2 versus sp^3-hybridized centers, susceptibility to substitution, 113–114
tautomers in mechanisms, 159–160, 162
combination addition and substitution reactions, 146–148
overview, 105–106
proton abstraction preference versus substitution, 76–78
S_N2 reactions
alcohol protonation, 108–109, 152
features, 106–107
leaving groups, 107–108
neighboring group effect, 111
phenolic oxygen alkylation, 110, 152–153
reactivities of carbons, 107
stereochemistry, 109
writing of mechanisms, 110–111, 153–157

Occam's razor, simplicity in writing reaction mechanisms, 88, 161
Olefin, cationic polymerization, 200
Orthoester, hydrolysis and formation, 216–218
Oxygen
Baeyer–Villager rearrangement, 236, 238–239, 270–271
bond number, 5
positively-charged species, 81–82

Pericyclic reactions
concerted nature, 343–344
cycloadditions
allyl cations and 1,3-dipoles, 362–365, 402–403
atom number in classification, 359–360, 402
electron number in classification, 355–357, 400–401
frontier orbital theory, 393
orbital symmetry, 378, 413
overview, 344, 355
selection rules, 360–361
stereochemistry
allowed stereochemistry, 361
antarafacial process, 356–357
classification of reactions, 400–401
exo:endo ratio and secondary factors, 361–362
suprafacial process, 356–357
electrocyclic transformations
cyclooctatriene, thermal cyclization, 350–351
cyclopropyl cation reactions, 353–355, 400
frontier orbital theory, 392–393
intermediates, 412
overview, 343–344, 346
selection rules, 346–347, 350
stereochemistry
conrotatory process, 347–348, 397–399, 405, 414
disrotatory process, 349, 354–355, 396–399
effect of reaction conditions, 399
ene reactions
intramolecular reactions, 374–375, 407
overview, 344, 373–374

molecular orbital theory
C_2 symmetry, 382, 386–387, 415
correlation diagrams
classification of relevant orbitals, 389
orbital phase correlations, 391–392
principle, 388
symmetry characteristics of reaction, 390–391
symmetry correlations between bonding orbitals of reactants and products, 389–390
frontier orbital theory, 345, 392–396
highest occupied molecular orbital, 387–388, 394–395, 416
lowest unoccupied molecular orbital, 387–388, 394–395, 416
mirror plane, 381–382
nodes, 380–381, 383, 385, 387, 415
π orbitals
allyl system, 386, 415
basis set, 384
bonding system in chemical reactivity, 381
energy levels, 384–387, 415
ethylene, 383–384
types, 384
wavefunctions, 379–380, 383
selection rules, theory, 345
sigmatropic rearrangements
Claisen rearrangement, 344, 403, 410–411
Cope rearrangement, 344, 366–367, 404, 410–411
frontier orbital theory, 393–396
overview, 344, 366
selection rules
alkyl shifts, 371–373
hydrogen shifts, 368–371, 405–406
terminology, 366–368, 403–404, 406–407
symmetry-allowed reactions, 345, 388, 398–399
symmetry-forbidden reactions, 345, 388, 396
writing mechanisms, 375–377, 408–412
Phenylhydrazone, synthesis, 128–129

Phosphorous, bond number, 5
π bond
    electrophile addition, 197–198
    estimation of number, 2–3
π orbital
    allyl system, 386, 415
    basis set, 384
    bonding system in chemical reactivity, 381
    energy levels, 384–387, 415
    ethylene, 383–384
    types, 384
Pinacol rearrangement, 206–208, 244–249
pK_a
    approximation from related compounds, 145, 185, 437
    calculation, 36–37, 60–61
    carbonyl groups, 185
    definition, 33
    values for common functional groups, 33–35, 457–463
Proton removal
    epimerization of reactants, 148
    nucleophilic substitution, competition with, 76–78
    susceptibility of specific protons, 74–76, 92–95
    writing reaction mechanisms
        condensation reactions, 432–434, 436
        rationale, 73–74
        strong base, 70
        weak base, 71–72, 102
Protonation
    carbonyl groups, 198
    olefins, 197
    susceptibility of specific centers, 75, 96–97
    writing reaction mechanisms
        condensation reactions, 432–434
        rationale, 73–74
        strong acid, 71, 100–101
        weak acid, 71–72

Radical
    addition reactions
        intermolecular addition, 294–296, 317
        intramolecular cyclization, 296–298, 317–321
    Birch reduction, 310–312

bond dissociation energies in determining feasibility of reactions, 292–294, 316–317
chain process
    balancing of equations, 287–288
    coupling of radicals, 288–290, 316
    disproportionation, 289
    halogenation by t-butyl hypochlorite, 288–290
    initiation, 287–288, 295, 298–299, 317–320, 334, 336
    propagation, 287–289, 295, 298–300, 317–318, 320, 334, 336
    termination, 287, 289
    definition, 283
    depicting mechanism, 283–284
    formation from
        functional groups, 286–287
        homolytic bond cleavage, 284–285, 316, 341
        hydrogen abstraction, 285–286
    fragmentation reactions, loss of small molecules
        addition followed by fragmentation, 301–302
        CO, 300
        CO_2, 299–300
        ketone, 300
        N_2, 300
        writing of mechanisms, 303, 322–325
    inhibitors, 290–292
    rearrangements
        alkyl migration, 201–203, 241–244
        apparent alkyl migration, 306–307, 325–327
        aryl migration, 304–305
        halogen migration, 305
        mechanisms in anion rearrangement, 312–313, 330–331
        non-migrating groups, 303–304
        resonance stabilization, 316
    $S_{RN}1$ reaction
        enolate reaction with aromatic iodide, 308–310
        features, 307
        identification of reactions, 310, 327–329, 331–333
        initiation, 308, 337
        propagation, 307–308, 338

stability, 284, 286
stereochemistry of reactions, 284
Rearrangement, *see also* Sigmatropic rearrangement
Baeyer–Villager rearrangement of electron-deficient oxygen, 236, 238–239, 270–271
base-promoted rearrangements
    benzilic acid rearrangement, 142–144, 179–180
    Favorskii rearrangement, 141–144, 179–180
    nucleophilic addition followed by rearrangement, 144–146, 181–183
    stereochemistry, 143–144, 180–181
carbenes, 230–231
carbocation
    alkyl shift, 201–203, 241–244
    dienone–phenol rearrangement, 204–205
    hydride shift, 201
    overview of pathways, 200
    pinacol rearrangement, 206–208, 244–249
    stability, 200
electrophilic nitrogen
    Beckmann rearrangement, 233–235, 267–269
    Curtius rearrangement, 235–236
    Hofmann rearrangement, 235–238
    Schmidt rearrangement, 235–236, 238, 269–270
radicals
    apparent alkyl migration reactions, 306–307, 325–327
    aryl migration, 304–305
    halogen migration, 305
    mechanisms in anion rearrangement, 312–313, 330–331
    non-migrating groups, 303–304
Reimer–Tiemann reaction, 229, 265–266
Resonance effects
    basicity, 33, 35, 57–59
    carbocation stability, 196
    protonation, 75, 96–97
    proton removal, 74–76, 92–95
    radical stability, 316

Resonance structures
cyclooctatetraenyl anion, 22–23
definition, 18
distinguishing from tautomers,
30–32, 56–57, 186
drawing, 18–19, 23, 26, 45–54
naphthalene, 19–21
p-nitroanisole, 21–22
rules, 23–25
stability, 25–26, 50, 35–36, 59–60,
82, 100, 448
Ring number, estimation, 2

Schmidt rearrangement, electrophilic
nitrogen, 235–236, 238, 269–270
Sigmatropic rearrangements
Claisen rearrangement, 344, 403,
410–411
Cope rearrangement, 344, 366–367,
404, 410–411
frontier orbital theory, 393–396
overview, 344, 366
selection rules
alkyl shifts, 371–373
hydrogen shifts, 368–371,
405–406
terminology, 366–368, 403–404,
406–407
Simmons–Smith reagent, synthesis,
226–227
S_N2 reactions
alcohol protonation, 108–109, 152
features, 106–107
leaving groups, 107–108
neighboring group effect, 111
phenolic oxygen alkylation, 110,
152–153
reactivities of carbons, 107
stereochemistry, 109
writing of mechanisms, 110–111,
153–157
Solvent, effects
basicity, 87
leaving group ability, 107
mechanism of reaction, 86–88

nucleophilicity, 39–40
radical stability, 425
sp hybridization, overview, 14–15, 45
sp^2 hybridization
nucleophilic substitution at
aliphatic carbon, 112–116,
157–161
overview, 14–16, 45
sp^3 hybridization, overview, 14–16,
44–45
$S_{RN}1$ reaction
enolate reaction with aromatic iodide, 308–310
features, 307
identification of reactions, 310,
327–329, 331–333
initiation, 308, 337
propagation, 307–308, 338
Stereochemistry
base-promoted rearrangements,
143–144, 180–181
cycloaddition reactions
allowed stereochemistry, 361
antarafacial process, 356–357
classification of reactions,
400–401
exo:endo ratio and secondary factors, 361–362
suprafacial process, 356–357
E2 elimination, 120–121
electrocyclic transformations
conrotatory process, 347–348,
397–399, 405, 414
disrotatory process, 349, 354–355,
396–399
effect of reaction conditions, 399
electrophilic addition
anti addition, 209–210
nonstereospecific addition,
211–212
syn addition, 210–211
epimerization of reactants, 148
pinacol rearrangement, 207,
244–245

radical reactions, 284
S_N2 reactions, 109
Styrene, bromination in acetic acid,
211–212
Sulfur, bond number, 5
Swain–Scott equation, nucleophilicity
calculation, 37–38
Symbols, chemical notation, 455–456

Tautomer
acid–base equilibrium, 30
definition, 29
distinguishing from resonance
structures, 30–32, 56–57, 186
drawing of structures, 31, 56
enolization, strong acids or bases as
intermediates, 72–73
equilibrium, 29–30
ketones, 30
stabilization of intermediates, 430
Toluene, electrophilic substitution by
sulfur trioxide, 220–221
Triflate, spontaneous ionization, 197
Trifluoroiodomethane, photochemical
addition to allyl alcohol, 294–295
Trimolecular reaction
breaking down into several bimolecular steps, 78–79, 89, 99,
193
rarity, 78

Unproductive step, 76

Verrucosidin, degradation products,
421, 440–448
Vitamin D$_2$, synthesis, 371

Wolff rearrangement
nitrogen analogues, 235–238
overview, 230–231

Zinc, reducing agent in acid catalysis,
421, 437–439